Philipp Wendler
Worked to the Bone

Philipp Wendler

Worked to the Bone

A Study of Gilded-Age Transatlantic Scientific Networks in Paleontology

Hamburg University Press
Verlag der Staats- und Universitätsbibliothek Hamburg
Carl von Ossietzky

Zugl.: Dissertation, Universität Hamburg, 2020 u. d. T.
Worked to the Bone. O. C. Marsh, his German Assistants, and their Contributions to the Genesis of US-American Paleontology. A Study of Gilded-Age Transatlantic Scientific Networks.

BIBLIOGRAFISCHE INFORMATION DER DEUTSCHEN NATIONALBIBLIOTHEK
Die Deutsche Nationalbibliothek verzeichnet diese Publikation in der Deutschen Nationalbibliografie; detaillierte bibliografische Daten sind im Internet über https://portal.dnb.de abrufbar.

ONLINE-AUSGABE
Die Online-Ausgabe dieses Werkes ist eine Open-Access-Publikation und ist auf den Verlagswebseiten frei verfügbar. Die Deutsche Nationalbibliothek hat die Online-Ausgabe archiviert. Diese ist dauerhaft auf dem Archivserver der Deutschen Nationalbibliothek (https://portal.dnb.de) verfügbar.
DOI https://doi.org/10.15460/HUP.HHD.009.216

PRINTAUSGABE
ISBN 978-3-943423-95-2

GRUNDGESTALTUNG Merle Papenfuß

DRUCK UND BINDUNG Books on Demand - Norderstedt

VERLAG Hamburg University Press, Verlag der Staats- und Universitätsbibliothek Hamburg Carl von Ossietzky, Hamburg (Deutschland), 2022
https://hup.sub.uni-hamburg.de

ACKNOWLEDGEMENTS

Great thanks to Fenja Heisig, Karin Heisig, and Dr. Alan Lena van Beek who tirelessly scrutinized and corrected the manuscript, and without whom the text would have taken a very different (undoubtably lesser) form. I would also like to thank Sebastian Prinz and Christina Urbanek who, with an impeccable eye for detail, helped in the early stages of the project.

Of course, I also want to thank my parents: my mother for always doing her best to support me and my father, who always expected better of me.

I thank Prof. Dr. Claudia Schnurmann, my PhD adviser, who supervised this thesis. She was always happy to answer any questions, provide a homecooked meal at a dinner party or organize an edutaining piracy lecture, complete with costumes and a parrot. I also want to thank Prof. Dr. Thorsten Logge and Prof. Dr. Markus Friedrich for co-assessing thesis and disputation.

I also thank Marianne Weis for her indispensable help in overcoming many a bureaucratic hurdle.

My colleagues at the department for history, especially Katja Reuther, Nils Schliehe, Felix Matheis and Yves Müller I thank for many entertaining lunch breaks and many a comment on academic work.

I thank the German Historical Institute in Washington D.C. for inviting me to a summer school program in 2014 and paying for my travel across the Atlantic, which I could not have afforded otherwise at the time. I also want to thank Prof. Dr. Mischa Honeck for organizing the summer school and guiding us through various archives, libraries, and pubs.

Another thank you to the tireless staff of the Sterling Memorial Library at the Yale University. My gratitude goes to the American Museum of Natural History in New York, and especially Susan K. Bell, who guided me through the labyrinthian innards of the Museum and made sure I had timely access to some of the most important sources of this thesis.

Hamburg, January 11, 2022

CONTENT

LIST OF FIGURES

ABBREVIATIONS

ANS	Academy of Natural Sciences of Philadelphia
AAAS	American Association for the Advancement of Science
AMNH	American Museum of Natural History
APS	American Philosophical Society
BAAS British	Association for the Advancement of Science
BIA	Bureau of Indian Affairs
NAS National	Academy of Sciences
USGS	United States Geological Survey

1 Introduction

1.1 Research Questions

In February 2020 a small scientific sensation was announced in the city of Hamburg, Germany. Sometime in the foreseeable future four dinosaur skeletons would be displayed at the zoo in Hamburg. The zoo, Hagenbecks Tierpark, is named after its founder Claus Gottfried Carl Hagenbeck (1810–1887), whose descendants still own the private wildlife park. The dinosaur skeletons will be the first ones to be exhibited in a German zoo,[1] and they might give answers to some long-standing questions concerning the social behaviorisms of sauropod dinosaurs. The four skeletons all belong to the genus *suuwassea*, and some 150 million years ago they inhabited the region now known as the Morrison Formation in Wyoming. Among the four specimens is a young individual and it is possible that the group consisted of a family unit, prompting the assumption that sauropods cared for their young and did not abandon their offspring entirely after a nest was built, a behavior exhibited by modern-day sea turtles and attributed to dinosaurs by some paleontologists.[2]

The skeletons were exhumed in 2009, bought by an anonymous "private investor," and shipped to Hamburg. Here the Hagenbeck Foundation (Stiftung Hagenbeck) purchased the dinosaurs to be displayed at the zoo and studied at the Center of Natural History (Centrum für Naturkunde) of the Hamburg University. Even the city's cultural minister (Kultursenator) chimed in, calling the acquisition a "great benefit to the city," adding that the supposed dinosaur family would surely prove to be "a special attraction for the zoo," and thanking the Hagenbeck Foundation for the dedication with which it had secured the dinosaurs for Hamburg and for science.[3] Cord Casselt of the Hagenbeck Foundation assessed that the dinosaurs would be a major attraction at the zoo, contributing significantly to its future funding.[4] When and where exactly the

1 Note that the Hagenbeck Zoo already displays a prehistoric landscape, including dinosaurs fashioned from concrete. The prehistoric panorama was designed by Josef Pallenberg (1882–1946) in 1909 and depicts reconstructions of various species according to the paleontological standards of that time. This mirrors another group of paleo-reconstructions, which were erected in England in the middle of the nineteenth century (see chapter 2. 5.).

2 N.N.: Dinoforschung am CeNak, https://hamburg.leibniz-lib.de/aktuelles/news/news-archiv/2020-newsarchiv/2020-04-09-news.html, as consulted online on May 14, 2020.

3 "Diese Dino-Familie ist ein großer Gewinn für Hamburg. Es ist beeindruckend neben diesen 150 Millionen Jahre alten Zeitzeugen zu stehen, die zu einer gan z besonderen Attraktion für den Tierpark werden können. Ich danke der Hagenbeck Stiftung für das Engagement, mit dem sie die Dinosaurier für Hamburg und für die Wissenschaft gesichert hat." Quoted after N.N.: Stiftung Hagenbeck sichert versteinerte Dinos für den Tierpark, https://www.hamburg-magazin.de/artikel/stiftung-hagenbeck-dinosaurier-für-hamburg, as consulted online on May 14, 2020.

4 "Die Hamburger Dino-Familie kann für den Tierpark Hagenbeck einen starken Publikums-Magneten darstellen und daher einen erheblichen Anteil zum Fortbestand des Tierparks beitragen." Quoted after N.N.: Stiftung Hagenbeck sichert versteinerte Dinos für den Tierpark.

skeletons will be mounted is currently still uncertain, but certain is that the skeletons do not only constitute a small scientific sensation but are also of great public interest for the city, if only to stimulate tourism.

This episode illustrates that dinosaur skeletons are of more than just scientific interest, that the products of paleontological discovery and research can become sources of local pride, and that the conduct of science and research is linked to finances and funding in a capitalist society. It is also exemplary for how paleontology is conducted today: it is not surprising that complete and impressive dinosaur skeletons come from the North American west, it is more the rule than an exception. These trends developed during the nineteenth century, when US-American paleontology came into its own and became a source of national pride:

> Paleontology had long since been proving historical narratives and American icons. This was true not only for the deep histories and animal as the mastodon that were reconstructed from fossil remains, but also for the history of American paleontology and its pioneers. The traces of the American deep past were national treasures; they were also bones of contention between men and institutions devoted to paleontology, a natural history that was strongly associated with the westward movement and the resulting territorial conflicts. In 'epic efforts' and public feuds, Marsh and Cope spearheaded the discovery of many Dinosaur species in the 1870s. When Osborn later organized museum expeditions to the American western states and territories, this triggered the 'second dinosaur rush', in which the Carnegie Museum of Natural History in Pittsburgh and all the other major museums followed with collecting expeditions.[5]

This study outlines the development of US-American paleontology during the nineteenth century with a special focus on the transatlantic influences on said discipline. The thesis depicts how paleontology came to be US-American, when, why, and by what means that Americanization happened, who its protagonists were, and how they interacted with each other. It touches on the genesis of said discipline in Europe, and traces how European, and especially German, know-how and techniques were imported to the US and implemented there. How US-American paleontology came into its own, how it was culturally propagated, and how it became a building block of rising nationalism are also subjects of this study.

This thesis is in part a history of the culture(s) that shaped the rise of paleontology in North America and argues that the influence these cultures exerted on the evolution of paleontology rivaled that of the fossil findings that constitute the foundation

---

5 Marianne Sommer: History Within. The Science, Culture, and Politics of Bones, Organisms, and Molecules, Chicago 2016, p. 27.

for this branch of science. As the emerging discipline of paleontology was imbued with nationalistic symbolism, and since the rise of US-American paleontology was intertwined with rising nationalistic sentiments during the long nineteenth century, this study situates itself also as a history of mentalities. Therefore, this study can be understood as part of what Keith Parsons calls the "Science Wars" (more on that below). To enable the study of said mentalities and culture, media and popular culture were examined as well, and the scope of this study also encompasses trends and developments in media during the second half of the nineteenth century.

The scientific discipline of US-American paleontology developed within the context of a greater effort to reform US-American higher education, heavily inspired by the German system of higher education. The development of scientific institutions is of special interest to the historian, because mutual, communal efforts would advance knowledge far more rapidly than the enterprise of isolated individuals. The US-American scientific institutions evolved in contact with the European institutions, foreign ideas and traditions were adapted from France, Britain and Germany and integrated by US-American institutions, as science historian Kohlstedt attests.[6] She further says that it is hardly imaginable to write about the history of a science without engaging at least a little in biographical work:

> Because many of the early leaders established and then sustained organizations from which their lives and work seemed almost indistinguishable, the line between biography and institutional history has not been precisely drawn in the history of science [...] The study of individuals and individual institutions provide, moreover, an intellectual coherence: they have a specific origin and identity, and, often equally important, their activities are likely to be systematically recorded and preserved in a major repository.[7]

All paleontologists were of course individuals. They communicated with each other and, besides professional information, exchanged pleasantries of friendship, complained about intimate hardships, or were bitter rivals. Exploring these interpersonal relationships necessarily gives the analyses biographical tendencies. Furthermore, as Jane Davidson notes in her monograph on the importance of government support for the science of paleontology: "Paleontology was an expensive business."[8] The quote shows how external, non-scientific factors such as the necessity to acquire funding

---

6 Sally Gregory Kohlstedt: Institutional History, in: Osiris, 2nd ser., vol. 1 (1985), pp. 17–36.

7 Kohlstedt: Institutional History, pp. 17, 18.

8 Jane Pierce Davidson: Patrons of Paleontology. How Government Support Shaped a Science, Bloomington, IN 2017, p. xv.

through rich patrons or a government also shaped the development of paleontology. These external factors of influence will therefore also be a crucial part of this study.

To sum it up, this is not science-history in the strictest sense, although the analysis of scientific networks and knowledge transfer will be a part of this thesis. The focus will not be on the individual discoveries, descriptions, and publications, within the discipline of paleontology. Instead, an analysis of interpersonal networks and an exploration of the genesis of US-American paleontology within the context of various (inter)cultural and (inter)personal influences will be the focal point of this study.

Dinosaurs play their special part in the history of paleontology. They were in the nineteenth century, as they are today, the most spectacular and most popular products of paleontology, with a vast pop-cultural legacy. And dinosaurs are US-American. Paleontologists like Othniel Charles Marsh (1831–1899) and his contemporaries laid the foundation for an US-American pop idol, which inspired and was cultivated by the likes of Michael Crichton, and Steven Spielberg.

O. C. Marsh, who "became, in the eyes of the public, perhaps the most well-known scientist in America"[9] is of special interest for this thesis, his scientific career and network is a most conclusive example showcasing how US-American paleontologists operated, received their education, and contributed to the rise of sciences in the United States.

The situation of US-American paleontology before paleontologists Edward Drinker Cope (1840–1897) and O. C. Marsh began their careers is depicted by Merrill as follows:

> Prior to the advent of Cope and Marsh a very large proportion of the work in vertebrate paleontology in America had been performed by Leidy, and that, too, on fragmental material that had weathered out of the matric and been gathered in many cases without an exact knowledge of the beds from which they were derived, during the haste and hurry of reconnaissance surveys. It remained for these men to take the field for themselves and for Marsh in particular to adopt new methods, train collectors, and, in short, to change entirely the mode of procedure. The results became shortly the wonder of the scientific world. The material was no longer collected haphazard and in form of weathered fragments, but actually shipped in the matrix in which it was embedded, to the laboratories in the east where proper time and facilities could be devoted

9 Url Lanham: The Bone Hunters. The Heroic Age of Paleontology in the American West, New York 1991 (orig. publ. 1973), p. 146.

> to it. In this way it became possible to restore entire skeletons and gain an idea of external form, before approximated by guessing.[10]

Marsh, Cope, and their colleagues doubtlessly made some very significant discoveries in the West. But those will not be the focus of this thesis as these discoveries were already duly honored in countless textbooks and other paleontological publications. Instead, the interplay between US-American nationalism, exemplified by the "conquest" of the West, and science will be examined here.

Unfortunately, next to no women will be mentioned throughout this thesis, as they prove to be largely absent from the sources examined here to trace the development of US paleontology.[11] During the period evaluated in this thesis women were not yet able to receive a scientific education, and they were socially barred from participating in the field of science, which was, in Europe and North America, quite literally dominated by (mostly rich) white men. Still, throughout the nineteenth century women contributed to paleontological publications by drawing paleontological plates, depicting the fossils described by the male paleontologists. They also enabled their husband's, son's, or relative's scientific work in other ways, such as through domestic and care work. Some women collected fossils for state surveys, and at the end of the century the first scientific descriptions of fossils were published by women.[12] However, the field of paleontology did only really open for women at a broader scale during the twentieth century. In 1974 Halszka Osmólska (1930–2008) became the first female scientist to craft the initial description of a dinosaur and consequently name it. She also participated in a Polish-Mongolian expedition to the Gobi Desert, which produced various new dinosaur findings between 1965 and 1971. The expedition was led by Zofia Kielan-Jaworowska (1925–2015) and all leading scientists were women.[13]

This thesis endeavors to answer some key questions about the genesis and further development of US-American paleontology during the second half of the nineteenth century.

How did US-American paleontology develop as a scientific discipline, especially in exchange with German higher education and through international scientific networks and knowledge transfer? Why did US paleontology (and science in gener-

10 George P. Merrill: The First One Hundred Years of American Geology, New Haven, CT 1924, pp. 528–529.

11 There is one notable exception at the beginning of the nineteenth century, when Mary Anning exhumed fossils at the South Coast of England (see chapter 2. 5.).

12 Michelle L. Aldrich: Women in Paleontology in the United States 1840–1960, in: Earth Sciences History, vol. 1, no. 1 (1982), pp. 14–22.

13 Peter Dodson: The Horned Dinosaurs. A Natural History, Princeton, NJ 1996, p. 9.

al) initially lag behind Europe? How did the situation of US-American science then change in the second half of the nineteenth century? Why did this change occur at that time? How could the US catch up with Europe so quickly, and arguably even overtake it? Where did the influences and inspirations for the change in US science and higher education come from? Who were the people who brought the change? How did they interact with each other? What made US-American paleontology, and dinosaur paleontology in particular, US-American? How did US paleontology and its findings shape US-American identity, the US-American national consciousness, and US nationalism? How, in turn, did US-American nationalism and self-understanding influence the way paleontology developed as a scientific discipline? Does paleontology serve to exemplify the broader changes happening within the US-American system of higher education?

Note that the achievements of science and its ranking within an international scientific field are highly debatable and subjective. When this thesis supposes that US-American paleontology rose to overshadow its European counterpart, this assessment may not be objectively measurable, but follows the evaluations of most paleontologists, living then and now. It certainly means that US-American paleontology became very productive, described many animals, and contributed greatly to scientific theories about life on this planet.

1.2 Theory and Methods

"Paleontology is a human endeavor, and like all human endeavors, ideas have changed as the context in which those ideas developed has changed."[14]

It appears that paleontologists are generally more interested in the history of their discipline than scientists who work in other fields. Maybe some paleontologists have been more open to the historic development of their discipline because paleontology is itself in part a historic science.[15] According to John Horner "[p]aleontology is not an experimental science; it's a historical science. This means that paleontologists are seldom able to test their hypotheses by laboratory experiments, but they can still test them."[16]

The methods used to investigate the genesis and early history of US-American paleontology in this thesis are of a strictly constructivist nature. This study is focused

14 David E. Fastovsky; David B. Weishampel: Dinosaurs. A Concise Natural History, Cambridge 2009, p. 315.

15 Lorraine Daston: The Sciences of the Archive, in: Osiris, 2nd ser., vol. 27 (2012), pp. 156–187.

16 James Gorman; John R. Horner: Digging Dinosaurs, New York 1988, p. 168.

on the actors, their training, their experience, and their interactions within their networks. In the past there have been numerous misunderstandings and disagreements between natural and cultural/social scientists. On the one hand, some scientists claimed that the history of their discipline, constructed by a sociological approach, would not further their science at all, and even worse, could undermine their scientific field in its presumed search for the objective truth.[17] On the other hand, one should avoid the danger of writing a whiggish history of science: a story of linear progress towards a lofty goal, complete with heroes (who furthered said progress) and villains (who sought to impede the noble quest). Even if empirical data could be collected impartially, one needs to interpret that data within some theoretical framework.[18] Stephen Jay Gould implores his fellow scientists to be open to a constructivist history of science as follows: "[...] objective minds do not exist outside of culture, so we must make the best of our ineluctable embedding. It is important that we, as working scientists, combat the myths of our profession as something superior and apart."[19]

Thomas Adam and Charlotte Lerg define the "academic diplomacy" in the following terms:

> The concept of 'academic diplomacy' captures how international relations played out within the academic world. This can literally mean making diplomacy visible on campus by hosting representatives of foreign governments, establishing research centres and museums, or actively pursuing exchange programmes in line with the nation's foreign policy. It also refers to the many roles academics played when they travelled abroad and became representatives of their university, their discipline but also of their country and sometimes even of their government. The notion of academic diplomacy is based on the premise that academia is institutionally tied in with nation-states and at the same time linked to an international and transnational community of scholars – be it real or idealised. As national institutions, universities could play a key role in cultural diplomacy and comparable policies, however, as academic institutions, universities followed their own agenda that included scholarly pursuits as well as the need and desire to secure funding, prestige, and influence. [...] It could result in intercultural transfer of ideas and concepts and create a trans-

17 Trevor Pinch: Does Science Studies Undermine Science? Wittgenstein, Turing, and Polanyi as Precursors for Science Studies and the Science Wars, in: Jay A. Labinger; Harry Collins (eds.): The One Culture? A Conversation about Science, Chicago 2001, pp. 13–26, see pp. 13–21.

18 Andreas Pacholski: Wahrheit in Gestalt. Sprachbedingungen der Wissenschaft. Die Ansätze T. S. Kuhns und M. Merleau-Pontys, Marburg 2009, pp. 74–76.

19 Stephen Jay Gould: Time's Arrow, Time's Cycle. Myth and Metaphor in the Discovery of Geological Time, Cambridge, MA 1987, p. 7.

> national discourse. However, it could also result in the rejection of ideas presented to a particular culture. Personal and professional networks of scholars and universities sometimes served as channels for diplomatic communication on a sub-governmental level, while a place on the diplomatic protocol provided universities and scholars with international visibility and access to power.[20]

While most of the examples which Lerg and Adam provide for their concept of "academic diplomacy" occurred during the twentieth century, the concept is also very applicable to the transatlantic exchange within Marsh`s paleontological network (see chapter 3).

They also assess that

> Transatlantic historians seek to evaluate the role of the state and of the individual in the historical process that led to the making of the transatlantic space, by combining social, cultural, political, and diplomatic history approaches. [...] Agency is also the key to analysing international relations within the academic world. Using merely a state-actor-driven cultural diplomacy approach fails to acknowledge the initiative of institutions and scholars who direct the political attention their transnational networks may have generated according to their own needs and circumstances.[21]

This study, too, works with the framework of the "transatlantic space" and focuses on "social, cultural and political history approaches." Most of the sources for this study were not produced by "state actors," but by individual scholars.

1.2.1 The Cultural Construction of Science

Parsons claims that for decades a "culture war" was waged in academia. One side argued that objective knowledge and research was possible, the other side argued that all knowledge and scientific conduct was political, heavily influenced by societal forces. He calls the first group "rationalists," the second "constructivists." These are the two battle lines in the "science wars" he writes about.[22] He states that:

---

20 Thomas Adam; Charlotte A. Lerg: Introductory Remarks. Diplomacy on Campus. The Political Dimensions of Academic Exchange in the North Atlantic, in: Journal of Transatlantic Studies, vol. 13, no. 4 (Dec. 2015), pp. 299–310, https://doi.org/10.1080/14794012.2015.1088327. Quote on page 302.

21 Adam; Lerg: Introductory Remarks, pp. 300–301.

22 Keith M. Parsons: Drawing Out Leviathan. Dinosaurs and the Science Wars, Bloomington, IN 2001, pp. 80–105.

> Rationalists think that physical reality ultimately drives consensus. They might admit that theories start off, as Einstein said, as free creations of the human intellect. [...] But rationalists hold that in the long run (and sometimes it is a rather long and circuitous run) science can transcend ideology and politics and achieve the rigorous constraint of theory by careful observation of or interaction with the natural world. [...] Constructivists radically oppose this rationalist image of science. They insist that science, like everything else, is governed by rhetoric, ideology, politics, vested interests, and other social factors.[23]

He himself falls more on the side of the "rationalists," and the former argument and his reasons for writing this book are rather personal ones for him.[24] This thesis, however, leans towards the more "constructivist" side of the argument (see above).

Parsons finally finds middle ground for himself:

> The lesson to draw about science is that science is a very complex and multifaceted process, a process not reducible to *any* stereotype. Like all human endeavors, science is subject to social influences at every level. However, to a greater degree than the vast majority of human enterprises, science incorporates methods and standards that permit the objective constraint of hypotheses to a very high level of credibility.[25]

And then even makes a point that underlines the whole approach of this study:

> Dinosaurs may not be cultural constructs, but the scientists that study them definitely are. [...] The history, social organization, and epistemological ideals of scientific communities are fair game for sociologists, historians, and philosophers.[26]

---

23 Parsons: Drawing Out Leviathan, p. 81.

24 He writes: "As one whose education began in the immediate post-Sputnik ear, these ideological assaults [by people he calls "postmodernists, Marxists, feminists, literary critics, radical ecologists, sociologists."] on science seemed odd and disturbing to me, especially the ones from the left. I had been taught that science was good, a force for progress and enlightenment and the most effective foil for obscurantism. From growing up in the Deep South, I knew all about fundamentalists antipathy to science; sweaty evangelists fulminating against godless 'evil-lution' were nothing new. The defection of many leftist scholars to trendy schools of anti-science was an unexpected betrayal." See Parsons: Drawing Out Leviathan, p. xiii.

25 Parsons: Drawing Out Leviathan, p. 101.

26 Parsons: Drawing Out Leviathan, p. 151.

Parsons criticizes Mitchell and the others, whom he refers to as "postmodernists," harshly and polemically when he states:[27]

> An initial reaction to the postmodernists is that they should turn off their televisions and get outside for some fresh air. Mitchell seems to have read a great deal about dinosaurs, but he never indicates what he has *done*. To really understand what paleontologists do, reading is not enough; you have to go to the badlands and actually dig up some fossils. Shoveling off a meter-thick layer of overburden in the desert sun gives one a robust sense of reality. [...] When you dig something out of the ground, remove the encrustation of a million centuries, and recognize a jaw or femur, there is a palpable sense of connection to a very real, very deep past.[28]

Parsons never delivers any further argument on why this hands-on approach is necessary to evaluate a science, or the scientists. He merely presents this non-argument and states it as fact to disprove the observations of "postmodernists," which seem to offend him personally.

Thomas Kuhn popularized the theoretical and methodological groundwork used in this thesis in "The Structure of Scientific Revolutions".[29] But was US-American paleontology actually revolutionary, or part of a scientific revolution? The paleontological work done in the United States in the second half of the nineteenth century may have been brilliant, matching, if not surpassing, anything that was done in Europe, but it was not necessarily revolutionary.[30] Yet O. C. Marsh's reconstruction of the equine lineage and his monograph regarding toothed birds were an important contribution to another ongoing scientific revolution: the theory of evolution by natural selection (see Chapter 7. 2.). However, Kuhn not only thought of an explanation for revolutionary processes in science, but also designed a theoretical framework that allowed scholars to analyze the production of knowledge.

Kuhn describes how external (non-scientific) impulses and scientists' worldviews have influenced the production of knowledge, and must therefore be part of any history of science.[31] He explains how a catalog of presumed facts becomes a broadly ac-

27 Note that Parsons' point of contention is Mitchell's monography on the cultural and symbolical meaning of dinosaurs: William John Thomas Mitchell: The Last Dinosaur Book. The Life and Times of a Cultural Icon, Chicago 1998.

28 Parsons: Drawing Out Leviathan, p. 108.

29 Thomas S. Kuhn: The Structure of Scientific Revolutions, Chicago 1962.

30 See Cohen for a historic contextualization of scientific and political revolutions, beginning in the seventeenth century: I. Bernard Cohen: Revolution in Science, Cambridge, MA 1985, pp. 26–101.

31 Kuhn: The Structure of Scientific Revolutions, pp. 1–9.

cepted "paradigm" and how the conversion of one such catalog to another (a "paradigm shift") might be interpreted as a revolution.[32] Such a shift is often accompanied by great upheaval and resistance, fueled by the clash of the participants' worldviews. Cultural prejudices may lead to two scientists observing the same phenomenon but interpreting it in radically different ways, according to the different paradigms they subscribe to.[33] According to this philosophy of constructivism it is necessary to study cultural influences in order to understand the genesis and evolution of knowledge and be able to write a history of science. This theory contrasts with the logical-positivistic position in which there is such a thing as an objective reality, which in turn can be observed and studied through the analysis of empirical data. In such a positivistic train of thought there should be a universal, scientific language (based on the empirical data), spoken by all scientists.[34] Kuhn disputes the existence of that universal language of objective empiricism and argues that scientists would still interpret the data in the context of their paradigms; the resulting theories would sometimes be "incommensurable".[35]

This, however, must not necessarily lead to relativism[36] (as Kuhn feared): two incommensurable theories concerning the same subject matter could be compared, even though they would come to radically different solutions. In that case the theory

32 One such shift, relevant to the history of paleontology, is the abandonment of the theory that fossils were merely games, or tricks, of nature ("lusus naturae"), and the realization that some colossal, petrified bones were not the remains of biblical giants, or angels, but belonged to extinct but very real mammals (see chapter 2. 2.).

33 Kuhn: The Structure of Scientific Revolutions, pp. 66–67, 110–134. See also: Barry Barnes: Scientific Knowledge and Sociological Theory, London 1974, pp. 45–68.

34 Edwin H.-C. Hung: Beyond Kuhn. Scientific Explanation, Theory Structure, Incommensurability and Physical Necessity, Aldershot 2006, pp. 3–5.

35 Hung: Beyond Kuhn, pp. 5–6.

36 Hans Lenk, however, postulates that everything a human being experiences is constructed, altered, and interpreted by the inner workings of the mind. All human activity is conducted in the interpretative context of society and the human brain: "Leistung und Eigenleistung sind immer interpretatorisch konstruiert und nur so zu erfassen." (p. 18). All production of knowledge happens within a cycle of cognition: the real world alters the thoughts and expectations which in turn alter the way the world is perceived. See Hans Lenk: Einführung in die Erkenntnistheorie. Interpretation – Interaktion – Intervention, Munich 1998. Another example of inescapable subjectivism is the argument that the way one studies his or her surroundings is itself dictated by cultural predispositions. One learns to group similar phenomena into wider categories: if a duck is a bird with a rounded beak, living in the water, it is therefore easy to assume that it also possesses webbed feet. The presumption does not necessarily mean that one knows through personal experience that ducks have webbed feet but that it is assumed that water dwelling birds have webbing between their toes. These conceptual frames are convenient but may be formed by paradigms and not necessarily by objective observation. See Hanne Andersen et al.: The Cognitive Structure of Scientific Revolutions, Cambridge 2006, https://doi.org/10.1017/CBO9780511498404, pp. 19–33.

that predicts future, or not yet studied phenomena, more accurately would be the superior one.[37]

The constructivist turn that had developed since the 1970s runs contrary to the methods and self-image of natural scientists, who seek to investigate an objective truth and reality. In the 1980s, Latour followed and expanded on this trend to which scientists often responded in a dismissive or hostile manner.[38] The internal pressures of scientific conventions can alter the production of science. Traditional conventions and accepted rhetoric and stylistic devices are often prerequisite to granting a new scientific thesis some authority.[39]

If someone dedicates their life to scientific pursuits, it is of great importance for that person to be able to present a great résumé: it helps to have been taught by the most brilliant or at least most prevalent professors, to have enrolled at the most prestigious institutions, and to publish. These prerequisites exemplify some of the external social pressures that impacted the work of scientists of the nineteenth century, and still impact scientific work today.[40] Besides economic necessities, a successful scientist is also rewarded with professional (and at least in this sense social) authority in the scientific community. This influence can in turn be reinvested into the production of science.[41] Katrin Knorr-Cetina even compares these social pressures to natural selection since the theories are constructed in accordance with the requirements of the scientific community: "Like adaptation, [professional] acceptance can be seen as the result of contextual pressures which come to bear on the scientists' selections in the environmental niches provided by the laboratories."[42]

Not only do external factors influence the production of science, new scientific discoveries change worldviews and social dynamics. Chemists might literally change the face of the earth (through the invention of new fertilizers or explosives, for example); astronomers (or paleontologists) might change belief systems and destroy old

37 "There are common measures between incommensurable theories. They can be compared in at least two ways: Given a set of phenomena, which of the two theories explains more members of the set, and which of the two explains these members better? […] As incommensurable theories can share phenomena as their explananda, it can be seen that interpretation ladders need not be linear: they can have branches." Hung: Beyond Kuhn, p. 134.

38 Matthias Wieser: Das Netzwerk von Bruno Latour. Die Akteur-Netzwerk-Theorie zwischen Science & Technology Studies und poststrukturalistischer Soziologie, Bielefeld 2012, pp. 17–26.

39 Wieser: Das Netzwerk von Bruno Latour, pp. 26–29.

40 See Latour for further and detailed elaboration: Bruno Latour; Steve Woolgar: Laboratory Life. The Construction of Scientific Facts, 2nd ed., Princeton, NJ 1986, pp. 187–233. See also: Barnes: Scientific Knowledge, pp. 117–120.

41 Karin D. Knorr-Cetina: The Manufacture of Knowledge. An Essay on the Constructivist and Contextual Nature of Science, Oxford 1981, pp. 68–87.

42 Knorr-Cetina: The Manufacture of Knowledge, p. 9.

dogmas through their discoveries and theories. Both aspects do in return change the way scientific knowledge is manufactured by altering the scientists' frames of reference.[43]

The scientific community upholds certain social and professional standards and is organized into communication networks. Foucault may have called these networks discourse communities: hierarchically organized and with their own rules, vocabulary, and communication rites.[44] These networks organize into schools of thought and compete for reputation and influence over the general public and potential patronage. Therefore, the exploration of governmental science policies is an integral part of the history of science.[45] Some of this prestige can be measured in the prizes and medals issued by governments or scientific associations. This system of non-scientific rewards often leads to the evaluation of a scientific school or even discipline with little – if any – consideration of the scientific achievements of said school or discipline. It is only dictated by outside perceptions, practical applicability of theories, or political calculation. This allows outside forces a certain control over the scientific process.[46] The distinct character of the scholar (determined by genetics and environment) also influences the direction of the research. Brilliant scholars might be held back or driven in other directions of inquiry through political or economic pressure.[47]

Another revolutionary aspect of the production of science, as explained by Kuhn's revolutions, is the separation of the natural sciences into independent disciplines: when two paradigms concerning the same subject grow too far apart for any meaningful scientific exchange, specialization can set in and a wholly new discipline can be born. Biology, for example, was spawned when medicine and paleontology diverged in parts from geology.[48] Laitko and Guntau describe this process not as a simple addendum to a preexisting system but as a radical restructuring of the system of scientific

---

43 Ian Hacking: Working in a New World. The Taxonomic Solution, in: Paul Horwich (ed.): World Changes. Thomas Kuhn and the Nature of Science, Cambridge, MA 1993, pp. 276–310, see pp. 280–283.

44 Michel Foucault: Orders of Discourse. Inaugural Lecture Delivered at the Collège de France, transl. Rupert Swyer, in: Social Science Information, vol. 10, no. 2 (Apr. 1971), pp. 7–30.

45 Ina-S. Spiegel-Rösing: Wissenschaftsentwicklung und Wissenschaftssteuerung. Einführung und Material zur Wissenschaftsforschung, Frankfurt/Main 1973, pp. 13–24.

46 Spiegel-Rösing: Wissenschaftsentwicklung und Wissenschaftssteuerung, pp. 37–51.

47 See Mohr, who is a compassionate defender of this constructivist approach: Hans Mohr: Subjektivität in den Naturwissenschaften, in: Hans Radermacher (ed.): Aktuelle Probleme der Subjektivität, Bern 1983, pp. 75–90.

48 Jouni-Matti Kuukkanen: Revolution as Evolution. The Concept of Evolution in Kuhn's Philosophy, in: Vasso Kindi; Theodore Arabatzis (eds.): Kuhn's The Structure of Scientific Revolutions Revisited, New York 2012, pp. 134–152, see pp. 134–139.

disciplines itself: a revolutionary act.[49] They call scientific disciplines self-revolutionizing systems,[50] and introduce another social factor that motivates the production of science: the sourcing of raw materials. In the eighteenth and nineteenth century there was great interest in the exploration and exploitation of natural resources like coal and metals. In turn, the interest in geology and mineralogy began to soar, and with it the public reputation of experts in these fields. This social recognition meant a boost in attention, funds, and new potential scholars. Not the scientific search for knowledge initiated this boost, but practical, economic, and ideological interests.[51] In Victorian England the fascination with geology grew so strong that fossil-collecting became some sort of national pastime. Many of the hobby-geologists became big players in this "heroic age" of geology (see Chapter 2.5.).

To give another example and argument for the constructivist analysis of a scientific process: Baconian science is in itself a prominent example for a scientific revolution, influenced by the cultural parameters of its time. Francis Bacon thought that human progress would be achieved through learning and technical advancement. His god was in the details; one could observe aforesaid detail and deduct the true composition of the big picture (maybe one early example for what Foucault calls "will to truth"). The observation should be conducted objectively and through experimentation and, if possible, unrestricted by policy and religious dogma.[52] This school of thought was probably influenced by the unsettling times Frances Bacon lived in; the early seventeenth century was a time of great religious and social upheaval. One could find security in the study of the minor parts, the details. Nature seemed to be a fine-tuned machine when reduced to its individual components. Even before Bacon, René Descartes proposed that the universe and all its beings might work like a machine. Plants and animals were sheer automatons, as was the human body (all of them created by a most genius god). The self-reflective human mind however, capable of abstract thought, made humans something more than functional robots. This was the foundation for a dualistic worldview in which (human) mind and body could be separated, a paradigm

49 "Die Entstehung einer neuen Disziplin darf man sich nicht als äußerliches Hinzufügen eines Bausteins zu einem Gebäude vorstellen, sondern als Resultate des Strukturwandels eines ganzen Systems bereits vorhandener Disziplinen." Martin Guntau; Hubert Laitko: Entstehung und Wesen wissenschaftlicher Disziplinen, in: Martin Guntau; Hubert Laitko (eds.): Der Ursprung der modernen Wissenschaften. Studien zur Entstehung wissenschaftlicher Disziplinen, Berlin 1987, pp. 17–89, see p. 25.

50 "Disziplinen sind selbstevolutionierende gegenstandsorientierte Systeme wissenschaftlicher Erkenntnistätigkeiten." Guntau; Laitko: Entstehung und Wesen wissenschaftlicher Disziplinen, p. 44.

51 Guntau; Laitko: Entstehung und Wesen wissenschaftlicher Disziplinen, p. 61.

52 Richard G. Olson: Science Deified & Science Defied. The Historical Significance of Science in Western Culture, vol. 1: From the Bronze Age to the Beginnings of the Modern Era, ca. 3500 B.C. to ca. A.D. 1640, Berkeley, CA 1982, pp. 278–290.

shift that enabled the scientific revolution.[53] But abstractions are not automatically scientific: a cat hunts a leaf, blowing in the wind, presuming it is some sort of prey. Humans frequently project their own emotions onto animals or objects; this might lead to animism, the idea that objects have agency of their own (or possess "souls"). This projection often happens automatically and unconsciously. The abstractions are hierarchically organized by the mind: one might mistake a rock formation for a bear because over the course of human evolution bears generally posed a greater and more immediate danger than rock formations. Because most humans are social beings, self-organized into societies, the ability to interpret facial expressions seems to be of crucial importance and happens instantly and instinctively. This is not only the reason for seeing faces in the clouds (or on a piece of toast), but also the basis for numerous jump scares in movies.[54]

1.2.2 Scientific Networks

One way of approaching the circumstances that may have influenced the production of science is to analyze the letters of scientists. From the late sixteenth century onwards, scholarly exchange was mostly conducted via mail. It was not only a prerequisite to discussing new findings and theories but also often the only way to obtain data.[55] Science was in large parts produced within these networks and therefore their importance for this thesis cannot be overstated; or as Browne puts it: "Letters exchanged between contemporaries were not just a daily fact of life but comprised a great deal of what it meant to be a man of science."[56]

Nicoline Scheidegger defines such a network as follows: a network is composed of independent participants, who interact to achieve certain objectives. The cooperation of these contributors is mostly decentralized; the decision-making-powers are often shared by more than one of the participants.[57] The network furthermore con-

53 Richard G. Olson: Science Deified & Science Defied. The Historical Significance of Science in Western Culture, vol. 2: From the Early Modern Age through the Early Romantic Era, ca. 1640 to ca. 1820, Berkeley, CA 1990, pp. 15–41.

54 Stewart Elliott Guthrie: Faces in the Clouds. A New Theory of Religion, New York 1993, pp. 39–48.

55 Janet Browne: Corresponding Naturalists, in: Bernard V. Lightman; Michael S. Reidy (eds.): The Age of Scientific Naturalism. Tyndall and His Contemporaries, London 2014, pp. 157–170, see pp. 157–160.

56 Browne: Corresponding Naturalists, p. 159.

57 „Netzwerke bilden Konfigurationen mit weitgehend autonomen Komponenten, die aber in selektiver Weise dauerhafte Beziehungen eingehen, um beispielsweise gemeinsame Projekte zu koordinieren. Ein Netzwerk stellt somit eine spezifische Form dezentraler und horizontaler Kooperation dar, bei der die organisationalen Entscheidungsrechte und Eigentumsrechte über die beteiligten Netzwerkpartner verteilt sind." Nicoline Scheidegger: Der Netzwerkbegriff zwischen einem Konzept für Han-

sists of social interactions which are the basis for the network-analysis. This analysis presumes that the relationships of the participants to one another and their position within the network in turn influence the decision-making and even perceptions of the participants.[58] In his Actor-Network-Theory, Bruno Latour divides scientific networks into "intermediaries," which can be persons or committees. The input of these intermediaries constitutes the output of the network. A "mediator" is an actor defined by their ability to radically transform the output, or even identity of the network.[59] Even inanimate objects can be interpreted as participants in networks: scientific instruments, for example, influence the perception of the studied subject (in the case of microscopes literally) and can greatly impact the production of knowledge. Because the instrument has no agenda of its own, it can be seen as an expression of the other intermediaries (a hammer, for example, has no interest in hammering a nail, it is an expression of the desire of the craftsman). Still, instruments and tools play an important part in the manufacture of science, and should not be overlooked.[60] Objects themselves can be imbued with meaning or vary in significance with changing scientific and social evaluation: a fossil is just an unusual rock until one realizes that it is the petrified remains of a life form.[61] The actions and positions of the intermediaries are always subject to outside influences, although these are often hard to identify due to their subtle nature.[62] The intermediaries are not bound together by some mystical force but by a common interest, and often by pure necessity.[63]

Bernd Kortländer studies the transfer of culture between nations (via private individuals); he subdivides the transfer into three distinct steps: the selection, the transportation, and finally the integration. The selection of what is to be transferred derives from an interest. This interest can be academically, practically, or politically motivated.[64] The transportation of knowledge is often hard to reconstruct, it happens on a personal level via written or verbal (and then usually undocumented) communication. Certain hindrances of communication, like cultural misconceptions or lan-

dlungskoordination und einer Methode zur Untersuchung relationaler Phänomene, in: Sabrina Kulin et al. (eds.): Soziale Netzwerkanalyse. Theorie, Methoden, Praxis, Münster 2012, pp. 41–51, see p. 43.

58 Scheidegger: Der Netzwerkbegriff zwischen einem Konzept für Handlungskoordination und einer Methode zur Untersuchung relationaler Phänomene, p. 48.

59 Bruno Latour: Reassembling the Social. An Introduction to Actor-Network-Theory, Oxford 2005, pp. 39–42.

60 Latour: Reassembling the Social, pp. 70–74.

61 Latour: Reassembling the Social, pp. 106–115.

62 Latour: Reassembling the Social, pp. 46–50.

63 Latour: Reassembling the Social, pp. 64–70.

64 Bernd Kortländer: Begrenzung – Entgrenzung. Kultur- und Wissenschaftstransfer in Europa, in: Lothar Jordan; Bernd Kortländer (eds.): Nationale Grenzen und internationaler Austausch. Studien zum Kultur- und Wissenschaftstransfer in Europa, Tübingen 1995, pp. 1–19, see pp. 6–7.

guage-skills, may alter the transfer and must be analyzed as well.[65] The integration can be split into three categories. In the first category, the transferred culture is imitated without alteration; in the second it is emulated but still distinctly foreign; and finally, it is transformed through native cultural norms. The last modus is the most common, but also the toughest to spot.[66] Because science is, at least in part, subject to cultural influences and expectations, Kortländer's methods may be used to analyze some aspects of the scientific exchange. The "German University" (e.g., German scientific methods, instruments, and institutions) of the nineteenth century was a role model for the reformation of higher learning in the United States during the late nineteenth century. The adaptation of new methods and institutions might be interpreted as the import of some aspects of German culture.[67]

In summary: In accordance with Kuhn and his successors, the analysis of the social and cultural premises that the production of paleontological knowledge is based on will serve as the foundation for this study. Because from the very beginning of the earth sciences geologists and paleontologists were linked with each other in international communication networks, some elements of Latour's Actor-Network-Theory are frequently used as additional methodological tools in this thesis.

1.3 State of Research

The conflict termed "Bone Wars" which erupted between the two most prolific and well-known US paleontologists during the late nineteenth century is one of the reasons why the history of US paleontology is as popular as it is. The term "Battle of the Bones" dates at least as far back as 1964, when Nathan Reingold used it to describe the conflict between Othniel Charles Marsh and Edward Drinker Cope.[68]

The first comprehensive Cope biography was written by Henry Fairfield Osborn (1857–1935) in 1931.[69] Osborn was a disciple and personal friend of Cope's. He fought side by side with Cope and Marsh's former assistants against Marsh during the most public battles of the "Bone Wars." His view on the conflict and on Marsh are very par-

65 Kortländer: Begrenzung – Entgrenzung, pp. 7–8.

66 Kortländer: Begrenzung – Entgrenzung, p. 8.

67 Kortländer: Begrenzung – Entgrenzung, pp. 16–17.

68 Nathan Reingold: Science in Nineteenth-Century America. A Documentary History, New York 1964.

69 Henry Fairfield Osborn: Cope. Master Naturalist. The Life and Letters of Edward Drinker Cope, Princeton, NJ 1931.

tisan, and favor Cope's position. The only other comprehensive work on Cope's life was written by Jane Davidson in 1997.[70]

The only comprehensive biography of Marsh to date was written by Clara Mae LeVene and Charles Schuchert (1858–1942) in the late 1930s. Schuchert knew Marsh personally, he worked preparing fossils and as a curator in the Yale Museum, held the chair for vertebrate paleontology in 1904, and was director of the Peabody Museum at Yale between 1904 and 1922; both positions were first held (and established) by Marsh.[71] Schuchert states that he decided to write the book after reading Osborn's "Cope: Master Naturalist," and calls his book a labor of love, expressing his wish that Marsh may claim his rightful place in the history of paleontology.[72] LeVene's and Schuchert's book might not be very objective, or neutral, but it is more objective then Osborn's book on Cope. According to Ronald Rainger, Charles Schuchert was "highly critical of Osborn, his ambitions, and his efforts to dominate work in paleontology."[73] Rainger judges that Osborn had the personal objectives of "undermining Marsh and sustaining the legacy of Cope"[74] Davidson writes on Schuchert and his relationship with Osborn:

> Schuchert's biography was for the most part pretty evenhanded. He had completed it, no doubt, in partial response to the biography of Cope written by Osborn and published in 1930 [sic!], Cope: *Master Naturalist*. The two protégés, now both important paleontologists in the twentieth century, took it upon themselves to defend their mentors.[75]

Jane Davidson, whose works are cited frequently throughout this thesis, is the only historian researching and publishing about Marsh, Cope, and this chapter of US-American paleontology in an academic manner. This thesis endeavors to expand this field of study by incorporating the perspectives of Marsh's so-far neglected German assistants. The analysis of their correspondence also adds nuance to Marsh's scientific network and the evolution of US paleontology. Almost all the information on them has been reconstructed from the primary sources or consist of a few scattered passages in the secondary literature. There is little scholarly work on Karl Alfred von

70 Jane Pierce Davidson: The Bone Sharp. The Life of Edward Drinker Cope, Philadelphia 1997.

71 For more information about Schuchert, see: Adolph Knopf: Charles Schuchert 1858–1942, Washington, DC 1952.

72 Charles Schuchert; Clara Mae LeVene: O. C. Marsh. Pioneer in Paleontology, reprint, New York 1978 (orig. publ. 1940), pp. xiii–xiv.

73 Ronald Rainger: An Agenda for Antiquity. Henry Fairfield Osborn & Vertebrate Paleontology at the American Museum of Natural History, 1890–1935, Tuscaloosa, AL 1991, p. 87.

74 Rainger: An Agenda for Antiquity, p. 242.

75 Davidson: Patrons of Paleontology, p. 133.

Zittel (1839–1904), a German paleontologist and correspondent of Marsh's. His role within Marsh's scientific network will also be subject of this thesis.

Other books covering the history of US paleontology have also been published and are often addressed to a larger audience,[76] some are purely entertainment literature.[77] Finally, David Rains Wallace's "The Bonehunter's Revenge"[78] and Mark Jaffe's "The Gilded Dinosaur"[79] are very comprehensive and detailed accounts of the "Bone Wars" and served as inspirations for this thesis. They are accessible and entertaining popular literature on the topic, yet they are not adequate secondary sources for this thesis.

To this date no publications shed more light on the relationship between Marsh and his assistants. This thesis suggests that the aforementioned relationship influenced both the "Bone Wars" and the development of US-American paleontology (Marsh specifically looked for German assistants).

1.4 Archives and Sources

Besides various articles published in (mostly US-American) scientific journals, correspondence between the scientists and some other ego documents comprise the sources for this thesis.

The main bulk of said ego documents is archived at Yale University, at the Sterling Memorial Library in the Othniel Charles Marsh papers (MS 343). This collection was

76 See for example: Thom Holmes: Fossil Feud, Parsippany, NJ 1998; Elizabeth Cody Kimmel: Dinosaur Bone Wars. Cope and Marsh's Fossil Feud, New York 2006. Noticeable examples for the popular scientific impact the "Bone Wars" have had are also documentary films, see for example: Mark Davis (director): Dinosaur Wars (film), in: American Experience (series), PBS 2011, http://www.pbs.org/wgbh/americanexperience/films/dinosaur/, as consulted online on April 10, 2020.

77 See for example: Sharon N. Farber: The Last Thunder Horse West of the Mississippi, in: Isaac Asimov's Science Fiction Magazine, vol. 12, no. 11 (Nov. 1988), pp. 20–44; Michael Crichton: Dragon Teeth, New York 2017; the "Bone Wars" even inspired a graphic novel: Jim Ottaviani: Bone Sharps, Cowboys, and Thunder Lizards: Edward Drinker Cope, Othniel Charles Marsh, and the Gilded Age of Paleontology, Ann Arbor 2005, MI. Another very interesting example for the popularity of the "Bone Wars" is a card game, which casts the players in the roles of rivaling paleontologists, competing for fossils in the "Wild West." The advertising text on the publisher's website reveals that the game was inspired by the history of US paleontology: "Most of the wild events in BONE WARS are based on things which actually happened. Edward Drinker Cope really did mount an elasmosaur's head on the wrong end of the skeleton, and then tried to buy up and destroy all the copies of the journal describing it. O. C. Marsh really did use his personal fortune to hire away all the best fossil collectors from rivals", http://www.zygotegames.com/bw.html, as consulted online on April 10, 2020.

78 Davis Rains Wallace: The Bonehunter's Revenge. Dinosaurs, Greed, and the Greatest Scientific Feud of the Gilded Age, New York 1999.

79 Mark Jaffe: The Gilded Dinosaur. The Fossil War Between E. D. Cope and O. C. Marsh and the Rise of American Science, New York 2000.

also used by LeVene and Schuchert for their biographical project. This thesis focuses on Marsh's correspondence with his German assistants and with German paleontologist Karl Alfred von Zittel, see chapter 6. Physical copies of Marsh's correspondence, his unfinished autobiography, notebooks, and some memorabilia are stored at the Sterling Library. Between 1940 and 1970 the Marsh papers were divided between the Peabody Museum and the library. In 1970 they were reunited in the library and photographed on microfilm.[80] Said microfilm as well as the digitalization of the correspondence, uploaded on the website of the Peabody Museum,[81] were used for this thesis.

Other extensive sources are archived at the American Museum of Natural History in New York. The most relevant to this thesis, and the most extensively used of the collections, is the general correspondence of the museum (VPA 1/108, General Correspondence), in this case the correspondence between Henry Fairfield Osborn, Zittel, and Marsh's assistants. These letters have never been published by a historian before and therefore give new insight into the relationship between Marsh and his assistants from their perspective, as well as the relationship between Osborn and the German scientists, who played an important part in mustering them for the final battles of the "Bone Wars."

Others, albeit less comprehensive, sources were consulted in Philadelphia, at the American Philosophical Society (APS) and the Academy of Natural Sciences (ANS) at Drexel University (Collections #1 and #567). The Edward Drinker Cope Papers of Haverford College in Haverford, PA (Haverford College Quaker & Special Collections Edward Drinker Cope Papers, HC.MC-956) were consulted as well to flesh out some of the aspects of the professional correspondence within the paleontological network during the second half of the nineteenth century.

Various visits to archives in Germany proved far less fruitful. Some letters written by Marsh, Cope, Zittel, and Marsh's assistants that are archived at the Berlin State Library (Staatsbibliothek zu Berlin-Preußischer Kulturbesitz) and the Library of the Museum of Natural History in Berlin (Bibliothek des Museums für Naturkunde) provided some material (mostly copies of scientific publications) as well. Some supplementary material was found at the Archive of the Leipzig University (Universitätsarchiv Leipzig) and in the library of the Paleontological Museum in Munich (Bayerische Staatssammlung für Paläontologie und Geologie). Only a few sources concerning Cope, Marsh, and their professional network survive in Germany, a great number of correspondence, archived in Munch, was destroyed by fire during World War Two.

80 Irving N. Fisher: The O. C. Marsh Papers, in: The Yale University Library Gazette, vol. 46, no. 1 (Jul. 1971), pp. 35–40.

81 Cf. https://peabody.yale.edu/explore/collections/vertebrate-paleontology/o-c-marsh-papers, as consulted online on January 28, 2020.

Most of the scientific periodicals keep their own archives, which are conveniently accessible online.

1.5 Quotes and Terminology

The original capitalizations in all quotes from the primary sources were kept in their original state. Necessary comments by the author were added in square brackets. Passages that were originally composed in German were translated into English by the author of this thesis, the original terms are documented in brackets following the translated words. In some cases, usually when whole passages from letters were reworded and adopted for this thesis, the original German passages are listed in a corresponding footnote. Note that the orthography of certain words has changed in the last two centuries, this is especially true for the German language. Again, passages were left in the original and the original orthography has been left unaltered. The most common example of this are differences in capitalization. Common examples for this change in orthography in German words are words nowadays written with a "k" that were sometimes spelled with a "c" in the nineteenth century, for example, "Kollege" (colleague) and "College." Other words were written with a "th," and are now spelled with a simple "t," like "Wirbeltiere" (vertebrates) and "Wirbelthiere." These instances have not been annotated for the purpose of not disturbing the flow of the reading every other sentence. Other instances, frequently encountered but not pointed out in this thesis by "[sic!]" for the sake of readability, are words such as "everyone" spelled as "every one," or "today" and "tomorrow" spelled as "to day," "to-day," or even "to morrow."

The term Native American will be used in this dissertation to refer to a great number of very different cultures. It is preferable to outdated terms like "Indians" for various reasons, the first and foremost among these being the racist connotations. Robert Berkhofer Jr. delivers a thorough treatise on the subject and the history of the terminology in his book (and yet still continues to use the term throughout his book).[82] Or, as Gerald Viznor puts it:

> The *indian* is the absence, natives the presence, and an absence because the name is a discoverable, and a historical simulation of distinct native cultures. Columbus warred, scored, rocked, talked, and coveted the other, and so we come around five centuries later to say, You made a mistake, and how ironic

82 See: Robert F. Berkhofer: The White Man's Indian. Images of the American Indian from Columbus to the Present, New York 1978, pp. 3–31.

> your discovery. Surely, five centuries as a discoverable is enough of victimry. […] The point is that we are long past the colonial invention of the *indian*. […]
>
> The *indian* is the invention, and *indian* cultures are simulations, that is, the ethnographic construction of a model that replaces the real in most academic references. Natives are the real, the ironies of the real, and an unnamable sense of presence, but simulations are the absence, and so the Indian is an absence, not a presence. You see, Indians are simulations of the discoverable other, and only posers or the naïve dare stand with that ironic name. That is to say, the simulations of the other have no real origin, no original reference, and there is no real place on this continent that bears the meaning of that name. The *indian* was simulated to be an absence, to be without a place. The reference of the simulation is a weak metaphor of colonialism and, of course, manifest manners.[83]

Also note that the term "Germany" will be used throughout the thesis synonymous with "German speaking region," though that would be the technically correct term before the founding of the German Empire in 1871.

1.6 Thesis Outline

Chapter 2 will provide an overview of the origins of paleontology in Europe and North America until c. 1870. For millennia humans had been finding the fossilized remains of strange animals and interpreted them within their own cultural and societal frameworks. The chapter describes in all briefness how conceptions about the Earth's past and its inhabitants were formulated in ancient Greece and Rome. The next part of the chapter describes how fossilized bones were then interpreted as ancient heroes, giants, or unicorns in the Middle Ages. The chapter will also describe how this interpretation changed with the renaissance, depict how the first international scholarly correspondence circles emerged, and summarize the role the establishment of the so-called cabinets of curiosities played in making proto-scientific findings publicly available and in laying the foundation for the establishment of the museum. The next short subchapter will focus on the scientific revolution of the seventeenth century. During this period new and innovative interpretations of the fossils emerged, along with a whole new conception of the world and the universe. Although still very much in accordance with the Christian Bible and a generally religious outlook, the world and all processes therein were now interpreted to be of a more mechanical,

83 Gerald Vizenor; Arthur Robert Lee: Postindian Conversations, Lincoln, NE 1999, pp. 84–85.

less mythical manner. Exact observation and the gathering of seemingly objective facts would define scientific conduct. The existing correspondence networks were built upon and became more formalized, a Republic of Letters was established, and so were scientific societies like the Royal Society in London or the French Académie des sciences in Paris. Theses societies began publishing new insights and findings in their journals, establishing the scientific periodical. New theories about the origins of the planets were imagined, and fossils were, for the first time, recognized as the remains of (non-mythical) animals when Nicolaus Steno compared the heretofore enigmatic "tongue stones" to the teeth of a shark. The next subchapter focuses on the eighteenth century, when well-off gentlemen of great general knowledge studied fossils and rocks. The cold rational of the scientific revolution was – at least partially – replaced with a romantic view of what was then called natural history, imbuing nature with a divine beauty. Then again, all flora and fauna were categorized and organized into a logical system, propagated by Carl Linnaeus and his disciples. The early nineteenth century, then, will be examined in greater detail in the penultimate subchapter. During this time, the geosciences really came into their own. In France, Georges Cuvier (1769–1832) perfected the method of comparative anatomy, matching fossilized bones to those of still living animals. He also realized that at least some of the fossils must belong to animals that had gone extinct, and theorized that extinctions happened en masse in great extinction events, which he called – and this is rather telling of the societal circumstances of Cuvier – revolutions. It was during this time that the term paleontology was coined, and the geosciences subsequently split into geology and paleontology. In Victorian Great Britain, collecting fossils became a popular pastime. A wholly new order of animals was described by Gideon Mantell, Richard Owen, and others. These were (at least in Victorian imagination) fearsome, dragon-like creatures from a strange age and accordingly they were called the terrible lizards, or dinosaurs. This time was also characterized by the gentleman hobby scientist, who had received little formal education and concerned himself with a variety of scientific subjects. Public museums were built to present the newest scientific findings, and in 1854 life-like reconstructions of dinosaurs and other extinct creatures were publicly exhibited in London. The conduct of science and higher education was reformed, and newly established universities and laboratories in Germany would claim a pioneering role. The final subchapter will explain how paleontology was established in North America. It will describe how American fossils were discovered for the first time by Europeans (Native Americans had known about them for a long time before then, but that is the subject of chapter 2. 6.). These fossils were brought to Europe and studied there. The chapter will also outline how US-American paleontology was linked to nationalism from the very beginning, when Thomas Jefferson (1743–1826) employed fossils in his battle for the honor of the young Republic against European biases. It will also describe how Joseph Leidy (1823–1891) of Philadelphia became the first professional American paleontologist of international renown and how he erected the first dinosaur skeleton, which proved to be a major attraction to the public.

Chapter 3 will go into greater detail concerning the genesis of US-American paleontology, focusing on one of the most prolific paleontologists, and the first professor for paleontology in the United States, O. C. Marsh. The chapter opens with a short description of Marsh's upbringing, his early education, and the beginning of his scholarly career at Andover Academy and Yale. The next subchapter is dedicated to Marsh's journey through Europe, his experiences with various leading German scientists and universities, how he began building a professional scientific network of his own, and how he finally decided to focus his professional ambitions on paleontology. Then a subchapter will analyze his correspondence with his former teachers in the years to come, and how knowledge was transferred within this network. The final subchapter will focus on the importance of the patronage of Marsh's famously wealthy uncle George Peabody for his nephew's career, further illustrating how circumstance – not genius – decisively influences the conduct of science.

Chapter 4 will provide an overview of the "Bone Wars," a decade long conflict between Cope and Marsh that greatly affected the development of US-American paleontology, and also had many consequences for Marsh's relations to other scientists, including his own assistants. The rivalry will be sketched out in all briefness, for many other monographies and articles focus on the very well documented affair that inspired many scholarly and fictional works.

Chapter 5 analyses how paleontology was linked to US-American nationalism and one of the most formative periods for US nationalism, the "winning of the West." It was then that paleontology really was Americanized. First the "discovery" of the west of the North American continent by US-American explorers and scientists will be described and the creation of the "frontier myth" will be examined. The second subchapter describes Marsh's expeditions to the West and how he attempted to tie his own image to the "frontier myth." Then the importance of the "Bone Wars" in relation to the overarching theme of this chapter will be discussed. Another subchapter focuses on the contribution of Native Americans to the genesis of US-American paleontology. They knew about the existence of many of the fossils and had interpreted them for centuries within their own cultural frameworks. Now US-American explorers were employing Native Americans as scouts and guides on their own fossil-hunting expeditions. Simultaneously, Native Americans were also painted as dangerous obstacles in the way of scientific progress, a constant threat to the expeditions. Finally, Native Americans were likened to the extinct fauna itself, positioned as "creatures" from another more primitive and long-lost age, a people now doomed to become extinct themselves. A small subchapter details Marsh's relationship to chief Red Cloud (Maȟpíya Lúta in Lakota), a personal friend of Marsh's. It examines how Marsh used Red Cloud to further his own public image, and how Red Cloud used his contacts with

Marsh and other influential US-Americans for his own political agenda, and to improve the life of his people. The final subchapter focuses on Buffalo Bill, a self-made living legend of the "Wild West," and the guide of Marsh's 1871 expedition. It will detail Marsh's attempt to link himself to Buffalo Bill and focus on the rise of US-American popular culture, the emergence of the "frontier myth," and on how paleontology was linked to this grand national narrative.

Chapter 6 will go into more detail on how US-American paleontology came to overshadow its European counterpart by the end of the century. While the main reason for this can be found in the abundance of fossils in the west of the North American continent, the relevance of Marsh's fossil collection as proof of Charles Darwin's (1809–1882) theory of evolution constitutes another important factor. The chapter details how this involvement in a highly topical, controversial, international discourse launched the international renown of US-American paleontology. It also sheds some light on Marsh's correspondence with Darwin himself, but more so with Thomas Henry Huxley (1825–1895) and his family.

Chapter 7 is the heart of this thesis. Here the knowledge transfer within Marsh's personal correspondence and his work-related network is the focus. The first subchapter describes the relationship between Marsh and Karl Alfred von Zittel, one of Germany's leading paleontologists at the time. Zittel's opinion on the North American continent, its landscape, and its geological potential will be discussed. Furthermore, the correspondence between Marsh and Zittel will be analyzed, for Zittel and Marsh exchanged not only paleontological publications, but Zittel also tried to acquire some fossils for Marsh in Germany and put Marsh in contact with Max Schlosser (1854–1932) and Georg Baur (1859–1898), two German paleontologists and alumni of the University of Munich, where Zittel held his professorship. The two young German scientists embarked for New Haven and became Marsh's assistants. Additional light will be shed on the relationship between Zittel and Henry Fairfield Osborn, a pupil of Cope's and himself, and an important US-American scientist. The next subchapter will detail the working relationship and personal grievances between Marsh and Baur, for the latter remained within the US-American system of higher education and stayed at Yale for a few years, working for Marsh. But the relationship soured almost instantly, when Baur realized that his salary would hardly cover his expenses and soon saw himself indebted to his employer. Furthermore, he felt used by Marsh, who allegedly published Baur's work under his own name. This brought Baur into contact with Cope and Osborn, Marsh's bitter rivals in the "Bone Wars." Osborn and Cope employed the statements of Marsh's disgruntled German assistants in a public attack Cope waged against Marsh in a series of newspaper articles in 1890 (see below). The next subchapter writes in all briefness of Max Schlosser and his time at Yale, for he left soon after he had arrived and returned to Munich, where he became a renowned paleontologist and the successor of Zittel. The penultimate subchapter tells of Otto Meyer (born in

1856), a third German, who was hired by Marsh in 1884. He had heard about Marsh's desire to hire German scientists and applied for a position. He also left New Haven disappointed and enraged by Marsh's treatment. Meyer remained a paleontologist for a few years, but it appears that he did not continue to work in science after 1895. In 1890 Meyer stated his case against Marsh in the New York Herald, where Cope started his most public attack. This attack is the subject of the last subchapter of chapter six. These newspaper articles are analyzed and their consequences for Marsh summarized.

Chapter 8 puts the emergence of US-American paleontology within the context of the broader effort to reform the system of higher education in the United States. This education reform was propagated by many men who had studied in Germany and were familiar with the "German University" and the laboratories and techniques employed by German scientists. First the conception of a modern German system of higher education, which developed during the eighteenth and early nineteenth century is discussed. The University of Berlin is often cited as the best example for this new type of "German University." This development goes hand-in-hand with the so-called "Humboldtian Ideals," which were supposedly implemented in Berlin and served as a role model for would-be reformers of higher education around the world, and in the United States. Therefore, the following subchapter details the state of higher education in the US and its evolution and reform during the late nineteenth century. The final subchapters focus on the role of the natural sciences within this process, as well as on developments in the field of public education via institutions such as the museum.

A concluding look will be taken at the questions raised at the beginning of this thesis, summarizing the findings and providing a further outlook into the area of research.

2

Euro-American Paleontology before 1870

To provide context for the scientific achievements of the German and US-American paleontologists described in later chapters, this chapter entails a brief synopsis of the history of paleontology. The evolution of German, French, English/British, and American sciences is of particular interest for this study. The cultural (non-scientific) influences on the production of knowledge are highlighted to illustrate the long chain of external influences on the genesis of the discipline paleontology in the second half of the nineteenth century. The particulars of American paleontology after c. 1865 are discussed in chapter 2. 6.

In the following chapters the modern terms "scientist," "geologist," and "paleontologist" are used to describe protagonists who might have called themselves "natural philosopher," "naturalist," or simply "scholar," and might even have taken offence at being called "scientist." The umbrella term "scientist" was coined in 1834 and remained a matter of debate and controversy throughout the nineteenth century.[84] While the terms "geology" and "paleontology" can be dated back to 1657 and 1822 respectively, the subject matters of both disciplines were examined by scholars throughout human history. This study adapts the modern terminology for the sake of tangibility.

The word paleontology derives from the Greek words *palaios* (old, ancient), *on/ontos* (creature, being), and *logos* (study, thought), describing the subject matter of paleontology: the study of ancient beings. Because most lifeforms studied by paleontologists became extinct several million years ago, the fossilized remains of animals, plants, and fungi are the only means of reconstructing prehistoric life on earth. Some of the more intact fossils allow even the casual observer a glimpse of a spectacular and lost world, truly alien to modern humans. No wonder some traces of ancient life inspired the imagination of their observers, who interpreted the fossils long before there was a science called paleontology (or any scientific method for that matter). Note that before the nineteenth century nearly every object that had been excavated, all minerals, gems, and all sorts of curious rocks, were called fossils; this study uses the term "fossil" according to its modern meaning, exclusively describing the petrified remains of organisms.

84 Ursula DeYoung provides a short but concise conceptual history of the term "scientist," complete with the current state of research. Ursula DeYoung: A Vision of Modern Science. John Tyndall and the Role of the Scientist in Victorian Culture, New York 2011, pp. 6–10.

2.1 Antiquity: Theories about Earth's Past

Even prehistoric humans collected fossils. Archeological findings suggest that fossils have been traded for as long as 30,000 years, and over great distances.[85] During classical antiquity the petrified remains of huge extinct vertebrates were interpreted as the remains of mythic monsters or heroes of ancient Greece, and even might have inspired some aspects of ancient mythology. For example, science historian and folklorist Adrienne Mayor makes a case for how the skulls of *mammoths* might have been interpreted as the remains of legendary cyclopes.[86] Mayor elaborates on how Plutarch (c. 46–120) assumed bones of the extinct elephant *mastodon* to be the skeletons of fabled war elephants that fought the no less mythic Amazons in the service of the god Dionysus.[87]

While some cunning observers like Herodotus (c. 484–425 BCE) identified fossilized shark teeth correctly, others believed them to be the tongues of snakes or dragons (and consequently, for centuries those fossils were called tongue-stones, or *glossopetrae*) and medicinal properties were ascribed to them.[88] According to Mayor, fossils were also studied in ancient Rome: Emperor Augustus (63 BCE–14 CE) for example is said to have acquired a respectable fossil collection, consisting at least in part of the spoils of war from all around the world (or better: the parts of the world ancient Romans had knowledge of).[89] To this day, fossils acquired in this way can become political issues (similar to "Beutekunst," the pieces of art looted by German soldiers and officials during World War II), raising questions of cultural heritage and national ownership. During the Renaissance, the ancient Greek and Roman philosophers and their theories were held in high regard once again. As will be seen, some of the ancient ideas made a comeback in the fifteenth and sixteenth centuries.

Not only paleontological, but also geological theories, similar to those of the scholars of the eighteenth and nineteenth centuries, were formulated in ancient Greece: The "Meteorology" of Aristotle (384–322 BCE), which was held in high regard during the medieval and Renaissance periods, explained that the earth had a fiery core, responsible for the violent eruptions of volcanoes, and that land might have (at least in parts) originated in the ocean, being pushed to the surface by earthquakes. This also explained why some fossils of obviously maritime origin were to be found on dry

---

85 Adrienne Mayor: The First Fossil Hunters. Paleontology in Greek and Roman Times, Princeton, NJ 2001, pp. 154–156.

86 Mayor: The First Fossil Hunters, pp. 3–8.

87 Mayor: The First Fossil Hunters, pp. 54–61.

88 Eric Buffetaut: A Short History of Vertebrate Palaeontology, London 1987, pp. 1–5.

89 Mayor: The First Fossil Hunters, pp. 138–148.

land.[90] Aristotle furthermore theorized that all matter consisted of four elements (fire, air, water, and earth) and shared the qualities of said elements (hot, cold, wet, and dry) in varying compositions. Thus, Aristotle established one of the first chemical theories and contested the common belief that minerals would grow like plants. In the early sixteenth century this chemical theory was rediscovered and refined by the likes of Paracelsus (1493–1541).[91]

The scholars and philosophers of classical antiquity knew of fossils and had various theories about their origins. While the organic origins of fossils were often identified correctly, they were linked to mythic beings like cyclopes and dragons and presumed to have magical and medicinal qualities. Some scholars, first and foremost Aristotle, invented a quasi-scientific theory on how the face of the planet had evolved.

2.2 Middle Ages and Early Modernity: The Emergence of Geosciences

During Late Antiquity and the medieval period, the Christian religion grew to become the dominant cultural force in almost all European societies. The Bible was regarded as the absolute authority on all questions of day-to-day life, morality, and philosophy. For several centuries, all scientific discoveries and theories had to be in accordance with the so-called Holy Scripture. Geoscientific discoveries challenged this religious world view, whether in the observation that life had changed since the days of creation or through the claim that the planet's features had evolved over a tremendous amount of time, an assertion incompatible with the timeline described in the Bible. Due to the chaos and violence of Late Antiquity and the Early Middle Ages, the discoveries of the ancient Greek philosophers had mostly been forgotten in Western Europe. But their geological and paleontological knowledge survived thanks to the Arabian scientific tradition. This knowledge returned to Europe as part of a slow process of cultural exchange, starting at the time of the Crusades. Aristotle's geological notions and explanations for the existence of fossils were again recited in scholarly circles. After the religious dogmatism of the Middle Ages, the production of scientific knowledge soared in early modern times.

Brian Oglive subdivides early modern naturalists into four generations: Members of the first generation (c. 1490–1530s) were almost exclusively from Italy and interest-

90 Frank Dawson Adams: The Birth and Development of the Geological Sciences, New York 1954, pp. 8–28.

91 Rachel Laudan: From Mineralogy to Geology. The Foundations of a Science, 1650–1830, Chicago 1987, pp. 28–35.

ed in medicine and botany. Some of the second-generation naturalists (1530s–1560s) were Northern Europeans but had been educated in Italy. Members of the third generation (1560s–1590s) had commonly studied at a medical academy, built gardens in which they grew plants with practical applications, and were interested in natural history. The discoveries of the third generation were studied and built upon by the scientists of the fourth generation (1590s–1620s). By doing so, they created new taxonomies and frameworks, and determined the way science was conducted for generations to come.[92]

For centuries, the Bible was still held in high regard as a historical document. Because the Bible is filled with biographical timelines, believers used it to determine the exact age of the earth. The most prominent of these chronologists is Bishop James Usher (1581–1656) of Ireland. In 1650, Usher proposed the twenty-second day of October 4004 BCE as the beginning of creation in his "Annales Veteris Testamenti, a prima mundi origine deducti." Most humans tend to imbue all things and processes with meaning, and it appears that therefore the human psyche can hardly imagine a time prior to the existence of humans who give meaning to the surroundings. This tendency might partly explain the existence of religion in general, but it definitely explains why the Christian myth of Genesis begins with the creation of man (or very shortly before that, five days to be precise).[93] This meant that scholars of natural history restricted themselves to the Biblical timeline due to this cultural dogma and could not fully grasp the concept of geological time, spanning thousands or even millions of years.

Many of the ancient theories were summarized and built upon by Georgius Agricola (1494–1555) in his "De Re Metallica," published posthumously in 1556.[94] Agricola was no newcomer to the geosciences; in 1546 he had described six hundred minerals in his book "De Natura Fossilium". Agricola had arranged the minerals into four major categories. This was contrary to the common medieval gemology, which attributed medicinal and magical qualities to various gems and jewels. This gemological knowledge was handed down from generation to generation in extensive collections of voluminous tomes, known as "Lapidaries."[95] Agricola's texts were the first attempts at a scientific methodology within the fields later to be called mineralogy and geology.

Fossils were most commonly identified as *lusus naturae*: jests, or marvels of nature, which either grew on their own, or, in some cases, were placed there by divine

92 Brian W. Ogilvie: The Science of Describing. Natural History in Renaissance Europe, pbk. ed., Chicago 2008 (orig. publ. 2006), pp. 1–24.

93 Martin J. S. Rudwick: Earth's Deep History. How It Was Discovered and Why It Matters,Chicago 2014, DOI: 10.7208/chicago/9780226204093.001.0001, pp. 9–30.

94 Adams: The Birth and Development of the Geological Sciences, pp. 51–76.

95 Laudan: From Mineralogy to Geology, pp. 22–27. See also Adams: The Birth and Development of the Geological Sciences, pp. 137–169.

whim, imitating living beings, or at least parts of their anatomy.[96] Scholars speculated that nature tried to imitate the divine creation, or that fossils were the evidence for a first, flawed attempt at creation, abandoned and buried in rock. Even with the realization that fossils were the petrified remains of living organisms, their origins were explained by referring to the Holy Scripture: Noah's flood was said to be in some way responsible for the distribution of fossils.[97] Alternative theories described how all of creation could be ordered into three categories, namely animals, plants and minerals. Animals and plants grew out of seeds, so subterranean minerals might too, growing unnoticed by human eyes. A third theory transferred the principles of humorism from the human body to minerals.[98] According to this theory there were subterranean veins of fluids (or humors), which were composed of chemical solutions. These chemicals accumulated in the ground due to extreme cooling, or dried due to the immense heat of the planet's inner fire. Other observers noticed how fossils sometimes resembled the organs of other lifeforms and concluded that fossils might reproduce sexually.[99] In short, there was a treasure trove of ideas and (mostly uneducated) speculations about the nature of fossils. Most of these speculations missed the fact that fossils are of organic origin.

Some of the enormous, petrified bones were considered to be the remains of giants which were described in the Bible, legendary Goliath being the most prominent example. During the course of the seventeenth century, the belief in giants and other monsters diminished, paving the way for more reality-based interpretations.[100] Others believed some of the huge bones to be the remains of dragons. Twentieth-century paleontologist and science historian Othenio Abel (1875–1946) for example concluded that the skull of the renowned Dragon of Klagenfurt ("Klagenfurter Lindwurm") was modelled after the skull of a *woolly rhinoceros*, missing its horn. The rhino bones were found in 1335 and most likely kept at Klagenfurt, Austria, where they inspired the imaginative design of a water fountain, sculptured as a water-spewing dragon and mounted in 1593.[101] Later, during the seventeenth century, another dragon design was gaining in popularity: amongst others, Athanasius Kircher (1602–1680) depicted

---

96 Helmut Hölder: Kurze Geschichte der Geologie und Paläontologie. Ein Lesebuch, Berlin 1989, pp. 11–13.

97 Adams: The Birth and Development of the Geological Sciences, pp. 254–263.

98 Humorism is the idea that all bodily and mental functions are determined by a mixture of four body fluids (or humors). The theory originated in ancient Greece and stayed popular in Europe until the sixteenth century.

99 Adams: The Birth and Development of the Geological Sciences, pp. 77–102.

100 For a detailed discussion on the significance of giants in medieval texts see: Alan Lena van Beek: Riesen in der Literatur des Mittelalters. Diskursive Formationen im deutschen Sprachraum, Frankfurt/Main 2021, https://doi.org/10.25716/amad-85226.

101 Buffetaut: A Short History of Vertebrate Palaeontology, pp. 13–18.

dragons as creatures with long, slender necks and pointed wings.[102] The unicorn is yet another telling example of how fossils were understood and culturally reinterpreted. It seems that tales of unicorns resurfaced in Western Europe in the thirteenth century (Greek antique myths about single horned creatures living in India were by then all but forgotten). The fabled creature came to be associated with the Virgin Mary, for it was said that the creature could easily be tamed by a virgin. Due to the connection with Mary, the unicorn then grew to be a popular figure in religious art and heraldry; soon it adorned many a coat of arms, and, most notably, the unicorn became the symbolic figure of Scotland. The *alicorn*, the horn of a unicorn, was said to be an antidote to all poisons, and to have many more medicinal uses. Similar superstitions concerning rhinoceros' horns can still be found today, with most tragic consequences for already endangered rhino populations. Narwhale tusks were often mistaken for the fabulous (and undoubtedly very valuable) *alicorn*. Indian rhinos and sporadic discoveries of prehistoric and now extinct rhinos or the tusks of *mammoths* and other ancient elephants most likely inspired stories of unicorn sightings. In 1663 some fossils (most likely *mammoth* bones) were unearthed in a quarry near Quedlinburg in the German Harz region. The bones were then studied by Otto von Guericke (1602–1686), a politician and scientist of great renown (most famous for his work in vacuum physics). Guericke identified the Quedlinburg fossils as unicorn remains, a diagnosis reinforced two decades later by Gottfried Wilhelm Leibniz (1646–1716), whose reconstruction of Guericke's unicorn was published posthumously in 1749.[103]

The aforementioned Athanasius Kircher contributed in another important way to the evolution of the geosciences: as a Jesuit scholar he was versed in many sciences and became the creator of an extensive cabinet of curiosities ("Wunderkammer"). These cabinets were all the rage in the fourteenth to seventeenth centuries; they stored various oddities from all around the known world, including rare rocks, gems, and fossils (which did not spoil as easily as stuffed animals for example). For the wealthy noble owners, these cabinets were status symbols, and for scholars of natural history they were treasure troves filled with objects to study. Collecting curiosities became more and more popular among aristocrats, princes, and even emperors. This meant that the cabinets were now prestige-objects and had to be presented in a courtly manner. Furthermore, the simple act of collecting and displaying soon was not enough anymore.

102 Abel postulated that this new design was modeled after the skeleton of the long-necked prehistoric marine reptile plesiosaurus, its fins inspiring the pointed dragon wings. Othenio Abel: Vorzeitliche Tierreste im Deutschen Mythus, Brauchtum und Volksglauben, Jena 1939, pp. 180–200. But this seems more than unlikely for plesiosaur skeletons are exceedingly rare and there is no known specimen dating back to the seventeenth century. Erich Thenius; Norbert Vávra: Fossilien im Volksglauben und im Alltag. Bedeutung und Verwendung vorzeitlicher Tier- und Pflanzenreste von der Steinzeit bis heute, Frankfurt/Main 1996, pp. 23–29.

103 Thenius; Vávra: Fossilien im Volksglauben, pp. 29–32.

A collector (or at least the curator of a collection) had to be able to elaborate on his treasures and their supposed origins and implications. Thus, the cabinet of curiosity became a forerunner to the modern museum, another important step in the evolution of science, inspired by social convention.[104]

Before the sixteenth century science was seldom done as a group effort, scholarly endeavors were more or less literary exercises: the theories of ancient Greek philosophers, above all those of Aristotle, were to be learned by heart and internalized. Science was done regionally and in relative isolation, mostly by the humanists of the Italian Renaissance. In the 1530s, scholars started to concern themselves with the production of new knowledge. Plants and their medicinal properties became the focus of scholarly attention. Scholars began to correspond about their findings and to visit each other. They also began to venture into the countryside to study nature firsthand; observation became the primary tool of scientific activity. In the sixteenth century traveling was fraught with danger, for there were few good roads and many murderous brigands. The work of the scientist had become more perilous but also more adventurous. To this day scientific excursions are a popular topic in literature, at least in part due to this whiff of adventure and excitement.[105]

Sixteenth- and seventeenth-century England did not produce any geoscientists of world renown, like Agricola for example. Instead, experts from the continent had to be hired to oversee the mining business in Elizabethan England. The founding of the Royal Society in 1660 was the first step to remedying the rather bleak conditions of English science. Now the first substantial collections were established, and for the first time mineralogical knowledge was circulated in an English magazine, the "Philosophical Transactions of the Royal Society." It still took the better part of the century until mineralogical theories were produced in England.[106]

Enlightened scientists believed that true scientific observation began in Asia but then fell victim to barbarian invasions. The accomplishments of ancient Egyptians, Greeks, and Romans were revered all the more, but then almost eradicated through chaos and intolerance during Late Antiquity and the medieval period. These ideas and theories barely survived thanks to Arabian scholars and were only gradually reimported into Europe by way of the crusades.[107] Twentieth-century sociologist and science historian Joseph Ben-David might have seconded this assessment. He understands the scientific revolution of the seventeenth century as a continuation of the natural

---

104 Ogilvie: The Science of Describing, pp. 40–42.

105 Ogilvie: The Science of Describing, pp. 89–100, 139–150.

106 Roy Porter: The Making of Geology. Earth Science in Britain, 1660–1815, Cambridge 1977, pp. 14–31.

107 Dietrich von Engelhardt: Historisches Bewußtsein in der Naturwissenschaft. Von der Aufklärung bis zum Positivismus, Freiburg/Breisgau 1979, pp. 32–50.

philosophy of ancient Greece,[108] the difference being that seventeenth-century scholars no longer saw themselves as philosophers, but as scientists, or as specialized astronomers, or as physicians etc. According to Ben-David, this scientific tradition had been interrupted by Late Antiquity and the Early Middle Ages. He states that in this age curiosity was frowned upon; theoretical knowledge had to be put into practical use or had to give some insight into the divine genius. The purpose of thirteenth- and fourteenth-century universities was to train civil servants, not to further scientific knowledge.[109] The Protestant reformation brought a sense of individualism to many Christians. A good Protestant had an obligation to interpret the divine will (as stated in the Bible) to the best of their abilities for themself. Using scientific observation to reveal God's will and genius was a side effect of this attitude, especially in Protestant England where Francis Bacon (1561–1626) employed his method of empiricism as a tool to interpret the divine creation. By doing so he managed to forge an alliance between science and religion that remained popular to at least the 1800s.[110]

2.3 The Seventeenth Century: The Scientific Revolution

Medieval thinkers had rediscovered Aristotelian theories and tried to reconcile them with Christian scripture, as described in the previous chapter. For centuries, Aristotle remained the only legitimate authority on scientific ideas. Change was to come slowly at the end of the medieval period when it seemed that the old explanations had run their course. Late-medieval and early-modern maps reflect this change as well: in the past, most "Western" (e.g. Christian) maps were of metaphorical design, oriented towards the most important place on earth: Jerusalem. Other places (and sometimes persons) of interest were depicted accordingly in beautiful pictures which had little in common with their actual geographic positions. This rapidly changed when geographically accurate maps were required in order to keep up with the improved methods of navigation. Scientific thinking changed likewise, thanks to the development of special instruments (microscopes for example); this gradual process changed the

108 Still, the theories of the revered philosophers were no longer taken as gospel by Francis Bacon and his contemporaries. Ancient sources were no longer to be just memorized and recited, but to be challenged and rebutted or amended accordingly. Ogilvie: The Science of Describing, pp. 258–264.

109 Joseph Ben-David: The Scientist's Role in Society. A Comparative Study, Englewood Cliffs, NJ 1971, pp. 33–66.

110 Ben-David: The Scientist's Role in Society, pp. 66–74. For more information on the patronage of paleontology in the fifteenth to seventeenth centuries see: Davidson: Patrons of Paleontology, pp. 2–27.

mindset of first scholars, and then society as a whole.[111] This process, referred to as the scientific revolution, is the topic of this subchapter.

According to Thomas Kuhn there have been numerous revolutions in science, but the events most commonly referred to as *the* scientific revolution took place during the seventeenth century, a time of great change and upheaval. Religious wars ravaged various European countries and principalities. Colonies in the Americas flourished. The Iberian monopoly on the riches of the New World had been broken and its natural treasures were now to be exploited by all seafaring Europeans, causing not only many colonial wars but also allowing the ascension of a whole new socio-economic class, later called the middle class.[112] But also God's own creation, nature itself, was to be explored in more detail than ever before. Scientists set out to study nature, discovering its hidden mechanisms, determining the rules of nature, and thereby demystifying at least parts of the apparently divine creation (this was essentially what Max Weber called "die Entzauberung der Welt," the disenchantment of the world). The disenchantment of nature was first and foremost furthered by Francis Bacon and René Descartes (1596–1650), the true protagonists of the scientific revolution. Spirits and the mythical forces of nature had to make way for a more mechanical explanation of the world, a world full of gears and coil springs, which adhered to mathematical rules. Phenomena now had to be ordered and categorized rationally. At the beginning of the revolution old concepts died hard. At first, magic and wonder were to be found in nature, only later to be explained in a completely rational manner. Learning, for example, that a vase was not completely empty, but full of oxygen, was a slow process.[113] Still, despite the triumph of Bacon, Descartes & Co. a deistic interpretation of nature never completely vanished, not even among natural historians and scientist, and natural theologians, like John Ray (1627–1705) saw the study of nature as proof of the existence and brilliance of God.

Like most other revolutions, the scientific revolution created a whole gallery of champions, Isaac Newton (1643–1727) being the most revered one, at least in the English-speaking world. Newton is remembered for his genius; one would be hard pressed to find anyone else who has done more than Newton for the advancement of science and rationality. His mental capabilities are held in such high regard that he

---

111 David M. Knight: Voyaging in Strange Seas. The Great Revolution in Science, New Haven, CT 2014, pp. 20–34.

112 David Knight stresses the influence of society and culture in science. Circumstances such as wars, economic change, and the invention of new methods of transportation and communication are essential for the scientific process, underlining that there is no such thing as "pure science." He stresses the changing modes of transportation in particular, arguing that the voyages and discoveries of the sixteenth and seventeenth centuries were most crucial in the development of the scientific disciplines. Knight: Voyaging in Strange Seas, pp. 1–9.

113 Lawrence Lipking: What Galileo Saw. Imagining the Scientific Revolution, Ithaca, NY 2014, pp. 1–19.

has become somewhat enchanted, immortalized, almost deified, set apart from the rest of humanity. Therefore, Newton's personal flaws and scientific errors are all but forgotten, overshadowed by his bright intellect; or so it would seem, and was handled accordingly by generations of historians indulging in the "history of great men." This variant of historical writing is no longer in style; even so-called great men are studied as the products of society and circumstances. Nonetheless, the achievements of the heroes of science were often recognized in their own lifetimes; their prestige reflected back on their nation's reputation and inflated the egos of patriots for generations.[114] Lawrence Lipking warns his readers not to fall into the same trap as the patriots of old, reminding them that genius and the production of scholarly achievements are group efforts:

> Instead of reaching a peak with the giant who mounted highest, the progress of science would undergo many ups and downs, without any clear destination. [...] Or alternatively, a historian might imagine science as the product of any number of minds collaborating and entering into one another over time, until together they formed a single great mind or genius compared to which even the mind of Newton might seem no more than a drop in the ocean of truth.[115]

Bacon's methods were based on observation and the realization that underlying rules and mechanisms affected all natural processes.[116] These mechanisms could best be studied under controlled, and therefore reproducible, circumstances which should lead to reproducible results (the whole process of course being called an experiment). This constitutes the invention of the laboratory and the foundation of modern science. Thanks to Bacon, England became the frontrunner of science in the seventeenth century.[117] In Bacon's time the scientific process was almost a democratic one: anyone could participate. The scholarly contributors would be organized into a communication network, discoveries would be reviewed within this network, and the resulting

---

114 Lipking: What Galileo Saw, pp. 158–165.

115 Lipking: What Galileo Saw, p. 199.

116 Klaus Fischer compiled a list with the four essential factors of Baconian science and therefore of the scientific revolution of the seventeenth century: 1. The regressive method of deduction entails that every observation is to be traced back to its initial point of origin. If the origins can be explained in principle, the phenomenon can be elucidated and furthermore some generalizations can be made about similar phenomena. 2. Nature became mathematized, meaning the discovery of mathematical principles behind natural processes. 3. Experimentation, following a strict set of rules and using specialized scientific instruments to test one's hypothesis objectively. 4. Realism, meaning the neutral, objective execution of science without fixed expectations. Klaus Fischer: Die Neue Ordnung des Wissens. Experiment – Erfahrung – Beweis – Theorie, in: Richard van Dülmen; Sina Rauschenbach (eds.): Macht des Wissens. Die Entstehung der modernen Wissensgesellschaft, Cologne 2004, pp. 155–185.

117 Rhoda Rappaport: When Geologists Were Historians, 1665–1750, Ithaca, NY 1997, pp. 53–63.

discussion would inspire all members. To this day, science is by definition peer-reviewed and therefore inevitably a group effort. Such a network had to reflect the hierarchies that dominated all seventeenth-century European societies. Bacon illustrated his positions in a utopian novel, published posthumously in 1627 as "New Atlantis". This novel, though unfinished, illustrates Bacon's notion of a perfect society of which rationality and empiricism constitute the backbone,[118] and in which science is highly revered. The notion of science as a communal effort inspired the establishment of scholarly societies and communication networks, such as the Royal Society and its "Philosophical Transactions."[119] The "Leopoldina" constituted another scientific communication network; it was situated in the German-speaking world and established in 1652, eight years prior to the Royal Society. David Knight argues that the new middle class and its tendency to self-organize had been essential to the formation of the scholarly communication networks of the seventeenth century. Furthermore, efficient modes of transportation and communication had to exist, first and foremost the postal service, otherwise scholarly letters could never have reached their destinations. The abandonment of Latin as the scholarly lingua franca was a side-effect of these correspondences in which the scientists tended to use their native tongues. The scientific journals also described the discoveries not in complicated technical jargon but in simpler terms. This allowed a wider audience to educate themselves scientifically, especially because literacy was also on the rise during the late seventeenth century.[120] With the establishment of objective observation as the main tool of knowledge construction, an abundance of new knowledge had been made available. In order to be handled, these observations had to be structured; modes of observing and recording the observed became increasingly formulaic. Scholarly practice became more and more demanding; scientists often had to juggle being a family man and provider, making financial gain, and dedicating time to their scientific activities. Therefore, their reputation within the scientific community became ever more important.[121]

In the sixteenth and the early seventeenth century interest in the history of the planet grew rapidly. During this period, the biblical story of Genesis was regarded by most Christian scholars as a reliable and true source of information, almost as an eye-witness account of the first days of earth, as was the story of the biblical flood that destroyed most life on the planet. It was generally regarded as the most likely

118 Note that in Jonathan Swift's satirical "Gulliver's Travels," published one year prior to "New Atlantis," the fictional flying island Laputa constitutes another utopia whose society is centered on science. Knight: Voyaging in Strange Seas, p. 83.

119 Knight: Voyaging in Strange Seas, pp. 44–49.

120 Knight: Voyaging in Strange Seas, pp. 136–159.

121 Lorraine Daston: The Empire of Observation, 1600–1800, in: Lorraine Daston; Elizabeth Lunbeck (eds.): Histories of Scientific Observation, Chicago 2011, pp. 81–113, https://doi.org/10.1007/s11191-012-9515-z.

cause for the geological transformation of the earth's surface and for other strange phenomena, namely that the fossils of clearly aquatic origin could be encountered in mountainous rock, many miles from any body of water. This diluvial explanation for the existence of fossils remained popular until the middle of the nineteenth century, as seen below. One of the earliest scholars to supplement the diluvial theory with a chemical explanation was Niels Steensen (1638–1686) of Copenhagen. Later in his life the Danish scholar converted to Catholicism (in 1678 he was even made a bishop) and latinized his name to Nicolaus Stenonius, or Steno for short. Steno, who went on to become the father of modern geology and paleontology, introduced the concept of comparative anatomy as a tool for fossil study. In Florence Steno dissected the head of a huge shark which had been caught a short time before and published his findings in 1667. He recognized that the shark's teeth bore a striking resemblance to stony objects found in rock formations, called tongue-stones, or *glossopetrae*, for their likeness to the split tongues of snakes. The tongue-stones were usually identified as *lusus naturae*. An alternative explanation, dating back to Pliny the Elder (23–79), supposed that the tongue-stones fell from the moon or the sky. Steno found both explanations to be unsatisfactory and instead theorized that if a solid body was engulfed by another solid body (in this case shark teeth buried on the ocean floor), the first object would harden and might become a fossil.[122] In his 1669 "Dissertationis Prodromus," Steno furthermore devised the rule of superposition, determining the sequence of succession of layers of earth, recognizing that newer layers (or strata) would come to rest upon older ones. Therefore, older strata would generally rest deep within the earth, buried by their successors.[123] Gottfried Wilhelm Leibniz, another early scientist of enormous reputation, met Steno 1667 in Hannover. Leibniz was a great admirer of Steno's paleontological observations, which in turn inspired his own, published in his "Protogaea," which he composed from 1691 to 1693 (not published in full length until 1749). In 1685 Leibniz arrived in the Harz region of Northern Germany, where he was to invent a way of pumping dry flooded mineshafts. While he failed in his mining endeavors, Leibniz had ample opportunity to expand his fossil collection. In 1692 he studied a fossilized tooth, which had been discovered near Wolfenbüttel in Lower Saxony. Leibniz realized it belonged to a gargantuan aquatic organism; he deduced that in primeval times the whole planet must have been covered by water, which had since then mostly dried up. The remains of the inhabitants of said boundless ocean could now be found in places distant from any body of water. Leibniz furthermore speculated that the an-

---

122 He had also noticed that tongue-stones had been used with other stones as building materials since Etruscan times and therefore had to predate the ancient culture and maybe even the great deluge. Rudwick: Earth's Deep History, p. 45.

123 Hölder: Kurze Geschichte der Geologie und Paläontologie, pp. 5–10. See also Laudan: From Mineralogy to Geology, pp. 36–40.

cient marine animals had either retreated to the darkest depths of the ocean, or that they had changed their form radically and were now hardly recognizable as the species evidenced by the fossil record.[124]

For the first time, technical human inventions were perceived as part of the natural world and not as stemming from its polar opposite. Nature now seemed to follow certain laws and rules that also had to be abided by human inventions. The same underlying rules applied to everything. Nature could therefore even inspire new inventions and machines. The newfound truths and their application should better mankind.[125] During the seventeenth century, science became more and more socially acceptable. Good science was also required to be of relevance to society as a whole, and ideally it was to be produced independently and then be peer-reviewed. English scientists were not organized for the best part of the seventeenth century, as David Elliston Allen, author of one of the first extensive accounts of the history of British science, notes. In 1698 the Temple Coffee House Botanic Club, a loosely organized scientific community, was founded. After Newton's death in 1727, British science grew stagnant, but scholarly correspondence networks became increasingly important to English science especially.[126]

In England, the Royal Society had established itself as the leading scientific institution. Yet besides scientific matters, politics and social reform in general were also discussed, since not only scientists joined its ranks but politicians and interested laymen also. In contrast to the English society, the French Academy of Sciences ("L'Académie des sciences," founded in 1666 by Louis XIV) was more exclusive, only open to the wealthy social elite.[127] Due to its focus on "pure science," the French model was emulated by all of Europe. At the beginning of the eighteenth century, French science seemed to be outperforming its European counterparts.[128]

Another accomplishment of the late seventeenth century, that would greatly further paleontology and science in general, was the Republic of Letters.[129] Within this

---

124 Eric John Aiton: Leibniz. A Biography, Bristol 1985, pp. 75, 136, 208–209.

125 Peter Weingart et al.: Nachrichten aus der Wissensgesellschaft. Analysen zur Veränderung der Wissenschaft, Weilerswist 2007, pp. 11–33.

126 David Elliston Allen: The Naturalist in Britain. A Social History, 2nd ed., Princeton, NJ 1994, pp. 3–21.

127 Bacon had always asserted that science had to be pursued with public welfare in mind, that the government should see to it that scientific discoveries were used for a greater good. The Royal Society, although established by King Charles II, failed in this regard, mostly operating independently from the English government. John Gascoigne: Science in the Service of Empire. Joseph Banks, the British State and the Uses of Science in the Age of Revolution, Cambridge 1998, pp. 16–23.

128 Ben-David: The Scientist's Role in Society, pp. 75–88.

129 Anthony Grafton writes a comprehensive history of the Republic of Letters. He focuses on the early history of the Republic (c. 1500–1700), but also describes its ripple effects to this day, and the historiog-

communication network, this *Respublica literaria*, scientific knowledge would be exchanged and debated and thereby amended and expanded. The scientific discourses were then published in journals such as the "Philosophical Transactions of the Royal Society" (of 1665), which in turn became part of the discourse themselves.[130] During the bellicose seventeenth century, this scholarly network grew to be especially important for the survival of international science:

> After the devastation wrought by the wars of the seventeenth century, engendered by a powerful brew of dynastic, religious and civil conflict between centralizing monarchs and aristocratic intrigue, the notion of a Republic of Letters offered some consolation to an intellectual elite weary of strife.[131]

Peter Weingart also underlines the immense significance of the Republic, for it made the participants of the scholarly discourse a collective. This collective developed its own common identity, and the scholarly correspondence within this collective evolved into the scientific journal.[132]

Even though the term *Respublica literaria* can be dated back to 1417, Dena Goodman argues that the Republic of Letters was a product of the emerging national state in conjuncture with the Enlightenment, really taking off after the religious wars of the sixteenth and early seventeenth centuries. European scholars supposedly saw their postal conversations as a continuation of the intellectual exchanges of the revered philosophers of Classical Antiquity. King Louis XIV of France (1638–1715) ruled a world-spanning empire and aspired to increase the king's authority over the state. In doing so, he developed the postal service of his realm, facilitating the means for an international communication network. French gentlemen considered themselves the embodiment of enlightenment and civilization, therefore France had to be the center of the Republic of Letters. One could describe oneself as a French patriot but still be a member of this international society. Furthermore, women were encouraged to

---

raphy of the Republic. See: Anthony Grafton: Worlds Made by Words. Scholarship and Community in the Modern West, Cambridge, MA 2009.

130 Rappaport: When Geologists Were Historians, pp. 7–40.

131 Gascoigne: Science in the Service of Empire, p. 148.

132 "Die Geschichte der modernen Wissenschaft ist die Entwicklung der Idee der respublica literaria seit Mitte des 17. Jahrhunderts. In ihren Anfängen ist es die Ausdifferenzierung einer Idee des Kollektivs der Gelehrten, einer Identität des kollektiven intellektuellen Lebens in Unabhängigkeit gegenüber den religiösen und politischen Autoritäten. Das Medium dieser Identität wird im 18. Jahrhundert die gelehrte Korrespondenz, die Vorstufe zum wissenschaftlichen Artikel." Peter Weingart: Die Einheit der Wissenschaft – Mythos und Wunder, in: Peter Weingart (ed.): Grenzüberschreitungen in der Wissenschaft – Crossing Boundaries in Science, Baden-Baden 1995, pp. 11–28, see p. 13.

participate; it was believed that a mixture of gender-specific virtues would enrich the network, making it even more civilized.[133]

Peter Weingart argues that growing nationalist tendencies and the rise of patriotism at the end of the eighteenth century led to a stagnation of international scientific communication. The Republic of Letters did not see its resurgence until the 1820s and the end of the Napoleonic Wars. Furthermore, he states that one of the most important tools for international communication is a common language. In the case of the Republic this happened to be Latin during the seventeenth and French during the eighteenth century. There were attempts to establish German as the scientific lingua franca in the nineteenth and early twentieth centuries, but there can be no doubt that English has assumed this position in the last one hundred years.[134]

2.4 The Eighteenth Century: European Natural History

In the eighteenth century the idea that knowledge had to be proven by empirical study and not just by invoking a scientific tradition, thus making an argument purely based on the authority of one's scientific forerunners, had firmly established itself. Furthermore, the sciences were no longer just studied and categorized by their respective subject matters; scholars began to think about the chronological development of their art.[135] Alongside the promotion of empirical science, there arose a notion of "good" and "bad" curiosity. Good curiosity inspired the observation and explanation of real and attestable phenomena, an occupation suiting virtuous gentlemen. The lust for sensation and distraction exploited commercially in showrooms constituted a bad style of curiosity, ill-suited for gentlemen but a diversion for the lower classes.[136] Some of the scholars who were part of the scientific revolution (Copernicus, Kepler, Galilei, and Newton for instance) possessed an acute historical awareness. On the one hand they knew how much their predecessors had accomplished (harking back to Greek and Roman times); on the other hand, they realized the future potential of their respective

---

133 Dena Goodman: The Republic of Letters. A Cultural History of the French Enlightenment, Ithaca, NY 1994, pp. 1–23.

134 Weingart: Die Einheit der Wissenschaft, pp. 11–28.

135 Wolf Lepenies: Das Ende der Naturgeschichte. Wandel kultureller Selbstverständlichkeiten in den Wissenschaften des 18. und 19. Jahrhunderts, Munich 1976, pp. 16–20.

136 Fabian Krämer: Why There Was No Centaur in Eighteenth-Century London. The Vulgar as a Cognitive Category in Enlightenment Europe, in: Kaspar von Greyerz et al. (eds.): Wissenschaftsgeschichte und Geschichte des Wissens im Dialog – Connecting Science and Knowledge. Schauplätze der Forschung – Scenes of Research, Göttingen 2013, pp. 317–345, https://doi.org/10.14220/9783737001717.317.

scientific disciplines.[137] It is no coincidence that the history of life on earth (paleontology) and the evolution of terrestrial features (geology) experienced a burst of growth at the same time that human history was established as an academic discipline. Rudwick refers to Edward Gibbon's (1737–1794) epic "The History of the Decline and Fall of the Roman Empire" (published in six volumes between 1776 and 1789) as a milestone in the development of historical science and an achievement to be envied and emulated by geologists and their accounts of the earth's history.[138] Similar to human history, the history of nature could be arranged into epochs and ages, and it seemed that most, if not all, known species had predecessors similar to but not entirely identical with them. This observation posed one important question: had life slowly and gradually changed, or had some species vanished completely, only to make room for some new and similar creations; had God stopped creating life on the sixth day, or was He still conjuring it up?[139] At the same time the idea was born that certain ancient myths were just poetically embellished retellings of actual historical events. Fossils were part of this embellishment, as the petrified bones of extinct animals were interpreted as those of mythical heroes and legendary monsters. Furthermore, archeological sources such as ancient monuments and coins were now recognized to be historical, factual sources, lessening the overreliance of historians on (inevitably subjective) written evidence. The same thing happened with fossils; petrified bones, too, became sources of (natural) history.[140] Most scholars came to believe that the earth was much older than religious chronologists like Bishop Usher had calculated. While the precise determination of the planet's age was still impossible, processes could be observed that had to have taken at least a few dozen millennia. In order not to come into conflict with the Bible and the still mostly religious establishment of eighteenth-century societies, the days of creation as described in the book of Genesis were reinterpreted into metaphorical days that could have lasted for thousands or maybe even millions of years.[141] Furthermore, romantic tendencies led to an equation of the humanities with natural sciences; all were part of the same spiritual process, so one could apply the same underlying rules to human history as to natural history. In the long term this led to the application of scientific rules to human societies, the most infamous example being Social Darwinism (see chapter 7).[142] Natural history was established as a scientific discipline at German universities in the second half of the eighteenth century. It divided nature into three distinct realms: animals, plants, and minerals. Natural history was

---

137 Engelhardt: Historisches Bewußtsein, pp. 24–31.

138 Rudwick: Earth's Deep History, p. 93.

139 Engelhardt: Historisches Bewußtsein, pp. 81–93.

140 Rappaport: When Geologists Were Historians, pp. 64–70, 83–94.

141 Rudwick: Earth's Deep History, pp. 99–102.

142 Engelhardt: Historisches Bewußtsein, pp. 201–210.

taught in Göttingen beginning in 1755; in 1765 a permanent professorship was established for this subject. Various other German universities followed suit, textbooks and journals were published; at the beginning of the nineteenth century natural history had run its course and dissolved into botany, zoology, and mineralogy.[143]

In this age of empires science became increasingly globalized. Botanical gardens for example, although hardly a new invention as they had been used since Roman times to study nature and explore the healing powers of plants and herbs, now housed exotic flora originating in the colonies. Consequently, the gardens were not only of scientific and medicinal use but also became figureheads of the globalized colonial empires.[144]

One of the most important contributions to modern science came from eighteenth-century Sweden, where Carl Linnaeus (1707–1778) invented a system of biological nomenclature, thereby giving a meaningful order to life itself.[145] Linnaeus laid the foundation for modern biology and provided a most useful tool for many other sciences, including paleontology, in his "Systema Naturae," published in 1735. Linnaeus even claimed that the scientific expedition was a Swedish invention, a grandiose claim, and, if true, another most important Scandinavian contribution to the development of modern science.[146] Linnaeus furthermore organized geological phenomena into *Petrae, Minerae,* and *Fossilia,* similarly to his classification of lifeforms. In contrast to the lifeforms, the geological phenomena were not ordered by their sexual but by their chemical properties; later Johann Gottschalk Wallerius (1709–1785) built upon

143 Kai Torsten Kanz: Von der BIOLOGIA zur Biologie. Zur Begriffsentwicklung und Disziplingenese vom 17. bis zum 20. Jahrhundert, in: Uwe Hoßfeld; Thomas Junker (eds.): Die Entstehung biologischer Disziplinen II. Beiträge zur 10. Jahrestagung der DGGTB in Berlin 2001, Berlin 2002, pp. 9–30, see pp. 20–22.

144 Richard Harry Drayton: Nature's Government. Science, Imperial Britain, and the 'Improvement' of the World, New Haven, CT 2000, pp. 3–25.

145 Kenneth Nyberg provides an overview of the propagation of Linnaeus' new system. Nyberg emphasizes the influx of new species of animals and plants to Europe due to global European expansion and colonization, beginning in the late fifteenth century. This influx necessitated a new system to categorize nature because many of the newly discovered species would not fit in with traditional European systems of classification. He then describes a group of seventeen or eighteen young naturalists who traveled the world between 1746 and 1799, spreading the word about the new taxonomic system. This group is commonly called "Linnaeus' apostles." Linnaeus himself used the term apostles and considered himself a "reformer," similar to Martin Luther. Indeed, his discovery revolutionized botany and could be counted as a scientific revolution in accordance with Kuhn. Linnaeus furthermore was convinced of the importance of traveling for science – not only would his taxonomy spread throughout the scholarly world this way, but his apostles were to collect new specimens for study. Kenneth Nyberg: Linnaeus's Apostles and the Globalization of Knowledge, 1729–1756, in: Patrick Manning; Daniel Rood (eds.): Global Scientific Practice in an Age of Revolutions, 1750–1850, Pittsburgh, PA 2016, pp. 73–89.

146 Lisbet Koerner: Daedalus Hyperboreus: Baltic Natural History and Mineralogy in the Enlightenment, in: William Clark et al. (eds.): The Sciences in Enlightened Europe, Chicago 1999, pp. 389–422, see pp. 399–404.

this taxonomy and refined it significantly.[147] This new scientific framework would alter the significance of fossils; in the seventeenth century fossils were largely regarded as unique occurrences, generated by some whim of nature; in the eighteenth century they were sorted into scientific categories. This enabled their objective study and allowed for some more general conclusions.[148]

An accelerating factor for the emergence of the geosciences was of a more economic nature: the industrialization of the eighteenth century produced a growing demand for minerals and metals, which in turn spurted the growth of the mining industry. To keep up with the need for personnel educated in both mineralogy and geology, scientific academies were founded across Europe. European governments suspected that the established universities would provide a mostly academic education, and not the practical know-how required by the mining industry. In contrast, some contemporary scholars criticized the new academies, supposing that this newfangled and ultimately profit-oriented mode of education would never promote true science and could only lead to the stagnation of scholarly endeavors.[149] Nevertheless, this new class of specialists was also taught the theoretical foundations of the craft. For example, one of the most brilliant and popular geologists of his time, Abraham Gottlob Werner (1749–1817), was employed by the Freiberg Mining Academy ("Bergakademie zu Freiberg") in Saxony to teach his theories.[150] Werner postulated that the earth's surface was changing gradually, and that all solid land originated from minerals found in ocean water and had separated slowly from the watery components. Werner's school of thinking ("Freiberger Schule") dominated the geological education at least up to the 1820s. Some of his ideas harkened back to Steno's law of superposition, but he expanded on the law by introducing index fossils ("Leitfossilien"), used to define or identify the geological period of the stratum in which they were found: Werner noticed that deeper, that is older, strata contained only the fossils of primitive life, but that life became more complex in younger strata, and that the two never mix.[151] He further theorized that all minerals encountered today were once dissolved in a global ocean fully covering the planet's surface. Later, sinking sea levels unveiled mountains that grew into landmasses and finally continents, composed of the minerals that were once dissolved in the primeval ocean. Werner's theory became known as Neptunism and stood in

147 Martin Guntau: The Natural History of the Earth, in: Nicholas Jardine et al. (eds.): Cultures of Natural History, Cambridge 1996, pp. 211–229.

148 Heesen, Anke te; Spary, Emma C.: Sammeln als Wissen, in: Heesen, Anke te; Spary, Emma C. (eds.): Sammeln als Wissen. Das Sammeln und seine wissenschaftsgeschichtliche Bedeutung, Göttingen 2001, pp. 7–21.

149 Engelhardt: Historisches Bewußtsein, pp. 192–193.

150 Laudan: From Mineralogy to Geology, pp. 47–55.

151 Laudan: From Mineralogy to Geology, pp. 87–100.

stark contrast to Plutonism, which emphasized the influence of magmatic activity on the creation of the features of the earth.[152] One of the chief proponents of Plutonism was James Hutton (1726–1797), a Scotsman educated in Leyden and another "Father of Geology." Hutton supposed that the processes that had formed the earth's features were still active, never ceasing their formative activity, whereas Neptunists believed the creative process had been concluded a long time ago. As a deist not bound to the literal interpretation of the Bible, Hutton did not believe that the earth was created and was someday to end. He envisioned an endless and perpetual circle of creation and destruction of the earthly features. Hutton introduced the concept of deep time and hypothesized that the planet was much older than most of his contemporaries presumed (most of whom dated the age of the earth by studying the time data found in the Bible).[153] Hutton, who after his return to Scotland operated within the relatively modern and liberal scholarly climate of Edinburgh, was less beholden to the clerical view, which still dominated the English universities.[154] Even though there was some effort made to reform dated curricula and strengthen scientific learning at Oxford and Cambridge, it took the better part of the nineteenth century to remedy the situation.[155] At the beginning of the nineteenth century, scriptural evidence was still interchangeable with historical or scientific evidence in Oxford.[156]

Johann Friedrich Blumenbach (1752–1840), who practiced anthropology as part of natural history and therefore had a great interest in fossils, noticed that some forms of fossilized life were nowhere to be found in the present day; therefore, he deduced that some lifeforms had been extinct during the course of history.[157] The Paris-based

---

152 Wolf von Engelhardt: Die Entwicklung der Geologischen Ideen seit der Goethe-Zeit, in: Karl Heinrich Olsen (ed.): Beiträge zur Geowissenschaft. Vorträge Anläßlich der Wwissenschaftlichen Vortragsveranstaltung und der Gauß-Gedenkfeier vom 30. 4. 1979, Göttingen 1979, pp. 1–23, see p. 5–10.

153 Laudan: From Mineralogy to Geology, pp. 113–122; Michael Freeman: Victorians and the Prehistoric. Tracks to a Lost World, New Haven, CT 2004, pp. 54–57.

154 Still encountering some religious opposition to his concept of geological time, he came to the conclusion that science and religion should not intermingle. Dennis R. Dean: James Hutton and the History of Geology, Ithaca, NY 1992, pp. 13–29.

155 Heather Ellis confirms this and writes that when George of Hanover inherited the British throne, the German universities became an inspiration for the English university reform of the nineteenth century. Heather Ellis: Enlightened Networks. Anglo-German Collaboration in Classical Scholarship, in: Heather Ellis; Ulrike Kirchberger (eds.): Anglo-German Scholarly Networks in the Long Nineteenth Century, Leiden 2014, pp. 23–38, https://doi.org/10.1163/9789004253117_004.

156 Nicolaas A. Rupke: The Great Chain of History. William Buckland and the English School of Geology (1814–1849), Oxford 1983, pp. 21–26.

157 Blumenbach specifically sorted the fossils into three classes: the first comprised animals that could still be encountered more or less unaltered and had changed little in the course of earth's history. These, he assumed, were the most recent fossils. The second and much older category encompassed those animals that had perished in Noah's flood. These animals were more foreign, more exotic, obviously displaced by the cataclysmic deluge. Still, they bore some resemblance to modern day animals

Georges Cuvier elaborated upon Blumenbach's design and proposed that innumerable species had gone extinct in a series of catastrophic events he called "revolutions."[158] This is rather telling as to the changing meaning of the term "revolution."[159] As other scientists had done before him, specifically the French naturalist and scientific celebrity Georges-Louis Leclerc, Comte de Buffon (1707–1788) had used the term to describe the movement of planets; now, with the bloodshed and confusion of the French Revolution in recent memory, Cuvier linked revolutions to catastrophic occurrences; yet another example for the ways in which science is culturally influenced.[160] The ambitious nobleman Buffon quickly rose through the ranks of the French scientific community and became the intendant of the royal garden ("Jardin du roi") in 1739. He wrote *the* book on natural history; after the first volume of his "Histoire naturelle, générale et particulière" was published in 1749, thirty-five volumes were to follow, the last one published posthumously in 1789. The books were products of a joint effort, and many French naturalists (and hundreds of amateur correspondents) contributed to them. Still, it was above all Buffon's name that came to be associated with the "Histoire Naturelle." Within his lifetime, the books were translated into German, Dutch, English, Italian, and Spanish. They were greatly appreciated by non-scientific readers and became bestsellers. Therefore, Buffon not only became one of the most prominent scientists of his time but is also considered one of the great literary figures of the French Enlightenment. Though Buffon's natural history was criticized by many of his educated contemporaries, the "Histoire Naturelle," at least in part, set the precedent on how natural history was to be done for the next generation.[161] Buffon was a representative of the *Ancien Régime*, the class of French noblemen-scientists that was diminished for the most part by the French Revolution (even though he died one year

---

to which they were related. Animals of the third class were even more alien. Blumenbach reasoned that they had migrated from their far-off habitats, forced by catastrophic climate change unrecorded by any human. Buffetaut: A Short History of Vertebrate Palaeontology, pp. 50–52.

158 Laudan: From Mineralogy to Geology, pp. 147–162.

159 While the association of the word "revolution" with a truly negative occurrence, that is the extinction of a species, might lead one to believe that Cuvier was opposed to the French Revolution, he owed his personal freedom to the radical changes and opportunities that it brought. Aristocratic patronage had allowed him to be educated and travel through France and Germany, but the revolution enabled him to leave the rural Normandy province and move to Paris in 1795. The Academy of Sciences and many of its aristocratic members had fallen victim to the terror of the revolution in 1793. In 1795, Paris recovered as the center of French science and an access hub to the international Republic of Letters. While Cuvier quickly rose to international acclaim thanks to his scientific capabilities, the political and social revolution facilitated his ascendance. Dorinda Outram: Georges Cuvier. Vocation, Science and Authority in Post-Revolutionary France, Manchester 1984, pp. 40–48. See also Toby A. Appel: The Cuvier–Geoffroy Debate. French Biology in the Decades before Darwin, New York 1987, pp. 34–39.

160 Cohen: Revolution in Science, pp. 276–277.

161 Jeff Loveland: Rhetoric and Natural History. Buffon in Polemical and Literary Context, Oxford 2001, pp. 8–14.

before the storming of the Bastille). Their passing made way for a new generation of less aristocratic French scientists, like Cuvier.

French geologists had suggested that a great flood had exterminated the ancient fauna, or that parts of the earth's surface had collapsed and sunk beneath the waves of ancient oceans. Cuvier thought that Noah's Flood (various accounts of catastrophic floods, originating from non-Christian folklore, apparently confirmed the biblical story) might have been the most recent of all extinction events. His geological exploration of the Paris Basin revealed many fossils of maritime animals, further confirming Cuvier's vision of a stable ecological system abruptly devastated by a catastrophic event.[162] He postulated that all great extinction events had taken place before modern man had emerged, or that mankind had survived the events rather unscathed, for there was yet to be found fossil evidence putting humans next to ancient and extinct animals. This statement incensed some scholars who tried to use geology to prove the scientific validity of the Bible. Most English academics, for example, still linked Noah's Flood, as described in the book of Genesis, to the distribution of fossils. Since said flood was caused by human sin, the fossil-producing extinction event could not predate human history.[163] Furthermore, Cuvier did not believe in a gradual transformation of animal species, as did his contemporary colleague and bitter rival Jean-Baptiste Lamarck (1744–1829), who established a theory of evolution by the means of gradual transformation. Cuvier had studied the bones of ancient elephant-like creatures and had compared them to those of modern-day elephants. Until then, most scholars had thought the old remains were those of elephants, probably of Hannibal's war elephants who had died crossing the Alps. But now Cuvier, being the most prolific and most celebrated anatomist of his time, realized that the old bones differed too much from those of modern elephants. While they were similar, they had to belong to a different species now extinct (they were *mammoth* bones).[164] He also employed Linnaean nomenclature in his extensive studies of natural history. His main tool remained comparative anatomy, which he taught at the "Cabinet d'anatomie compare," established in 1806. His teachings at this museum constitute Cuvier's contribution to the development of museums. His occupation became part of the international Republic of Letters, for Cuvier corresponded frequently with the British Royal College of Surgeons, which had acquired an extensive collection of anatomical specimens.[165]

162 Martin J. S. Rudwick: Worlds before Adam. The Reconstruction of Geohistory in the Age of Reform, Chicago 2008, https://doi.org/10.4000/miranda.1393, pp. 11–16.

163 Marjorie Grene; David Depew: The Philosophy of Biology. An Episodic History, Cambridge 2004, pp. 158–159.

164 Buffetaut: A Short History of Vertebrate Palaeontology, pp. 52–71.

165 Philippe Taquet: Establishing the Paradigmatic Museum. Georges Cuvier's Cabinet d'Anatomie Comparée in Paris, in: Simon J. Knell et al. (eds.): Museum Revolutions. How Museums Change and Are

Before Lamarck's aforementioned theory, no scientist could really inquire into the origins of minerals or animal species. It appeared to be pointless because a lion was and always had been a lion, since the day of its creation, a mineral had always been a mineral, and so forth. Theologians and philosophers could question the origin of this divine creation, but a scientist could not. This changed with the conception of transmutation over time. Étienne Geoffroy Saint-Hilaire (1772–1844), a French naturalist and colleague of Lamarck's, elaborated upon the theory of evolution: he reasoned that animal species did not change, but that sometimes drastic changes happened to individuals of said species, creating a whole new kind of animal. He interpreted sporadic birth deformations as (unfortunate) variations of this phenomenon but suggested that, under the right circumstances, a mutation could lead to the emergence of a new species. Saint-Hilaire's ideas could not explain why species seemed to be adjusted to their environment, a major flaw later corrected by the Darwinian theory of evolution.[166] Now scientists could observe how species had evolved; the empirical study of fossils was a crucial tool in documenting this change. Therefore, theories of evolution and transmutation were closely linked to the science of paleontology. Because of Cuvier's professional authority, scientists trying to prove that such a gradual transformation was happening had to fight an uphill battle.[167] This further illustrates how the production of scientific knowledge is subject to social factors, rather than a pure quest for truth. In the aftermath of the French Revolution pragmatic academies were founded all over the up-and-coming empire. These academies were no longer interested in research but in practical education. Scientific research continued to be the business of mostly independent scholars. The academies were extensively sponsored by the French government and flourished between 1800 and 1830. Yet this led to the stagnation of French science, as the French system remained mostly unchanged for the remainder of the nineteenth century, whereas Britain and Germany reformed their universities over the course of the century. The reformation of Prussia's entire educational system began in the Napoleonic Age and, by cause of the establishment of modern laboratories, led to a boom of German sciences.[168]

During the eighteenth century, scientific knowledge made great headway in all fields. For the first time in modern history, scholars had the feeling they had caught

---

Changed, London 2007, pp. 3–14.

166 Rudwick: Earth's Deep History, pp. 195–197.

167 Dirk Backenköhler: CUVIERs langer Schatten. "Il n'y a point d'os humains fossiles", in: Uwe Hoßfeld; Thomas Junker (eds.): Die Entstehung biologischer Disziplinen II. Beiträge zur 10. Jahrestagung der DGGTB in Berlin 2001, Berlin 2002, pp. 134–147; see pp. 133–141.

168 Ben-David: The Scientist's Role in Society, pp. 94–126.

up to and even surpassed the philosopher-scientists of classical antiquity. Yet they optimistically expected that greater accomplishments were still to come.[169]

2.5 The Nineteenth Century: The Rise of European Paleontology

At the beginning of the nineteenth century no professional (in the literal sense) geologists or paleontologists existed anywhere yet.[170] Instead, interested (and well-off) gentlemen furthered their understanding of the earth sciences through attentive observation and rigorous collecting. And while it can be argued that collecting data is the principal task of all scientists, a less abstract style of collecting is the basis for the geosciences.[171] Jeff Loveland further underlines the heterogeneous background of seventeenth- and early eighteenth-century scientists, incorporating all kinds of knowledge to the study of nature: "Natural history was the business of academic experts, but also, just as importantly, of writers, collectors, teachers, curators, patrons, explorers, and thousands of readers of books on the subject."[172]

Due to the enormous popularity of natural history and the size of its audience, the production of said science was subject to public scrutiny, and the linguistic finesse of scientific publications was of tremendous importance. Lacking significant institutional control, almost anybody could contribute. Simon Knell argues that this laissez-faire approach to science brought forth the true brilliance of some individuals and led to a fierce competitive struggle within the "free market of geology." Therefore, scientific publications not only had to further geological understanding, but also the reputation of their authors in their struggle to stay relevant within their scientific community. Most of the enterprising hobby geologists joined the Geological Society of London (founded in 1807), which was to become one of the first institutions to bring some semblance of order to the free market of geology, and to substantially advance

169 Engelhardt: Historisches Bewußtsein, pp. 75–80. Davidson provides insights into the funding of paleontology in the eighteenth and early nineteenth century with a special focus on the Royal Society and the Geological Society of London, see: Davidson: Patrons of Paleontology, pp. 28–43.

170 Martin Rudwick points out that the early British geologists were no amateurs, meaning their methods and education were far from amateurish. Although they were not paid for their geological work, they engaged in intense competition with each other. Marin J.S. Rudwick: The Great Devonian Controversy. The Shaping of Scientific Knowledge among Gentlemanly Specialists, Chicago 1985, pp. 17–18.

171 For further reading and for a substantial historical overview of this practice see: Bruno J. Strasser: Collecting Nature. Practices, Styles, and Narratives, in: Osiris, 2nd ser., vol. 27 (2012), pp. 303–340.

172 Loveland: Rhetoric and Natural History, p. 1.

the progress of the emerging discipline.[173] This marks the first important step of British geology on its way to becoming an established and self-reliant scientific discipline:

> In the early 19th century, then, science was becoming more of a profession in England: that is, a person who spent most of his energy on this activity might initiate, maintain, or improve his middle-class status; might devote his life to this activity without needing the justification of being a doctor, a lawyer or clergyman in addition; and might feel that his success was determined by the reputation he gained among his peers, not by monetary returns.[174]

Mines, quarries, cliff sides, and, above all, recently dug canals and railroad beds were treasure troves for fossil hunters and rock collectors. During the construction of a canal, some fossils and other unusual rocks piqued the interest of the British engineer William Smith (1769–1839). Inspired by his discoveries Smith created the first geological map of Great Britain in 1815; a milestone which made Smith one of the many "Fathers of Geology." Smith continued his work, and in his "Strata Identified by Organized Fossils" (published in four volumes between 1816 and 1819) he underlined the importance of fossils for the classification of strata, in accordance with Werner's "Leitfossilien." The construction of the British railroads, beginning in the 1830s, enabled a whole generation of hobby geologists to gather fossils to their hearts' content. Collecting fossils became a national pastime.[175] Furthermore, the development of the British railroads allowed for faster communication, transportation, and travel within the United Kingdom; the British science community also profited from this improved infrastructure.[176] By taking the railroad one could study the geological cross section of England simply by looking out the window, no digging required, for the railroads were cut deeply into the rocky countryside. William Buckland (1784–1856),[177] one of the most prominent paleontologists of all time, held some of his lectures in a railroad car.[178] A keen eye often was of more use in geologic endeavors than an academic education. Special equipment and laboratories were not required, but hours of dedication

173 Simon J. Knell: The Culture of English Geology, 1815–1851. A Science Revealed through Its Collecting, Aldershot 2000, pp. 7–12.

174 Susan Faye Cannon: Science in Culture. The Early Victorian Period, Folkestone 1978, p. 146.

175 Freeman: Victorians and the Prehistoric, pp. 9–10, 29–41.

176 Allen: The Naturalist in Britain, pp. 110–111.

177 William Buckland was born into a family with a long-standing religious tradition. He received a classical education at the Corpus Christi College in Oxford from 1801 to 1805, but his true passion was science. He became one of the most distinguished geologists of the nineteenth century but remained faithful to his Anglican convictions, which explains Buckland's attachment to diluvialism. See Rupke: The Great Chain of History, pp. 5–15.

178 Freeman: Victorians and the Prehistoric, pp. 42–51.

and some manual labor were. No wonder fossil-collecting became a beloved diversion for all social classes.[179] The beauty of nature as revealed by the study of fossils appealed to the romantic sentiments of Victorians. Books on geology were eloquently written and beautifully illustrated, as authors like Buckland blurred the line between natural history and natural theology.[180] The association of science with religion might also have been a reason for the growing acceptance and even appreciation of scientific theories by the majority of Victorian society. This changed gradually in the second half of the nineteenth century, thanks to the work of popularizers such as the outspoken agnostic Thomas Henry Huxley, "Darwin's Bulldog." Thus, British scientists at the end of the nineteenth century had little in common with the romantic theologians of nature.[181]

The Isle of Wight, off the south coast of England, proved to be a rich source to some of the most prominent geological scholars of Great Britain. First and foremost among his colleagues was Charles Lyell (1797–1875), who drew heavily on the island's features and fossils as inspiration and proof for his "Principles of Geology" (published in three volumes between 1830 and 1833), a work of monumental importance for the establishment of modern geology as a scientific discipline.[182] Lyell imagined that rainfall and river water had formed the countryside by slowly washing away some solid components of the ground, and that said components could congregate and form new landmasses and even continents, when given enough time. Lyell's deliberations were supported by Louis Agassiz (1807–1873), a Swiss scientist whose concept of the ice age reformed geology, making Agassiz an international celebrity.[183] He imagined that the whole northern hemisphere had once been covered by ice. Agassiz thus explained the huge boulders located far away from mountain ranges, transported to their current location by glacial ice. This ice age also could be imagined as one of Cuvier's extinction events.[184]

The new scientific findings sometimes challenged Christian beliefs concerning the age of the planet and the finite nature of time (starting with divine Creation, ending with Apocalypse). While Lyell saw no problem with this challenge,[185] other scholars,

179 Rudwick: The Great Devonian Controversy, pp. 37–41.

180 Knell: The Culture of English Geology, pp. 33–40.

181 Cannon: Science in Culture, pp. 2–3, 29.

182 Freeman: Victorians and the Prehistoric, pp. 10–25.

183 Freeman: Victorians and the Prehistoric, pp. 65–79.

184 Rudwick: Worlds before Adam, pp. 517–533.

185 Still, he did not want to stir up any public commotion about his publications. Even though Lyell was a liberal and a reformer himself, he wrote his publications for a conservative and well-off audience. He knew that some conservatives believed that science and particularly geology inspired materialism and atheism. He attacked radical ideas like Lamarck's theory of transmutation and managed not to offend his mostly Christian readership. Thus, he made his science socially acceptable. If Galileo has

like Buckland for example, tried to reconcile their scientific findings with the Christian scripture. Especially in Great Britain scientific institutions were associated with the church; the scientific establishment remained attached to Christianity.[186] For centuries, scholars had linked the occurrence of fossils in apparently strange places to Noah's flood. Depictions of imposing tidal waves were a popular motive for paintings and murals. In his best-selling "Reliquiae Diluvianae" of 1823, Buckland supposed that some animal remains were covered in mud by the great flood, but later had to rethink this statement, accepting that the biblical flood could not be confirmed by geological evidence. This exemplifies how the tide slowly began to turn; rational explanations based on science began to replace the literal interpretation of the Bible.[187] Buckland acknowledged that the biblical flood was an untenable explanation for the geological phenomena in his "Bridgewater Treatise" of 1836,[188] a major factor in making rational and secular geological ideas socially acceptable:

> Buckland's accomplishments led to something of a personality cult. His geological knowledge became proverbial, and his personal habits as well as his discoveries and theories were made the subject of jocular verse and cartoons.[189]

Geology became an important part of the meetings of the British Association for the Advancement of Science (BAAS); geological societies and clubs emerged all over the country.[190] This was a clear indication that geology was on its way to becoming an independent scientific discipline, similar to natural history in the 1750s. When William Smith sold his extensive collection to the British Museum in 1816, it became apparent that sorting and displaying the pieces was as important as collecting them. Arranging the collections became a fulltime job, curators now had to be employed and paid (whereas before mostly honorary curators tended to the fossils). Although these curators were not paid especially well (indeed they could be likened to struggling artists), this constitutes another important step in the professionalization of the geosciences; the cabinets of curiosities had come a long way.[191] The foundation of the British Geo-

---

founded astronomy and Newton physics, Lyell has done as much for the science of geology. James A. Secord: Visions of Science. Books and Readers at the Dawn of the Victorian Age, Oxford 2014, pp. 138–172.

186 Freeman: Victorians and the Prehistoric, pp. 85–91.

187 Freeman: Victorians and the Prehistoric, pp. 164–178, 181–189.

188 Rupke: The Great Chain of History, pp. 81–88.

189 Rupke: The Great Chain of History, p. 71.

190 Freeman: Victorians and the Prehistoric, pp. 118–121. Or as Marjorie Grene and David Depew put it: "During the period in question, no science was developing as fast or provoking more intense debates than geology." See Grene; Depew: The Philosophy of Biology, p. 156.

191 Knell: The Culture of English Geology, pp. 99–111.

logical Survey in 1835 marks the end of the hands-off approach of the British government to geology.[192]

According to historian Kai Torsten Kanz, for a scientific discipline to become independent the following milestones have to be accomplished: there must be a stable and self-reproducing communication network of experts specialized in said discipline. The new scholarly insights are published in scientific and discipline-specific periodicals and textbooks and the scientists organize in societies and educational institutions.[193] At the beginning of the nineteenth century, geology had established itself as a discipline. In 1822 Henri Marie Ducrotay de Blainville (1777–1850), who would later succeed Cuvier as professor of comparative anatomy, further specified the disciplines formally collectively known as natural history: in the "Journal de Physique" he coined the term "paléontologie," meaning the study of ancient living organisms through fossils.[194] It still took some decades until paleontology found its own identity apart from biology or geology. At first, a paleontologist was perceived as a geologist meddling in biology, or vice versa as a biologist meddling in geology. In some respects, this prejudice lived on even after there were paleontological journals, textbooks, museums, and professorships. Paleontology did not find its place as a truly independent and valued discipline until the 1950s, due to the accomplishments of scientists like George Gaylord Simpson (1902–1984).[195]

The formation of scientific disciplines is of monumental importance to the history of science. The discipline dictates which aspect of nature is studied and, by doing so, becomes part of the process of knowledge generation. The real world is observed through the filter of a discipline; methods, instruments, and customs are predetermined by its conventions. Furthermore, the scientific discoveries are primarily discussed within the framework of the discipline and with other members of that specific community, abiding by the same rules and operating within a social framework.[196]

---

192 Knell: The Culture of English Geology, pp. 226–227.

193 Kanz: Von der BIOLOGIA zur Biologie, p. 20.

194 Rudwick: Worlds before Adam, pp. 47–58.

195 David Sepkoski; Michael Ruse: Introduction. Paleontology at the High Table, in: David Sepkoski; Michael Ruse (eds.): The Paleobiological Revolution. Essays on the Growth of Modern Paleontology, Chicago 2009, pp. 1–11, DOI: 10.7208/chicago/9780226748597.001.0001, see pp. 1–5.

196 "Jede Analyse von Veränderungen der Wissensproduktion muss die Struktur der wissenschaftlichen Disziplinen in den Blick nehmen, Die Disziplinen repräsentieren die Gegenstände (und das heißt die Inhalte) der Forschung. Man könnte auch sagen, dass die Gesellschaft ihre natürliche Umwelt und sich selbst durch die Brille der Disziplinen sieht. Insofern sind die Disziplinen auch ein zentrales Element der Wissensordnung. Zugleich sind Disziplinen auch die sozialen Organisationen, innerhalb derer die Forschungsfragen generiert werden, durch die also die Richtung der akademischen Wissensproduktion bestimmt wird." Weingart: Nachrichten aus der Wissensgesellschaft, p. 41.

As a result of the Napoleonic Wars, British geologists were somewhat isolated from the rest of the European scientific community, but after 1815 it began to blossom as communication (and competition) with the rest of Europe was re-established. Joseph Banks (1743–1820), president of the Royal Society from 1778 to 1820, endeavored to keep the Republic of Letters between Britain and France alive. He ensured that this essential communication network was re-established as soon as possible following the Ages of Revolution and Napoleon had ended; he had done the same thing with regards to the United States during and after the War for Independence.[197] As it had been since the sixteen hundreds, in the eighteenth and early nineteenth century science was still a product of collective discussion. Scientific discoveries were in large still spread verbally within the scientific network. This discussion within the peer group was of more importance than the written publication of scientific discoveries.[198] Alongside scientific discoveries, private news was exchanged within the network, contributing to its longevity. The aforementioned president Banks, for example, was a figure of enormous reputation and significance within the scientific community, yet published very little. While the community was open to all interested gentlemen (due to cultural practice, persons of other genders were rarely admitted) at the beginning of the nineteenth century, increasing professionalization of the scientists in the 1850s led to the establishment of "shop talk" within scholarly circles. The laymen and hobbyists, who had contributed most of the geological discoveries at the beginning of the century, could no longer easily follow the scientific discussions. Scientific circles became more exclusive.[199]

Thanks to Cuvier and the National Museum of Natural History ("Muséum national d'histoire naturelle"), France had taken the pole position in the study of natural sciences. Which is not to say that there were no German paleontologists' contributions to this new science. Paleontologists like Christian Keferstein (1784–1866) and Johann Jakob Kaup (1803–1873) also contributed to the rapid expansion of paleontology. Fossils became fundamentally important in geological practice, not only for determining the age of strata, but also because some fundamental truths about the history of life

197 Gascoigne: Science in the Service of Empire, pp. 151–157.

198 Though according to Martin Gierl science had to be published in a book or journal to be of any importance, and above all to be accessible: "So blieb denn, was nicht in den Büchern ist, nicht in der gelehrten Welt. Und der, der Wissen schafft, hatte Autor zu sein." This may be true for modern day scientific practices, but not for the scientific networks of the sixteenth through (early) nineteenth centuries. Martin Gierl: Korrespondenzen, Disputationen, Zeitschriften. Wissensorganisation und die Entwicklung der gelehrten Medienrepublik zwischen 1670 und 1730, in: Richard van Dülmen; Sina Rauschenbach (eds.): Macht des Wissens. Die Entstehung der modernen Wissensgesellschaft, Cologne 2004, pp. 417–438. Quote on page 438.

199 James A. Secord: How Scientific Conversation Became Shop Talk, in: Aileen Fyfe; Bernard V. Lightman (eds.): Science in the Marketplace. Nineteenth-Century Sites and Experiences, Chicago 2007, pp. 23–59, DOI:10.1017/S0080440107000564.

could be learned from them. Continental European fossils were sold at record prices in Great Britain. Because of their importance, fossils were treated like national treasures and British scientists tried to limit the exports of British fossils to be able to compete with their French and German counterparts.[200] Nonetheless, geology and science in general not only became more institutionalized but also more and more international: Cuvier and Werner for example joined the Geological Society of London. After Napoleon's wars and the Continental Blockade had ended, scientific exchange flourished.[201]

In the opinion of science historian and Buckland biographer Nicolaas Rupke, the importance of fossils can hardly be overstated: "A paleontologist was to some extent as good as his fossil collection. Civic or even national pride could be based on the possession of particular specimens or of collections, especially those of large vertebrate fossils."[202]

Images of gigantic, fire-breathing, and distempered dragons were firmly established in the imaginations of British Victorians. They were the stuff of beloved fairy tales and Christian mythology. Now, at the beginning of the nineteenth century, the fossilized remains of enormous, spectacular, dragon-like creatures were discovered not in some far off mysterious and exotic land, but in the seemingly mundane British countryside: In 1811, Mary Anning (1799–1847) of Lyme discovered parts of the skeleton of a large marine reptile in South England, which was later christened *ichthyosaurus*. Anning went on to discover other remarkable fossils but lacked the education (and social status) to describe them in a scientifically relevant way. Instead, Anning sold her fossils to the emerging paleontologists of England, such as the aforementioned William Buckland.[203] Anning represents a group of professional fossils collectors who did not study their discoveries themselves but sold them to the educated elite. Scientific demand had created a whole new occupation for rather uneducated subcontractors.

> Thus, geology became a mirror of Victorian industrial society. Fed by a proletariat of fossil gatherers, middlemen supplied the leisured classes who in turn manufactured thoughts on the subject at society meetings or adorned their palatial homes with a new kind of intellectual wallpaper.[204]

---

200 Knell: The Culture of English Geology, pp. 28–30.

201 Rudwick: Worlds before Adam, pp. 25–34.

202 Rupke: The Great Chain of History, p. 133.

203 Freeman: Victorians and the Prehistoric, pp. 132–136.

204 Knell: The Culture of English Geology, p. 6.

At the same time a medical doctor turned paleontologist,[205] Gideon Mantell (1790–1852), used fossils to determine that the British Isles were once covered by primeval tropical vegetation completely different from the fauna of nineteenth-century England. Even more startling was the discovery of some alien looking bone fragments and teeth. Elaborating on his findings, Mantell crafted the second science-based description of a dinosaur in 1825 and named it *iguanodon* (for the shape of the discovered teeth resembles those of a modern iguana). The first non-avian dinosaur had been discovered and named one year earlier, in 1824, by William Buckland, who gave his discovery the rather unimaginative name *megalosaurus* (meaning great lizard).[206] *Megalosaurus* and *iguanodon* (and *hylaeosaurus*, discovered in 1832 by Mantell) baffled Victorian scientists. They seemed not to fit within any known taxonomic group and were therefore given their own group by paleontologist Richard Owen (1804–1892) when he coined the term Dinosauria (meaning terrible lizards) in 1842.[207] Presumably at least in part due to the immense popularity of his creation, Owen is to this day remembered as one of the most important naturalists of all time:

> His name was mentioned in one breath with Isaac Newton's, and he was idolized as Britain's answer to France's Georges Cuvier and Germany's Alexander von Humboldt. [...] It is fair to say that among Britain's Victorian naturalists Owen came second in importance only to Charles Darwin.[208]

Gideon Mantell, too, is remembered today for his dinosaur discoveries, or as Dennis Dean put it rather poetically: "Gideon had ridden on the back of his *iguanodon* into

205 As a physician Mantell knew a lot about anatomy; later he could use this anatomical knowledge to study fossils. From an early age Mantell was fascinated by fossils, he started his collection in 1803 and added to it by collecting and buying from other enthusiasts and professional fossils gatherers. He became a member of the Linnaean Society in London and corresponded with Buckland and Lyell. He also sent some iguanodon teeth to Cuvier in Paris, who misidentified them at first, thinking they were of mammalian origin (later he corrected his first diagnosis, stating that Mantell's teeth must have belonged to an herbivorous reptile). Cuvier's interest in Mantell's findings boosted the physician's credibility as a paleontologist, illustrating the reach and importance of scientific correspondence in the early nineteenth century. Dennis R. Dean: Gideon Mantell and the Discovery of Dinosaurs, Cambridge 1999, pp. 11–23, 41–43, 71–85.

206 Note that the megalosaurus bones were first described and depicted in Robert Plot's (1640–1694) "The Natural History of Oxfordshire" in 1677, it was the earliest British book in which fossils were illustrated. Plot's illustrations were reused in 1763 by Richard Brookes (1721–1763), who gave them the name scrotum humanum for their resemblance to the human body part with the same name. Lambert Beverly Halstead; William A. S Sarjeant: Scrotum Humanum Brookes – The Earliest Name for a Dinosaur?, in: William A. S. Sarjeant (ed.): Vertebrate Fossils and the Evolution of Scientific Concepts. Writings in Tribute to Beverly Halstead, by Some of His Many Friends, Amsterdam 1995, pp. 219–222.

207 Freeman: Victorians and the Prehistoric, pp. 137–140. For a more detailed analysis of the discovery and description of the first dinosaurs see: Justin B. Delair; William A. S. Sarjeant: The Earliest Discoveries of Dinosaurs, in: Isis, vol. 66, no. 1 (Mar. 1975), pp. 4–25.

208 Nicolaas A. Rupke: Richard Owen. Victorian Naturalist, New Haven, CT 1994, p. 1.

the Temple of Immortality."[209] There seems to be something extraordinarily fascinating about the "terrible lizards."

Along with petrified tracks and plant fossils, the study of *coprolites*, which are fossilized feces, allowed the reconstruction of the environments in which dinosaurs and other extinct animals had lived. These lost worlds were captured in countless popular drawings and illustrations.[210] With his "The Age of Reptiles," published in 1831, Mantell greatly influenced how the general public would perceive dinosaurs for the next decades.[211] The early dinosaur reconstructions of Mantell, Owen, Buckland, and others differ greatly from modern reconstructions. All then-known dinosaurs were imagined as slow-moving, quadruped lizards of enormous size. Buckland's *megalosaurus*, for example, had been calculated to have a length of at least 20 meters. He had consulted with Cuvier, who compared the dinosaur's teeth to the teeth of modern-day lizards and scaled *megalosaurus* up accordingly,[212] whereas modern reconstructions describe a more agile bipedal carnivore, just 6 or 7 meters in length.[213] A case can be made that this misrepresentation of dinosaurs was due to the influence of Richard Owen, who had begun to "correct" (if not to say belittle) the achievements of other British paleontologists. Mantell had long realized that *iguanodon* must have been a bipedal animal, but his objections to Owen's reconstructions were not heard due to the latter's status as "the English Cuvier." Again, social standing trumped scientific reasoning.[214] It stands to reason that if the production of scientific fact was greatly influenced by socio-political factors, the (often very artistic) depiction of dinosaurs was even more affected by cultural trends.

The scientific findings were published in beautifully illustrated magazines and books, inspiring the imagination of the romantically minded British middle class. Droves of citizens traveled to the beaches and bluffs to collect fossils on and of their own. This public enthusiasm culminated in the construction of a prehistoric amusement park, in which life-sized reconstructions of dinosaurs and other antediluvian lifeforms were exhibited. The exhibitions had great scientific merit; the extinct animals were crafted in concrete by Benjamin Waterhouse Hawkins (1807–1894) in

---

209 Dean: Gideon Mantell and the Discovery of Dinosaurs, p. 88.

210 Rupke: The Great Chain of History, pp. 139–148.

211 Dean: Gideon Mantell and the Discovery of Dinosaurs, pp. 106–110.

212 Rudwick: Worlds before Adam, pp. 59–69.

213 Gregory S. Paul: The Princeton Field Guide to Dinosaurs, Princeton, NJ 2010, p. 86.

214 Dean: Gideon Mantell and the Discovery of Dinosaurs, pp. 178–191. The discord between Owen and Mantell grew further, deteriorating into a rather personal feud. At its apex, Mantell even tried to discredit his rival before the Geological Society, hoping the institution would suspend the funding of Owen's projects. This chapter in British paleontology foreshadows the bitter rivalry of American professors Cope and Marsh in the second half of the nineteenth century. Dean: Gideon Mantell and the Discovery of Dinosaurs, pp. 243, 258.

collaboration with Richard Owen and in accordance with the latest scientific knowledge.[215] This Victorian Jurassic Park opened its gates in 1854 in Crystal Palace Park,[216] south of London.[217]

Charles Darwin brought Lyell's "Principles" with him on his journey on the Beagle, studied South American valleys and mountainsides knowing that they were the product of the slow geological processes described in Lyell's book. Darwin brought numerous geological specimens and fossils back to England, where he sent them to Richard Owen. Owen noticed that while the fossils bore some resemblance to contemporary South American wildlife, they all belonged to an extinct fauna. Darwin reasoned that species not only went extinct but slowly changed their form, that some small anatomic details were altered in a long chain of successive generations, until the descendants were almost incomparable to their ancestors. Inspired by Thomas Malthus' (1766–1834) demographic theory,[218] Darwin gradually developed his theory of evolution as a struggle for resources on a global scale and as a process spanning millennia.[219] When crafting his theory, Darwin frequently consulted with Richard Owen, whose paleontological findings seemed to prove Darwin's ideas about transmutation (Owen however never embraced the theory of evolution and later became one of the most outspoken opponents of Darwin's theory). In contrast to Lamarck, Darwin realized that species not only went extinct, but that the struggle for survival was the driving force for transmutation.[220] Still, while crafting his theory Darwin could not rely on the plethora of fossil evidence that was discovered in the second half of the nineteenth century (see chapter 7. 1.), instead he had to refer to more recent examples found in plant and animal breeding.[221] Simon Knell elaborates on how the cultural evaluation of fossils shifted due to new scientific discoveries and theories. He argues

215 According to Rupke, Owen was the natural choice to scientifically reconstruct the lost creatures: he had described many species after studying them himself in numerous British museums, meaning he had a knack for observation and was a somewhat talented illustrator. He drew the construction plans for Hawkins' models, which became a sought-after commodity for every modern Victorian museum. Rupke: Richard Owen, pp. 130–135.

216 Named after the Crystal Palace, which was part of the Great Exhibition of 1851, that took place in London's Hyde Park. After the closure of the exhibition, the Crystal Palace was moved to its new location in the south of the city, nowadays the suburb Crystal Palace.

217 Freeman: Victorians and the Prehistoric, pp. 154–161.

218 Best expressed in his "Essay on the Principle of Population" of 1798. See: Thomas Robert Malthus: An Essay on the Principle of Population, as it Affects the Future Improvement of Society. With Remarks on the Speculations of Mr. Godwin, Mr. Condorcet, and Other Writers, London 1798.

219 Freeman: Victorians and the Prehistoric, pp. 201–212.

220 Sandra Herbert: Charles Darwin, Geologist, Ithaca, NY 2005, pp. 320–331.

221 Herbert: Charles Darwin, Geologist, pp. 331–354.

that the changing appreciation can best be studied by looking at the reactions of contemporary individuals.[222]

In the 1830s, English geology caught up with French, German, and Scottish science as a result of focusing on fossils, or as English paleontologist William Conybeare (1787–1857) judged: "The study of fossils rather than of minerals had been a root cause of the success of English geology."[223] Grene judges (with some historical perspective): "British science came of age in the first half of the nineteenth century."[224] Paleontology and geology had established themselves as independent scientific disciplines. Paleontological discoveries, especially the fearsome dinosaurs, captured hearts and minds.

2.6 Geosciences in North America

"If the democratic principle does not on the one hand induce men to cultivate science for its own sake, on the other, it does enormously increase the number of those who do cultivate it."[225]

The enlightened scholars of the eighteenth century linked the success of science to a climate of liberty; therefore, great things were to be expected of US-American scientists after 1783. Wars and rigid religious interference were recognized as prohibitive to science, sponsorship by the state, or a sovereign beneficial.[226] Yet the absence of wealthy (aristocratic) sponsors was a point of concern for the future of science in the young republic, especially since nationalistic sentiments, sweeping the States following their recent independence, gave most endeavors a patriotic dimension:

---

222 Simon J. Knell: Museums, Fossils and the Cultural Revolution of Science. Mapping Change in the Politics of Knowledge in Early Nineteenth-Century Britain, in: Simon J. et Knell al. (eds.): Museum Revolutions. How Museums Change and Are Changed, London 2007, pp. 28–47.

223 Quoted from: Rupke: The Great Chain of History, p. 182.

224 Grene; Depew: The Philosophy of Biology, p. 154. Note that dinosaurs had also been discovered in France, maybe as early as the late eighteenth century. There, the first remains of undoubtedly dinosaurian origin were described by Cuvier in 1808, long before the name dinosaur was coined. Megalosaurus was discovered and described in France in 1828, but it seems that French dinosaur paleontology quickly fell behind its British and then US-American counterparts. Many of the French paleontologists did little field work themselves and had little interest in the dinosaur fossils, which were rather fragmentary, especially compared to the discoveries made in North America in the 1870s, and in later decades. Eric Buffetaut et al.: The Discovery of French Dinosaurs, in: William A. S. Sarjeant (ed.): Vertebrate Fossils and the Evolution of Scientific Concepts. Writings in Tribute to Beverly Halstead, by Some of His Many Friends, Amsterdam 1995, pp. 159–180.

225 Alexis de Tocqueville: Democracy in America, quoted from: Dirk J. Struik: Yankee Science in the Making, Boston, MA 1948, p. 199.

226 Engelhardt: Historisches Bewußtsein, pp. 186–190.

> To Americans living in this period of exploding scientific inquiry, the fundamental fact conditioning every thought and deed was the consciousness that they were now an independent nation. With respect to science this meant two things: as the example par excellence of useful knowledge, science must be cultivated to promote the interests, prosperity, and power of the rising American nation; and as the supreme example of the powers of the human mind, the successes of science challenged Americans to prove to the world that republican institutions were as favorable to intellectual achievement as they were to liberty.[227]

> Science transcends national boundaries and rivalries, but scientists are citizens and patriots as well as co-workers in the pursuit of truth. [...] Americans felt themselves to be a chosen people with a sacred duty to prove that republican institutions were at least as favorable to letters and science as monarchical ones.[228]

Furthermore, no hubs of knowledge, such as London or Paris, existed in the United States yet, and therefore the history of the geosciences in the US is regional.[229] In the early days of the republic, mineralogy was occasionally taught at medical colleges, but never properly incorporated at American institutions of higher education. Although the Philosophical Society of Philadelphia (founded in 1743) acquired a mineralogical collection early on, the science of mineralogy was not taught at American colleges up until the early nineteenth century, when it became part of the curriculum in Philadelphia. The City of Brotherly Love with its philosophical society, the university, a medical school, and a museum of natural history remained at the forefront of American science for decades. When Benjamin Silliman (1779–1864) became professor for natural history and chemistry at Yale College, he had to travel to Philadelphia to have a suitcase full of minerals classified at the university. He later traveled through Britain to further his knowledge concerning the geosciences. It is in large parts thanks to Silliman that Yale's teaching of these sciences caught up with Philadelphia's in the middle of the nineteenth century.[230]

---

227 John C. Greene: American Science in the Age of Jefferson, Ames, IA 1984, pp. 5–6.

228 Greene: American Science in the Age of Jefferson, p. 10.

229 John C. Greene; John G. Burke: The Science of Minerals in the Age of Jefferson, Philadelphia 1978, pp. 20–22.

230 John C. Greene: The Development of Mineralogy in Philadelphia, 1780–1820. A Summary, in: Cecil J. Schneer (ed.): Toward a History of Geology. Proceedings of the New Hampshire Inter-Disciplinary Conference on the History of Geology, September 7–12, 1967, Cambridge, MA 1969, pp. 184–185. Greene; Burke: The Science of Minerals in the Age of Jefferson, pp. 22–44, 93–107.

Nonetheless, it seems that even in the first half of the nineteenth century there must have existed at least some semblance of public interest in the geosciences, for Lyell's "Principles" was republished in Boston in 1841. On this occasion Lyell and his wife visited the Northeastern parts of the country, where the renowned geologist had the chance to study American fossils firsthand. The Lyells were invited by Prof. Silliman to stay at his house in New Haven, Connecticut. Silliman was the linchpin of the transatlantic network of geoscientists.[231] Back in England, Lyell published the findings of his travels in two volumes in 1845. The second volume included a detailed and colorized geological map of Canada and the US. Besides his scientific discoveries, Lyell also discussed some socio-economic subjects like slavery and the condition of the US-American universities (the latter of which he praised highly).[232] Nonetheless, science historian Mott Greene argues that geologists and paleontologists were the most crucial developers of US-American educational reforms, and therefore an interest in their activities seems very well-founded.[233]

One of the first known reports of American fossil findings dates back to the most prominent puritan author and theologian of Salem witch-hunting infamy: Cotton Mather (1663–1728).[234] When in 1705 a seventeen-foot thighbone was unearthed in Claverack, New York, Mather identified it as the bone of a giant. He knew that giants had existed because the Bible said so, therefore this was physical evidence for the existence of at least one biblical giant, who furthermore had lived in America. Decades later it turned out that the enormous bones belonged to a *mastodon,* nonetheless this episode marks the first major fossil discovery in North America. It exemplifies how the early transatlantic scientific networks operated, for Mather discussed this discovery and possible implications with his peers at the Royal Society in London. He did this at length and in detail in a series of letters sent across the Atlantic between 1712 and 1724. It is likely Mather also knew of Native American stories about giants but ignored them because to Mather the legends of heathens could hardly be taken seriously, or indeed might even be diabolical deceptions.[235] Whether Puritanism, which dictated life in the

231 Leonard Gilchrist Wilson: Lyell in America. Transatlantic Geology, 1841–1853, Baltimore, MD 1998, pp. 7–15.

232 Wilson: Lyell in America, pp. 141–146.

233 "Because geologists and paleontologists had a distinguished formative role in the history of American scientific institutions, public and private, this interest is well merited." Mott T. Greene: History of Geology, in: Osiris, 2nd ser., vol. 1 (1985), pp. 97–116. Quote on page 101.

234 Cotton Mather drew much of his scientific knowledge from German scholars, such as Athanasius Kircher and Otto von Guericke, see: Henry A. Pochmann: German Culture in America. Philosophical and Literary Influences 1600–1900, Madison, WI 1957, p. 84.

235 David Levin: Giants in the Earth. Science and the Occult in Cotton Mather's Letters to the Royal Society, in: The William and Mary Quarterly, vol. 45, no. 4 (Oct. 1988), pp. 751–770. Adrienne Mayor: Fossil Legends of the First Americans, Princeton, NJ 2005, pp. 36–37. Also see: Paul Semonin: American

young colonies in North America and still dominates their historiography, held back the development of scientific institutions or not is a controversial subject to this day. Some historians argue that it was in no way at odds with science; instead, new scientific discoveries were simply interpreted within a puritan framework to demonstrate the glory and genius of the divine creation. Others say that Puritanism and religious fervor caused American colleges to stick to a curriculum centered on classical education and moral instruction, and that this conservative influence discouraged the study of natural sciences until the Morrill Act of 1862.[236]

For the sake of completeness, it should be mentioned that the very first American fossils studied by Europeans were of South American origin. In a tradition dating back to Hernán Cortés (1485–1547), South American Natives told Spanish travelers legends of incredible creatures (presumably inspiring numerous folktales). The Aztecs used fossils as jewelry and (ground into powder) as medicine. Alexander von Humboldt (1769–1859) encountered some South American fossils on his travels and sent them to his friend Cuvier who recognized them as the teeth of prehistoric elephants.[237]

The first North American fossils that were identified correctly were found in 1725 near a plantation called Stono in South Carolina. A group of enslaved Africans discovered some huge teeth and identified them as elephant molars, with which the Africans (presumably they were Congolese or Angolan) were apparently familiar. The teeth indeed belonged to a *mammoth*, as Cuvier noted in 1806, and consequently the enslaved workers noticed the kinship of modern elephants to *mammoths* long before European scientists did. Adrienne Mayor suggests this episode in American paleontology might be as obscure as it is for racist reasons and was only begrudgingly incorporated in an otherwise heroic saga, dripping with national pride.[238]

In 1739, Baron Charles de Longueuil (1687–1755), a major in the French Army in Montreal, was ordered to lead his soldiers south to Louisiana to fight the native Chickasaws, who were allied with the British. The expedition made its way down the Ohio River. While passing through the lands that are now known as Kentucky, native scouts (Adrienne Mayor identifies them as Abenaki) brought numerous bones belonging to gargantuan animals back into camp. They had found the bones in a nearby marsh near a salt lick. Longueuil took some of the bones to Paris the next year, where they were stored and studied. In 1756 drawings of the North American fossils were published by Jean-Étienne Guettard (1715–1786). In 1821 the bones were identified by

Monster. How the Nation's First Prehistoric Creature Became a Symbol of National Identity, New York 2000, pp. 15–40.

236 Ronald L. Numbers: Science and Religion, in: Osiris, 2nd ser., vol. 1 (1985), pp. 59–80.

237 Mayor: Fossil Legends of the First Americans, pp. 73–105.

238 Mayor: Fossil Legends of the First Americans, p. 56.

Cuvier as those of an *American mammoth* ("mammouth d'Amérique" as he called it).[239] Before the American bones were correctly identified by Cuvier, they had been studied in London by William Hunter (1718–1783), fellow of the Royal Society and physician to Queen Charlotte. Hunter called the creature the unknown bones belonged to *American incognitum*; the name would stick until Cuvier's abovementioned rectification. Hunter also suggested that the *incognitum* might have been a carnivore. Consequently, the *incognitum* was imagined to be a fierce and bloodthirsty beast, was metaphorically linked to the revolting American colonies, and "became an emblem of the rebellion" (see below).[240] This illustrates how from the very start the history of American paleontology was linked to Europe's through colonial ties. Still lacking scientific institutions, equipment, and funding in general, the colonies offered an abundance of nature, providing European scientists with data to study. Therefore, it might be argued that the study of nature and its history has a longer tradition than any other science in America. However, the salt lick in what is nowadays Kentucky would provide many more assets for generations of American paleontologists to come; it is known to this day as the Big Bone Lick (designated as a State Park in 1960). It also demonstrates how Native Americans were crucial to the discovery of American fossils. They knew their way around the land and found some of the fossils that would fuel European and American theories, and it seems likely that giant petrified bones had motivated some Native American legends: legends about terrible monsters, slain by fabled heroes. The Irish trader George Croghan (c. 1718–1782) might have heard some of these stories since he acted as middleman between Native bone collectors and European scientists. He sold bones from the Big Bone Lick in great quantities to Europe in the middle of the eighteenth century.[241] Transatlantic communication was of the utmost importance to the development of American geosciences and science in general, from its inception all throughout the nineteenth century: "American geologists have always had close contact with their European counterparts [...]. Therefore, one cannot write the history of American geology without knowing the history of European geology."[242]

Native American legends, knowledge of the land and of the whereabouts of fossil hunting grounds would continue to be of enormous importance to American paleontology for decades to come, especially when the western territories were explored in detail after the Civil War (see chapter 5). "And as European and Euro-American nat-

---

239 Mayor: Fossil Legends of the First Americans, pp. 1–7. Buffetaut: A Short History of Vertebrate Palaeontology, pp. 37–39.

240 Semonin: American Monster, pp. 137–161. Quote on page 161.

241 Mayor: Fossil Legends of the First Americans, pp. 8–15.

242 Greene: History of Geology, p. 110.

uralists became aware of the significance of fossils in the New World, Native knowledge and guides actively contributed to the development of paleontological science."[243]

This resulted in legal issues concerning the ownership of some fossils, still fought to this day.[244]

Like Longueuil's discoveries, most North American fossils were brought to the Old World to enrich the collections of European paleontologists. The notion of natural space is of enormous importance to the self-conception of many generations of US citizens. But even before the United States were founded, and even before European colonies could gain a foothold in North America, the natural landscape of the continent was perceived as a utopic, almost sacred phenomenon. This New World promised to be a new Garden of Eden, the reverse of the old European spaces associated with the supposedly profane decadence of stalled societies and deadlocked kingdoms. On the one hand, the New World promised a return to a more natural, divinely inspired existence; on the other hand, the new Garden of Eden could prove itself to be a deadly wilderness threatening the survival of cultivated rural life. A compromise between the two extremes, manifesting itself in "frontier life," seemed to encourage a virtuous existence.[245] This notion, or variations of it, can be found all throughout Euro-American history. In the nineteenth century the railroad quickly came to symbolize the expansion of the United Sates along its "western frontier." Allegedly, the "frontier experience" permanently renewed democratic core values like self-reliance and self-determination, which would then trickle back to the east coast.[246] The sheer size of the continent and the natural conditions of the North American landscape, seemingly untouched by human hands, provided a plethora of fossils. Thus, American paleontology had always been intertwined with the geography of the land and the exploration of the continent: "Sensational discoveries of fossil vertebrates in North America were soon to bring American scientists to the forefront of research in this field [paleontology]."[247]

243 Mayor: Fossil Legends of the First Americans, pp. 296–297.

244 Mayor: Fossil Legends of the First Americans, pp. 303–313.

245 Robert Berkhofer Jr. highlights this ambivalence between the benevolent Garden of Eden with its abundance of natural resources, ready for the taking, on the one hand, and the fear of the untamed wilderness and its savage, if not diabolical, beasts on the other hand, which the (mostly puritan) Euro-Americans harbored during the seventeenth century. See: Berkhofer: The White Man's Indian, pp. 72–85. For a very comprehensive analysis of the evolving view English settlers had of the wilderness, sometimes finding paradise in nature, other times suffering degenerative influences upon their own civilization, see: Richard Slotkin: Regeneration Through Violence. The Mythology of the American Frontier; 1600–1860, 4th ed., Middletown, CT 1987.

246 Gerhard Strohmeier: Wild West Imagery. Landscape Perception in Nineteenth-Century America, in: Mikuláš Teich et al. (eds.): Nature and Society in Historical Context, Cambridge 1997, pp. 257–273.

247 Buffetaut: A Short History of Vertebrate Palaeontology, p. 121.

In his 1924 review of the "First One Hundred Years of American Geology" George Merrill summarizes the humble beginnings of geology in the United States:

> Not [in] a single university was geology taught as a science. There were no accurate maps, and topographic delineation was undreamt of. Neither were there railroad cut nor deep well borings to give a clue to the earth's structure beneath the immediate surface. The country was largely a wilderness, and the information with which the geologist of today begins his career was uncreated. Naturally such as was available was almost wholly of European derivation; indeed, many of the workers had received what training they may have had in European universities.[248]

Charles Miller writes that nature has been a source of American nationalism since the early days of the republic. Naturalists like Thomas Jefferson associated the raw, apparently untamed nature with the young nation. It becomes apparent that the study of nature and natural history has always been a political issue for the United States:

> Further, insofar as nature symbolized America in its entirety, nature *was* America for Jefferson. His interest in nature and his use of the word are therefore a form of nationalism. In Europe national sentiment was expressed through a common history, a royal family, a culture or a literature. In America and for Jefferson it was expressed through, and as, nature.[249]

At the dawn of the nineteenth century, some of America's brightest and most well-educated minds took an interest in the gigantic bones from the Big Bone Lick. Jefferson, for example, had convinced himself that some of the huge animals whose bones fired his imagination were still roaming the western parts of North America, a terrain mostly unknown to Euro-Americans at that time.[250] Jefferson had always been fascinated with nature: "He was always the farmer, always the natural historian, always interested in natural theology and natural morality, always ready to base the political order on natural law and natural right."[251] In Jefferson's mind scientific pursuits and American patriotism were inextricably linked.[252] Indeed, the self-evident truths, the

---

248 Merrill: The First One Hundred Years of American Geology, p. 1.

249 Charles Allen Miller: Jefferson and Nature. An Interpretation, Baltimore, MD 1988, p. 3.

250 Buffetaut: A Short History of Vertebrate Palaeontology, pp. 122–126. See also: Daniel Justin Herman: Hunting and the American Imagination, Washington, DC 2001, pp. 88–89.

251 Miller: Jefferson and Nature, p. 1.

252 "In most of its leading characteristics – patriotism; utilitarianism; antitheoretical bent; fascination with the geography, flora and fauna of the North American continent; and interest in the relations

unalienable right to life, liberty, and the pursuit of happiness as demanded by the Declaration of Independence, derive from the conception of natural (human) rights, a notion that can be dated back to Isaac Newton and John Locke (1632–1704).[253] Jefferson had collected numerous Native American stories about fossils and their origin. French fur traders shared rumors about massive animals still roaming the uncharted prairies, reinforcing Jefferson's hopes.[254] The discovery of these giants would come in handy when refuting Buffon's theories on New World degeneracy: Many eighteenth-century scientists linked appearances and attributes of organisms to their surroundings. It was believed that climate and resources dictated how a creature would develop; that too much rain and humidity would lead to degeneration, making species weak, lazy, and stupid. None other than French scientific superstar Buffon argued that the American climate led to such degeneration and that all animals imported from Europe had degenerated in size. Jefferson argued against that. He filled pages of his "Notes on the State of Virginia" (published in 1787) with tables comparing animals indigenous to Europe with their American counterparts. If climate actually caused life to degenerate, the abundance of land and resources in conjunction with the spirit of American civilization would more than make up for the climatic shortcomings. Native Americans, too, could be "civilized" in that manner.[255] Jefferson had used the

---

of science, politics and religion – American science found an appropriate spokesman and symbol in Thomas Jefferson." Greene: American Science in the Age of Jefferson, p. 27.

253 I. Bernard Cohen: Science and the Founding Fathers. Science in the Political Thought of Jefferson, Franklin, Adams, and Madison, New York 1995, pp. 114–121. Furthermore, at the age of 77 Jefferson wrote in his unpublished and unfinished autobiography: "Celebrated writers of France and England had already sketched good principles on the subject of government. Yet the American Revolution seems first to have awakened the thinking part of the French nation in general from the sleep of despotism in which they were sunk. The officers too who had been to America, were mostly young men, less shackled by habit and prejudice, and more ready to assent to the suggestions of common sense, and feeling of common rights. They came back with new ideas & impressions. The press, notwithstanding it's [sic!] shackles, began to disseminate them. Conversation assumed new freedoms. Politics became the theme of all societies, male and female, and a very extensive & zealous party was formed which acquired the appellation of the Patriotic party, who, sensible of the abusive government under which they lived, sighed for occasions of reforming it." Jefferson was of the opinion that the French revolution was directly inspired by the American Revolution. And if American politics and preconceptions of natural rights could transform and inspire the political and philosophical landscape in Europe, why should not American science fulfill the same function as well? See: Thomas Jefferson: Autobiography, in: The Avalon Project – Documents in Law, History and Diplomacy (URL: http://avalon.law.yale.edu/19th_century/jeffauto.asp, as consulted online on January 19, 2018).

254 On the other side of the Atlantic Cuvier had also heard the Native Americans' legends. He was more of a skeptic then Jefferson and thought the fabled animals had long gone extinct. Nonetheless it seems that the highly acclaimed French paleontologist held the Native American legends in high regard and did not simply ignore them like Cotton Mather had done one century before. Mayor: Fossil Legends of the First Americans, pp. 55–64.

255 Jenny Davidson: Breeding. A Partial History of the Eighteenth Century, New York 2009, pp. 152–159. Also see Miller: Jefferson and Nature, pp. 43, 61–63; Greene: American Science in the Age of Jeffer-

essential method of Baconian science, data collecting, to demonstrate how Buffon's theory proved to be fallacious.[256] Paul Semonin, who demonstrates that from its very beginning American paleontology was linked to regional and national identity, describes the Buffon-Jefferson conflict in the following drastic manner:

> Buffon's theory was an attack on the manhood of the American patriots and was particularly galling to those republicans who saw themselves as the Anglo-Saxon masters of a future empire extending across the continent to the Pacific Ocean. For the founding fathers, the discovery of the fossil remains of the American *incognitum* enabled them to refute Buffon's theory and to demonstrate the grandeur of the new republic's antiquity. The campaign to repudiate Buffon's humiliating theory actually began during the American Revolution itself with the first efforts by the founding fathers to celebrate the new nation's natural history.[257]

He furthermore links the desire to conquer the imagined American wilderness to the desire to master the knowledge of the truly ferocious ancient creatures, who had populated North America in the distant past. Such creatures were imagined as bloodthirsty carnivores:

> Despite their desire to create a universal society and to repudiate Buffon's view of American degeneracy, the founding fathers continued to see themselves as the dominant race, bringing civilization to the 'savages,' both Native Americans and African Slaves. The savagery of prehistoric nature, symbolized by the jaws of American *incognitum*, was linked to their own aspirations of empire over the natural world, which for them, included the heathen nations and races.[258]

---

son, pp. 27–35. For further reading on the ambivalent nature of Jefferson's conception of nature and "race" concerning Native and African Americans, see Miller: Jefferson and Nature, pp. 63–76. Also see: Berkhofer: The White Man's Indian, pp. 42–44.

256 Buffon was convinced by Jefferson's counterarguments and retracted. Cohen: Science and the Founding Fathers, pp. 86–87. Yet the style of environmentalism employed by Buffon was never fully abandoned and is yet another example for how the supposedly objective study of nature can be (mis-) used to further a political agenda: slaveholders used arguments similar to Buffon's in defense of their "peculiar institution." They claimed that enslaved Africans were created differently from Europeans to be able to cope with their "natural" environments and were best adapted to the humid climate of Dixie (even better than to the climate of their place of ancestral origin). Mart A. Steward: 'Let Us Begin With the Weather'. Climate, Race and Cultural Distinctiveness in the American South, in: Mikuláš Teich et al. (eds.): Nature and Society in Historical Context, Cambridge 1997, pp. 240–256.

257 Semonin: American Monster p. 6.

258 Semonin: American Monster p. 13.

Jefferson moved to Philadelphia in 1797 to assume his duties as vice-president of the United States and as president of the American Philosophical Society, where he held a lecture on *megalonyx*, a creature he imagined to be a lionlike predator (in reality the claws that had been described by Jefferson belonged to a giant sloth). He also imagined the American *incognitum* to have been a carnivorous elephant, as Hunter had done decades before (see above). This episode demonstrates how Jefferson's desire to construct the image of the American prehistoric fauna to have been as ferocious as possible interfered with his scientific conduct, and led to false conclusions in his research on fossils that belonged to herbivores and did not fit his gruesome presumptions about a savage past.[259]

Jefferson was interested in paleontology throughout his life. After the Louisiana Purchase of 1803, Jefferson, now president of the United States, commissioned Meriwether Lewis (1774–1809) and William Clark (1770–1838) to survey the recently acquired territory. A further objective of the celebrated Lewis and Clark Expedition was to bring back bones of the rumored animals or, even better, living specimens.[260] Or to say it in Jefferson's own words, as written in the "Notes on the State of Virginia":

> The bones of the mammoth which have been found in America, are as large as those found in the old world. It may be asked, why I insert the mammoth, as if it still existed? I ask in return, why I should omit it, as if it did not exist? Such is the economy of nature, that no instance can be produced of her having permitted any one race of her animals to become extinct; of her having formed any link in her great work so weak as to be broken. To add to this, the traditionary testimony of the Indians, that this animal still exists in the northern and western parts of America, would be adding the light of a taper to that of the meridian sun. Those parts still remain in their aboriginal state, unexplored and undisturbed by us, or by others for us. He may as well exist there now, as he did formerly where we find his bones.[261]

Cohen writes on the relationship between Jefferson and paleontology:

> Throughout his mature life, paleontology always remained for Jefferson a principal scientific interest. He was an avid collector of fossil bones and he even

259 Semonin: American Monster pp. 288–314.

260 Greene: American Science in the Age of Jefferson, pp. 196–197. See also Miller: Jefferson and Nature, p. 241.

261 Thomas Jefferson: Notes on the State of Virginia, Philadelphia 1788, p. 54.

> believed that some giant mammoths were still in existence somewhere in the wilds of America.[262]

The study of nature was interwoven with nationalism:

> Paleontology thus took a patriotic dimension and spurred an especial interest in two animals: the mammoth (also called mastodon in Jefferson's day), a hulking North American precursor of the elephant; and the megalonyx, an outsized sloth which Jefferson believed (or hoped) to be a massive lion.[263]

Jefferson sent William Clark to the Big Bone Lick to obtain fossilized bones; some he studied in the White House, others he sent to France for further examination.[264] The Lewis and Clark Expedition was one of the first instances in which the US-government directly financed a scientific excursion to gain strategic information and, in the process, advanced natural science. The United States geological surveys drew on this tradition and became a crucial tool in furthering American science and nationalism. In a letter to Silliman, published in the American Journal of Science and Arts in 1836, Charles Thomas Jackson (1805–1880) praised the involvement of the government in executing numerous geological surveys of the country:

> No other people in the world, I may safely affirm, have ever called on their governments, to furnish information of this kind; from which fact we may conclude that the American people are more enlightened respecting the application of science to the arts, than the people of any European state.[265]

Note that while Lewis and Clark did not encounter the fabled great beast Jefferson was hoping they would find, they might have brought back with them one of the first dinosaur bones discovered by Euro-Americans in North America. But because that bone has never been fully described and has since been lost, it is up to speculation if it really was of dinosaurian origin.[266]

---

262 Cohen: Science and the Founding Fathers, p. 290.

263 Miller: Jefferson and Nature, p. 51.

264 Greene: American Science in the Age of Jefferson, pp. 282–291. See also Cohen: Science and the Founding Fathers, p. 63; Miller: Jefferson and Nature, p. 50.

265 Charles T. Jackson: On the Collection of Geological Specimens and on Geological Surveys, in: The American Journal of Science and Arts, vol. 30, no. 1 (Jul. 1836), pp. 203–208. Quote on page 203.

266 Delair; Sarjeant: The Earliest Discoveries of Dinosaurs, pp. 10–11.

Note that after Jefferson's presidency ended in 1809, "federal support for natural history lapsed for nearly a decade."[267] The absence of good infrastructure hampered travel and communication and the transport of specimens, as "[t]he country lacked the resources and the audience for both scientific journals and the elaborately illustrated volumes that at the time seemed essential to natural history."[268] This demonstrates how crucial funding and infrastructure were and are to the conduct of science, and how important government funding and the expansion of the railroads during the second half of the century were to become for scientists (see chapter 5).

The bones of ancient elephants like *mastodon* and *mammoth* were studied in Europe; original American contributions to paleontology were of a sporadic nature up until the middle of the nineteenth century. The exploration of the American West after the Civil War came to change that; it opened a treasure trove of exciting fossils. Above all, dinosaur skeletons were to captivate the public's attention and direct it towards paleontology, as they had in Britain.[269]

At the beginning of the nineteenth century, American science was still regarded as having been severely lackluster: "America had seemed to be lagging behind in science, work done here was largely derivative, and at the beginning of the century it was virtually impossible to arouse either public or private support for any scientific enterprise."[270] It seems that George Daniel, the author of that quote, does not consider the Lewis and Clark Expedition a scientific enterprise (indeed it was not a purely scientific pursuit; if Congress was to sponsor the expedition it had to promise at least some practical gain). He further claims that American science really kicked into gear after the War of 1812, when American scientists began to organize nationwide and established lasting ties with their European colleagues. Advancements in transportation and communication furthered the establishment of national and then international scientific communication. The slow start of American science was, in the minds of most Americans, not due to an inferiority of the American mind or industry to their European counterparts. What the United States lacked initially was time for scientific development (in contrast to decades and centuries of history in Europe) and adequate funding that would allow bright Americans to focus on the production of knowledge and not just material profit. Scientific advancement began to pick up pace after 1815, when Americans began investing in their scientific institutions. The number of col-

267 Philip Pauly: Biologists and the Promise of American Life. From Meriwether Lewis to Alfred Kinsley, Princeton, NJ 2000, p. 20.

268 Pauly: Biologists and the Promise of American Life, p. 20.

269 Joseph T. Gregory: North American Vertebrate Paleontology, 1776–1976, in: Cecil J. Schneer (ed.): Two Hundred Years of Geology in America. Proceedings of the New Hampshire Bicentennial Conference on the History of Geology, Hanover, NH 1979, pp. 305–335, see pp. 305–307.

270 George H. Daniels: American Science in the Age of Jackson, Tuscaloosa, AL 1994 (orig. publ. 1968), p. 7.

leges, scientific societies, and journals was rapidly growing, the "American Journal of Science and Arts," co-founded by Silliman in 1818, being of the utmost importance to the development of American science: while most other periodicals had a very short lifespan, Silliman's journal prevailed. After some initial financial troubles (Yale would only lend moral support to the magazine) in 1829, the journal could afford to pay for original contributions.[271]

Leonard Warren aptly summarizes the American scientific environment in the first half of the nineteenth century as follows:

> Having achieved political independence, Americans cultivated the English naturalist tradition of direct observation and illustration adapted to a democratic society. The dependence of American scientists and intellectuals on current European ideas was accompanied with a defensive, nationalistic element in science. [...] Americans realized that the most effective way to achieve parity with Europeans was to borrow from and mimic them. And so, from the 1840s on, they looked to German, French, and English thinkers and scientists for guidance.[272]

Philip Pauly writes that the United States transformed between the American Revolution and the 1820s, and that one important aspect of that transformation was that "American nationalism became an established ideology."[273] He describes how this newfound nationalism and the desire to remove the young republic from its British roots altered the view on science and nature; both were now employed in the name of nationalism. He writes that the "interest in describing organisms specific to America"[274] were employed in the name of nationalism, and gives a vivid example of this practice:

> At his museum, opened in Philadelphia in 1786, Charles Willson Peale displayed together the first reconstructed skeleton of the 'American mastodon,' stuffed specimens of such nationally symbolic native animals as the bald eagle and wild turkey, and the portraits he had painted of Washington, Jefferson, and other Revolutionary leaders.[275]

---

271 Daniels: American Science in the Age of Jackson, pp. 7–26.

272 Leonard Warren: Joseph Leidy. The Last Man Who Knew Everything, New Haven, CT 1998, p. 41.

273 Pauly: Biologists and the Promise of American Life, p. 17.

274 Pauly: Biologists and the Promise of American Life, p. 18.

275 Pauly: Biologists and the Promise of American Life, pp. 18–19.

Indeed, Charles Willson Peale (1741–1827) had succeeded in linking the freshly unearthed skeleton of a *mastodon*, which he called *mammoth* and imagined to be a carnivore, to US-American patriotism. The bones of the giant creature were linked to the destiny of the young nation as the "conqueror" of a continent that was presumed to be a wilderness and that had been the home of such savage creatures as the *mastodon*, and still was the home to many human "savages." Peale displayed his skeleton in Philadelphia and toured with it through London and the English countryside in 1803. Later, some skins, skeletons, and Native American artifacts that had been acquired by the Lewis and Clark Expedition adorned Peale's museum alongside the *mastodon*. After his death, Peale's museum was sold to P. T. Barnum (1810–1891), continuing the tradition by sensationalizing the exhibition and further linking the bones and artifacts to US-American nationalism, self-imagination, and popular culture (see Chapter 5. 3.).[276] Irmsher describes Peale's museum "as a kind of secular temple," and claims that the museum reflected "his quiet confidence that the eyes of God rested perhaps a little more, favorably on the American portion of his creation." Peale even envisioned that an American museum harbored the potential of becoming "one of the first in the world" thanks to all the new specimens that would surely be found in the near future in the "vast territories" of North America. Peale not only aimed for patriotic reaffirmation but also for public education, and labelled his exhibits extensively, adopting the Linnean System in ordering his natural specimens.[277]

Louis Agassiz emigrating to the United States in 1846 constitutes another milestone in the development of transatlantic sciences. Agassiz had been educated by Humboldt and Cuvier, and the latter had bequeathed his extensive fossil collection to Agassiz. Humboldt had swayed the king of Prussia to fund Agassiz' journey to Boston, where Agassiz was to deliver a series of public lectures. Furthermore, Lyell encouraged his friend Agassiz to take the trip; he had been welcomed to America with open arms six years before. The lectures were a huge success and attracted more than 5,000 attendees, and each lesson had to be held twice to keep up with demand. Thanks to his attractive appearance and demeanor Agassiz made many friends and professional contacts, and in 1847 the Lawrence Scientific School, headed by Agassiz, was established at Harvard University. He fully integrated into his adopted home, his three children marrying into Bostonian high society. Agassiz taught science in a very modern European manner: practically independent from political and religious influences and with a focus on empirical observation, a revolutionary and most uncommon

---

276 Semonin: American Monster pp. 315–361. For more on the funding of the expedition by the US government see: Davidson: Patrons of Paleontology, pp. 43–46.

277 Christoph Irmscher: The Poetics of Natural History. From John Bartram to William James, New Brunswick, NJ 1999, pp. 56–100. Quote on page 57.

practice at American universities at that time (see chapter 8. 3).[278] Struik even postulates that "Agassiz also hoped to free American savants from their dependence upon Europe, from the subservient role in which they had voluntarily placed themselves."[279] Furthermore, it seems that radically new scientific theories from Europe were not embraced in the US until the 1840s and until transatlantic scholarly ties strengthened, as illustrated by the visits of Lyell and Agassiz.[280]

The first American dinosaur skeleton was discovered in 1858 by William Parker Foulke (1816–1865),[281] a law professional turned hobby paleontologist. Foulke had dug up the bones from a marl pit in Haddonfield, New Jersey, and gave them to Joseph Leidy of Philadelphia. The anatomy professor Leidy (yet again more evidence for Cuvier's enormous influence on paleontology through to his methodology of comparative anatomy) properly described the skeleton and in honor of his associate named it *hadrosaurus foulkii*.[282] Much like Mantell, Leidy was a physician by education, who struggled with the practice of medicine and harbored a true passion for natural science. Leidy had met Lyell in 1842 and the British geologist had encouraged him to become a paleontologist. Unlike Mantell, Leidy managed to leave his unloved profession and fully committed himself to science when he became the librarian and then the curator at the Academy of Natural Sciences in 1845, which allowed him to commit more time to his paleontological studies (yet in addition to being appointed curator of the Anatomical Museum of the University of Pennsylvania in the same year, he still had to practice medicine until 1847). Besides being a pioneer of American paleontology, Leidy was also one of the first American scientists who utilized the microscope extensively, which allowed him to contribute immensely to American parasitology. Most fossils, including some dinosaur bone fragments and teeth, were provided to Leidy by Ferdinand Vandeveer Hayden (1829–1887), who surveyed the Dakota and Nebraska territories in the 1850s (see chapter 5.1.). In 1868, in collaboration with Benjamin Waterhouse-Hawkins (and Cope), Leidy completely assembled the dinosaur skeleton and displayed it at the Philadelphia Academy of Natural Sciences. *Hadrosaurus* was the

278 Louis Menand: The Metaphysical Club, New York 2001, pp. 97–101.

279 Struik: Yankee Science in the Making, p. 279.

280 Struik: Yankee Science in the Making, pp. 158–160.

281 A few American dinosaurs were described by Leidy as early as 1856, namely the trachodon, troodon, deinodon, and palaeoscincus. His reconstructions of all aforementioned dinosaurs were based on a handful of teeth and bone fragments send by Ferdinand Vandeveer Hayden, who had gathered them in the western territories. Of the dinosaurs mentioned above only troodon is valid today, the others are regarded as dubious genera, lending all the more importance to the discovery of hadrosaurus. Dodson: The Horned Dinosaurs, pp. 123–125.

282 It is possible that Benjamin Franklin had examined some hadrosaurus bones found near Philadelphia as early as 1787 but never made head or tail of them. Mayor: Fossil Legends of the First Americans, p. 69.

first publicly mounted dinosaur skeleton and instantly drew enormous crowds to the Academy. Contrary to Owens lumbering quadrupedal Crystal Palace reconstructions, Leidy correctly imagined *hadrosaurus* as a more agile, mostly bipedal creature due to the small size of its forelimbs.[283] With Hawkins' help efforts were made to recreate an exhibition of prehistoric life in the Central Park of New York, very much inspired by the Crystal Palace exhibition, but with the addition of various examples of American extinct animals. The project failed due to the political corruption of New York's Tammany Hall. When Hawkins publicly complained about the corruption, hired goons of William "Boss" Tweed (1822–1873) stormed his workshop and smashed the sculptures. They were buried in Central Park, where they stay to this day.[284] Leidy is generally considered the real "father of US-American paleontology," who provided the basis for the meteoric rise of this discipline in America: "Leidy performed the same function in America as had Cuvier in France and Richard Owen in England, both founders of vertebrate paleontology in their respective countries."[285] John Strong Newberry (1822–1892) was another pioneer of American paleontology. He became professor of geology and paleontology at Columbia College (now Columbia University) in 1866 and specialized in the study of fossilized fish. Fish and many other fossils of the New World were very similar to European specimens; another reason why paleontology is considered an international science. Due to the shifts in plate tectonics, which changed the very face of the earth radically over the course of millions of years, paleontology has to be studied globally. Thanks to the public appeal of dinosaurs and the genius of individuals like Leidy, and later Cope and Marsh, American paleontology became internationally acclaimed long before other American branches of science did.[286]

> As the nineteenth century drew to a close, the field of paleontology was prominently represented by Americans. Leidy had helped to provide the indispensable foundation, but it was Cope and Marsh who raised American paleontology to a position of international eminence with their discoveries of immense, fossilized dinosaurs and mammals, their description of large numbers of new genera and species, and working out of their affinities and phylogenies.[287]

In 1850 Agassiz wrote a letter to Leidy in which he reflected on the state of US-American science. He told him that US scholars had made huge advancements in the last

---

283 Warren: Joseph Leidy, pp. 45–87. Also see Buffetaut: A Short History of Vertebrate Palaeontology pp. 126–127.

284 Davidson: Patrons of Paleontology, pp. 58–62.

285 Warren: Joseph Leidy, p. 123.

286 Gregory: North American Vertebrate Paleontology, pp. 308–316.

287 Warren: Joseph Leidy, p. 190.

years, but also cautioned against overstating the gains on the grounds of patriotism, simply because they were the achievements of US Americans:

> Your works in almost all special branches of Natural History were known to me before I came to this country; I may add that they are justly valued in Europe by all those who aim at knowing what is doing abroad in all departments of science; and I personally should feel very unhappy if the partiality shown to Europeans in this country should interfere in any way with your prospects and the credit you duly deserve for your extensive and valuable researches. On [?] this particular point I wish to add a remark, that [the] time has come when American scientific men should aim at establishing their respective standing without reference to the expression of opinion of Europeans respecting them, and at the same time to be cautious not to allow national feeling to exaggerate their value. I have been surprised to find American men of eminence value their correspondence with Europeans of no standing at home, and on the other hand seen things and characters praised beyond bounds, simply because they are American. Let us in [the] future make an effort to do right and to be what we can without the assistance of anybody, and let me include myself in the list, if I can be welcome.[288]

According to Bernard Cohen, no groundbreaking theoretical or "pure" science was done in the early nineteenth-century United States. Instead, more practical inventions like machine tools originated in the young republic. European science was envied, and imitated, young American scientists had to travel to Europe to round off their education. American scientific forebears like Franklin and Jefferson were remembered for their political ideas and seldom for their scholarly achievements. Only in the second half of the nineteenth century was this to change slowly. The National Academy of Sciences (NAS) had been founded in 1863, at the apex of the Civil War; the Smithsonian Institution, though established in 1846, was to gather its momentum only after the Civil War.[289] John Greene also paints a dire picture of early American scientific endeavors, but notes these shortcomings were more than made up in the second half of the nineteenth century:

> In the years from 1780 to 1830 American scientists ceased to be mere purveyors of the raw materials of science to Europe and became junior partners in the

---

288 Jean Louis Rodolphe Agassiz, Cambridge, MA to Joseph Leidy, Philadelphia, 2 May 1850, The Academy of Natural Sciences of Drexel University, Coll#1, Box#1, Folder#1.

289 I. Bernard Cohen: Franklin and Newton. An Inquiry into Speculative Newtonian Experimental Science and Franklin's Work in Electricity as an Example Thereof, Philadelphia 1956, pp. 27–39.

> Western scientific enterprise. In the succeeding century and a half they have become leaders in that endeavor.[290]

In his 1912 memoir Henry Fairfield Osborn called Joseph Leidy "the last great naturalist of the old, or eighteenth and early nineteenth century type," who possessed an "encyclopaedic knowledge, broad grasp of the whole field of natural history, precision and originality of observation in every field."[291] The next generation of American paleontologists was more professional in the sense that they were much more specialized in their studies than their forbears. The era of the gentleman polymath was coming to its end. O. C. Marsh, who would occupy the first American professorship for paleontology, was first and foremost a paleontologist, neither an anatomist, nor a mineralogist, or a natural historian. Furthermore, a contest for the discovery and scientific description of the extinct American fauna began. The rivalry of Cope and Marsh would further American paleontology in a completely unprecedented manner. Leidy could not and would not participate in this breakneck and vicious tournament. Osborn described this generational shift as follows:

> For the long period of twenty-one years (1847—1868) he [Leidy] had enjoyed a monopoly of vertebrate palaeontology in America. Now the situation is suddenly changed; two younger men, full of energy and enthusiasm and with ample means, render it impossible for him to compete in the collection of fossils or to continue his best loved work.[292]

2.7 Conclusion

All throughout human history fossils were gathered, collected, and traded. While the origin of fossils was much debated until the seventeenth century and Steno's observations, they fascinated countless generations for their resemblance to living beings. Fossils allegedly possessed various magical and medicinal qualities, the specific attributes depending on the respective cultural background of the interpreter. While some scientific theories about their origins (derived from objective observation and embedded in a greater theoretical framework) arose in ancient Greece thanks to philosophers like Aristotle, their true meaning and origin remained a subject of debate

290 Greene: American Science in the Age of Jefferson, p. 3.

291 Henry Fairfield Osborn: Biographical Memoir of Joseph Leidy, 1823–1891, in: National Academy of Sciences. Biographical Memoirs, vol. 7, Washington, DC 1913, pp. 335–396. Quote on page 339.

292 Osborn: Biographical Memoir of Joseph Leidy, p. 365.

for many centuries. Some of the more spectacular bones were thought to be the remains of mythical beasts or fabled heroes.

The Middle Ages saw the rise of Christianity in Europe, dominating cultures and societies. Therefore, the perception and interpretation of fossils was now predominately influenced by religious tradition and the Bible. Gargantuan bones were now commonly believed to be the remains of giants. Noah's flood was thought to be the reason why the fossils of aquatic animals could be found miles away from any modern sea. Still, the organic origin of fossils was debated, theories about abandoned attempts at creation or tricks or jests of nature were circulated as alternatives. Much of the ancient Greek knowledge had been forgotten during the turmoil of Late Antiquity and was now slowly rediscovered. Aristotelian theories were held in the highest regard, arguments about nature had to derive from authority and had to be in accordance with Christian scripture. The age of the planet, and therefore all lifeforms, was derived from biblical data. The biblical method of chronology remained the predominant one until the early eighteenth century, Bishop Usher being the most celebrated chronologist. Scientific work was ruled by cultural and religious presumptions. This only changed gradually, beginning in the late fifteenth century. The humanists of the Italian Renaissance and their successors began to question the old ideas, which appeared to be solely derived from arguments of authority and religious dogma. Data collecting, scholarly correspondence, and debate were incorporated as tools for the production of knowledge. As plants were gathered and studied in gardens, other natural oddities, like gems and fossils, were displayed in cabinets of curiosities; these cabinets became the forerunners of the modern museums.

Empirical observation, extensive data collecting, and experimenting became the tools of the seventeenth-century scientific revolution. The heroes of this revolution, above all others Bacon, Descartes, and Newton, are celebrated to this day. There was a boom of new scientific theories and discoveries. Nature itself seemed to follow subtle mathematical rules; the far corners of the earth were explored in more and more detail, and all manners of strange and exotic lifeforms were discovered by Europeans. As methods of transportation and communication were improved, the white spots on the maps grew smaller, and scholarly exchange was furthered along the growing global trade routes. Scientific correspondence within the so-called Republic of Letters grew to new heights. Still, this republic was hierarchically structured in accordance with early modern social norms; scientific discoveries were not solely discussed on their own merit, but the status of the discoverer was of importance too. Scientific societies like the British Royal Society further advanced the scientific exchange of ideas; ideas that were now circulated in journals sold internationally. Thanks to the efforts of scientists like Steno and the methodology of comparative anatomy, the organic origin of fossils was finally asserted as true.

During the eighteenth century, Paris became the center of European science and the Republic of Letters, thanks to generous funding by the French crown and the endeavors of well-off aristocrats like Buffon. Natural history developed as an independent scientific discipline, the study of fossils being one of the most important tools of this science. By the likes of Werner and Hutton fossils were also recognized to be an indicator of the age of strata, thereby becoming more important to geology, which was establishing itself as a discipline. The gradual progression of life on earth as documented by the fossil record became a point of interest, inspiring Lamarck's, and later Darwin's theories of evolution. Meanwhile, Cuvier was recognized as the most brilliant comparative anatomist of his time, interpreting many fossils, and thereby learning that some lifeforms had gone extinct. While several aristocratic scholars fell victim to the French Revolution, the ensuing chaos and rearrangements provided opportunities for the next (and less aristocratic) generation of French scientists. Again, practical events significantly influenced the production of knowledge. At the end of the eighteenth century, scholars in the United Sates began to formulate their own scientific theories and to build up their own collections. Until then, America had mostly provided raw scientific material for European minds to study. At the beginning of the nineteenth century scientists of the young republic began to organize and longed for European recognition of their labor; therefore, science became an astoundingly patriotic undertaking.

The construction of railroads and canals in nineteenth-century Britain provided natural historians with unexpected opportunities: as a byproduct of these lifelines of progress, fossils were literally lying by the wayside. This accessibility, coupled with a widespread romantic appreciation for nature's beauty, triggered public enthusiasm for geology in Victorian Britain. At this time, thanks to the increased attention to the study of fossils, paleontology slowly emerged as a self-regulating scientific discipline in Britain. Now British paleontologists took the lead, especially with the discovery of the fearsome dinosaurs. Thanks to the immense popularity of dinosaurs, best illustrated by the construction of the Crystal Palace Park, paleontology gained in popularity. Another important factor in the specialization and founding of scientific disciplines was the Prussian educational reform at the beginning of the century, also partly inspired by nationalism (see chapter 8.1.). During the nineteenth century, science became increasingly professionalized. Before, only wealthy gentlemen with ample leisure time could turn to scientific endeavors. Now, thanks to better funding for museums and professorships and the opportunity to easily publish scientific discoveries in professional journals, a scientist could earn a living just by doing science. Joseph Leidy of Philadelphia may have been the last gentleman polymath in the style of the eighteenth and early nineteenth centuries in America. The next generation and its two most celebrated representatives, Cope and Marsh, were of a new breed, their almost frantic dedication to science furthered American paleontology greatly. It came to surpass European paleontology at the end of the nineteenth century, in part due to the "Bone Wars," the intense rivalry of Cope and Marsh. On the other hand, this process can be understood as part of a more general effort to reform the American universities.

Many of the socio-cultural shaping the conduct of science, as discussed in this chapter, can be observed with regard to American paleontology in the second half of the nineteenth century: the public and professional prestige of the paleontologist remained a prime motivating factor (at least subconsciously) in the minds of scientists. Prestige was the prize to be gained in the race for the most promising fossil hunting grounds and the first scientific description of a new-found species. The need for public funding for expeditions and the acquisition of fossil collections might have motivated the patriotic embellishment and aggrandizement of the paleontological discoveries. Railroads remained to be of the utmost importance to American paleontology; the fossil beds of the western territories were reached by railroad, and the exhumed bones were sent back to the east by the same means. A fairly new invention, the telegraph, also contributed to paleontology. Some brand-new findings and deductions were wired back to the east coast by telegraph. These developments can be understood as a continuation of the improvements in transportation and communication during the Age of Discovery. Furthermore, the scientific survey of the American West might be linked to the investigation of the Americas in the fifteenth to seventeenth centuries and the plethora of new scientific data it brought. The public interest in dinosaurs did not decrease, as is demonstrated by various nineteenth-century newspaper articles covering the new breathtaking discoveries. This connection to dinosaurs might be the main reason why Cope and Marsh are reasonably well known even today, while scientists like Buffon, international celebrities in their lifetime, are all but forgotten. The Republic of Letters rose to be a fundamental instrument to produce science; within it, scientific information, material, and even personnel was traded across the Atlantic.

3

“They are a Nation of Scholars” – O. C. Marsh and the Necessity of Transatlantic Education

The subject of this chapter is O. C. Marsh's education and upbringing. Both will be described in all briefness, for LeVene and Schuchert have already described Marsh's early education in some detail. This thesis can add but a little more focus on Marsh's voyage to Europe. Still, his early education and most of all his experiences in Germany are essential to understanding Marsh's scientific network, his relationship with his German assistants (chapter 6), the paleontological exchange that happened during the last half of the nineteenth century, and also the genesis of paleontology as a scientific discipline in the US within the context of the education reform that took place at the same time (chapter 8. 3).

The first part of this chapter will briefly detail Marsh's childhood education, his time at school, at the Andover Academy, his studies at Yale, and his first scientific ventures abroad. The second subchapter will focus on Marsh's first travels through Europe with a focus on the scientific education at German universities. The third subchapter will analyze some of Marsh's correspondence with several people he met in Europe between 1863 and 1865. A small excursion into the importance of the extreme privilege Marsh gained through the financial support of his rich uncle follows.

Besides LeVene and Schuchert's detailed biographical descriptions, Marsh's correspondence, notebooks, and various memorabilia from the O. C. Marsh collection provide the basis for this chapter.

3.1 Childhood and Education in the United States

Othniel Charles Marsh was born to Mary Gaines Marsh nee Peabody (1807–1834) and Caleb Marsh (1800–1865) on October 29, 1831. The Marsh family lived on a farm near Lockport, New York. Othniel had an older sister, Mary (1829–1852), born in 1829, and a younger brother, George (1834–1835), who was born in 1834. Soon after giving birth O. C.'s mother died of cholera, throwing the family into deep crisis. Caleb Marsh sold the family home at Lockport and moved with Mary Jr. and O. C. to Danvers, Massachusetts. The newborn George was left with a nurse and died the next year. Even though his mother had died when O. C. was not even three years old, she had left him with family ties that would prove to be of the utmost importance for his education and scientific career. Mary Peabody's brother was the famously rich businessman, entrepreneur, and, most importantly, philanthropist George Peabody (1795–1869), the namesake of O. C.'s ill-fated little brother. When Caleb remarried in 1837 George Peabody kept close ties with the two surviving children of his beloved sister, promoting their education and wellbeing in general. After a business venture of Caleb's had failed, leaving him indebted, he, his new wife, and O. C.'s halfsiblings moved back to Lockport in 1839, leaving O. C. and Mary Jr. back in Danvers with an aunt. Caleb's financial situation worsened during the 1850s, throwing his family into never-ending financial predicament.[293] According to George Bird Grinnell, later colleague and friend of Marsh's, Caleb "was, however, both stern and impulsive, and not being always in sympathy with the tastes of his strong-willed son [O. C. Marsh], he occasionally inflicted severe punishment on the boy."[294]

LeVene and Schuchert describe Marsh's early childhood and his relationship to his father as follows:

> As he grew into a strong boyhood, the oldest son in a rapidly increasing household, he was expected to be his father's mainstay in the farm work, and his reluctance to do so was a source of friction between them. He preferred, instead, to roam the field and woods, hunting the small game then still abundant in the Lockport region.[295]

These early childhood experiences with hunting and the outdoors were undoubtedly good preparation for O. C.'s later fossil-hunting excursions to the West (see chapter 5. 4). In 1878 George Bird Grinnell wrote about Marsh's youthful sport exploits:

> As a boy he was passionately fond of field-sports, and devoted much of his time to fishing and shooting. The writer has heard him remark that he was a sportsman before he was a naturalist; and it cannot be doubted that the open-air life of his early years gave him the vigorous health he has since enjoyed, while to the habits of observation acquired in the woods and fields much of his subsequent success in science has been due. He is still a keen sportsman, and very hard to beat with rod or gun.[296]

Furthermore, the Marsh farm was situated roughly one mile from the Erie Canal. The earth and rocks, dug out during the construction of the canal, were full of fossils and, as in other places before and after (see chapter 2. 5.), these discarded obstacles to the march of civilization and transportation became hunting grounds for many fossil collectors, including young O. C. Marsh. When Colonel Ezekiel Jewett (1791–1876), a prolific paleontologist, came to Lockport, he took O. C. under his wing:[297]

---

293 Schuchert; LeVene: O. C. Marsh, pp. 12–16.

294 George Bird Grinnell: Othniel Charles Marsh. Paleontologist, in: David Starr Jordan [ed.]: Leading American Men of Science, New York 1910, pp. 283–312. Quote on page 284.

295 Schuchert; LeVene: O. C. Marsh, p. 16.

296 George Bird Grinnell: Sketch of Professor O. C. Marsh, in: Popular Science Monthly, vol. 13 (Sep. 1878), pp. 612–617. Quote on page 612.

297 Schuchert; LeVene: O. C. Marsh, pp. 17–18.

> He held a summer school in Geology at Lockport for four years, and he was visited there by many a noted geologist, both from this country and from abroad. Such a combination of vivid personality and collecting skills, we may assume, drew young Othniel like a magnet. Moreover, Jewett was as skilful [sic!] with a rifle as he was with a collecting hammer, and Marsh told Ray Stannard Baker many years later that his first great ambition had been to shoot as well as the colonel. [...] It is known that the boy came under Colonel Jewett's influence some time about 1845, and learned from him where and how to collect fossils and minerals; and from that time on he seems to have had even less taste for farming.[298]

Jewett and O. C. Marsh stayed in contact and thirteen letters from Jewett to Marsh are preserved in the Marsh papers. That Jewett had become a friend of the family is evidenced by the first archived letter, in which he expresses his condolences about the death of Caleb Marsh: "I have your note of yesterday giving me the sad inteligence [sic!] of the decease of your Father. Most sincerely do I condole with you in this great and irreparable loss."[299] All other letters between Jewett and Marsh are of a more professional nature, referring to the acquisitions of fossil collections and of potential fossil-hunting grounds. Starting in 1868, Jewett addressed Marsh as his "dear friend" (instead of addressing Marsh by his title, as he did in the letters sent before 1868) and unsuccessfully invited him several times to pay him a visit, at one time downright begging him to come: "I beg you to oblige me with a visit."[300]

As for Marsh's formal education: it seems that he visited school almost exclusively in winter terms. In 1847 he enrolled in the Collegiate Institute in Wilson, New York, and in 1850 changed to Lockport Union School. Afterwards he became a schoolteacher for a short period of time (he gave up on school teaching because of frequent headaches). After a short period of vacillation, Marsh enrolled in the Philips Academy at Andover, Massachusetts, in 1852. George Peabody was a patron of education and spent a portion of his ever-increasing wealth enabling various relatives to follow their scholarly pursuits. Motivated by the lack of education opportunities he had experienced in his youth, he was now driven by the urge to improve the situation of his younger relatives. He funded Marsh's higher education although his nephew was already nineteen years old, which made him almost two years older than most of his classmates. Marsh

298 Schuchert; LeVene: O. C. Marsh, p. 18.

299 Ezekiel Jewett, Albany, NY to Othniel Charles Marsh, New Haven, CT, 23 September 1865, Yale University, Sterling Memorial Library, Othniel Charles Marsh papers (MS 343), Series I. Correspondence, Box 18, Folder 724.

300 Ezekiel Jewett, Utica, NY to Othniel Charles Marsh, New Haven, CT, 26 October 1869, MS 343, Series I. Correspondence, Box 18, Folder 724.

began studying geology and mineralogy and spent part of his summer vacation of 1852 collecting minerals in New York and Massachusetts. Later he recalled that at this time he made up his mind to sincerely focus on studying, having spent a lot of time playing backgammon and duck hunting during his first year at Andover. Marsh really did improve, being top of his class every term from then on. In the summers of 1854 and 1855 Marsh went to Nova Scotia, vacationing, studying the geology of the land, and collecting fossils and minerals; following in the footsteps of Lyell and Silliman Sr., as he noted himself. Marsh's collection grew to notable size during this time and he became known for his extensive collection among his peers. In 1856 Marsh graduated from Andover.[301]

Yet again bankrolled by Uncle Peabody, Marsh enrolled at Yale College in 1856. The curriculum of the first years proved quite diverse: besides natural sciences various courses on Greek, Latin, Rhetoric, History, and other subjects were taught, for the holistic and moral education was still promoted in US higher education. This would change during the course of the nineteenth century, when the training of specialists became the focus. His seniority prompted his classmates to give him the nicknames "Captain" and "Daddy" and again his extensive fossil and mineral collections were distinguishing and noteworthy features to his classmates. In 1860 Marsh received his bachelor's degree, beginning graduate studies soon thereafter. It is of note that Marsh had to write to his Uncle Peabody and request more funding, for he constantly went over budget spending his uncle's money. This tendency, the inability to balance a budget and the callousness towards money would never change, giving many an employee of Marsh sincere grievance when he did not receive payment in time while being dependent on a steady income, unlike the well-off professor (see chapter 6. 2. 3).[302]

After receiving his bachelor's degree from Yale College, Marsh continued his higher education pursuing the goal to obtain a professorship in some discipline of natural science (in 1860 he had not decided in which). Only recently a scientific school (Sheffield Scientific School) had been established at Yale, and Marsh was one of the first students to attend it. He chose the Chemistry Course, which at that time also included studies in metallurgy, mineralogy, botany, French or German, and in the second year of the two-year Course included, amongst others, physics and geology. Marsh's instructors included James Dana (1813–1895) and the younger Silliman (1816–1885). In November 1861 Marsh's first scientific paper was published, the subject being the observation of the newly discovered gold fields in Nova Scotia[303] he had visited in 1860. According to himself as well as to LeVene and Schuchert, this first paper garnered

301 Schuchert; LeVene: O. C. Marsh, pp. 18–27.

302 Schuchert; LeVene: O. C. Marsh, pp. 28–41.

303 Othniel, Charles Marsh: The Gold of Nova Scotia, in: The American Journal of Science, ser. 2, vol. 32, no. 96, (Nov. 1861), pp. 395–400.

much attention and was often quoted in other publications of that time. Marsh followed it up with a description of the vertebrate remains he had found on an earlier trip to Nova Scotia.[304] This paper was also received warmly; it even caught the attention of Louis Agassiz, who was at that time one of the most celebrated US scientists (see chapter 2. 6.). LeVene and Schuchert write that all the attention and praise his second publication created might have been the deciding factor for Marsh to focus on vertebrate paleontology. LeVene and Schuchert further write that Marsh was offered the position of a major in a Connecticut regiment at the start of the Civil War but turned the offer down due to his flawed eyesight.[305] After he had received his master’s degree in 1862, Marsh was apparently still torn between joining the Army or focusing on his scientific career and furthering his education by traveling to Europe to visit the world’s most prolific and prestigious scientific institutions, learning from some of the most famous scientists of that time.[306] In the end Marsh chose the latter (after his father had virtually begged him not to “expose” his “valuable life on the field of battle”[307]). Thanks to the continued funding by George Peabody, he embarked for Europe in November 1862.[308] Grinnell writes that “Marsh refused the professorship offered him by his Alma Mater,”[309] meaning that he forewent a secure position in favor of a chance to really round out his scientific education, studying at the – then – leading institutions of higher education in Germany. That he was in fact offered a permanent professorship however is dubious, for even after his educational tour through Western Europe Yale had no paid professorships to offer (see below).

---

304 Othniel Charles Marsh: Description of the Remains of a New Enaliosaurian (Eosaurus Acadianus) from the Coal Formation of Nova Scotia, in: The American Journal of Science, ser. 2, vol. 34, no. 100, (Jul. 1862), pp. 1–16.

305 At the outbreak of the war Marsh received a letter from his half-sister, Martha, inquiring about his nearsightedness, and if it would be a factor in his decision whether to join the army or not, telling him that other men had not been rejected because of poor eyesight. See: Martha Marsh, South Danvers, MA, to Othniel Charles Marsh, New Haven, CT, 30 April 1861, MS 343, reel 23, frame 9.

306 In August 1862 his aunt Mary wrote Marsh, telling him about some local men enlisting in the army and asked him if he would go to Europe or enlist. She asked him the suggestive question: “I suppose you cannot go now, until after the draft, can you?” See: Mary Marsh, South Danvers, MA, to Othniel Charles Marsh, New Haven, CT, 18 August 1862, MS 343, reel 22, frame 362.

307 Caleb devoted the better part of one letter to telling his son that the fighting should be done by men more suitable: “I do not think my dear Son, that you are called upon to fight her [the United States’] battles, for certainly, there are thousands that are willing & anxious to engage in deadly strife, that are equally & perhaps better qualified than yourself for battle.” Later in the same letter he even evokes O. C.’s dead mother, urging him not to enlist: “In conclusion, my dear Son, by the memory of your dear Mother and the love I have ever bore you, not to expose your valuable life on the field of battle.” See: Caleb Marsh, Lockport, NY, to Othniel Charles Marsh, New Haven, CT, 08 August 1862, MS 343, reel 20, frame 735.

308 Schuchert; LeVene: O. C. Marsh, pp. 42–48.

309 Grinnell: Othniel Charles Marsh, p. 290.

3.2 "Many Pleasant Acquaintances" – O. C. Marsh in Europe

Equipped with a letter of introduction written by his instructor James Dana, Marsh crossed the Atlantic. The first stop on Marsh's nearly three-year tour of Europe was Great Britain, where he visited various museums. Albeit spending but little time in England, Marsh acquired a close circle of friends, who would provide him with letters of recommendation, opening the doors of European academia to the young US-American. At the end of November Marsh had arrived in Berlin, where he enrolled at the University and studied chemistry and mineralogy with Gustav (1798–1873) and Heinrich Rose (1795–1864). Both brothers held professorships at the University of Berlin, the former for mineralogy, the latter for chemistry. The Rose brothers numbered among the very first professors working at the University of Berlin, contributing to the eminence of the institution in the 1820s (see chapter 8. 1.). LeVene and Schuchert write that Marsh spent most of the winter of 1862/63 working on his German.[310]

In January 1863 Marsh told his father not to worry, that he had made many friends already and, due to their encouragement, had made up his mind to pursue a career in science. This underlines how important and formative the personal local network was for Marsh:

> Although in a foreign country I have already many pleasant acquaintances and friends, among whom are many very distinguished men, who have showed me a great deal of attention on account of the little I have already done in Science. This greatly encourages me to try to do much more, and if my life & health are spared, I intend to accomplish enough to satisfy a reasonable ambition.[311]

In the spring of 1863 Marsh moved to Heidelberg, where he worked with Robert Wilhelm Bunsen (1811–1899), professor of chemistry and eponym of the Bunsen burner, Johann Reinhard Blum (1802–1883), professor for mineralogy, and Gustav Robert Kirchhoff (1824–1887), physicist.[312]

LeVene and Schuchert reproduce a letter that Marsh wrote to Silliman Jr. on May 10, 1863, which gives insight into Marsh's thought process concerning his future in science:

---

310 Schuchert; LeVene: O. C. Marsh, p. 48.

311 Othniel Charles Marsh, Berlin to Caleb Marsh, Lockport, NY, 26 January 1863, MS 343, reel 20, frame 750.

312 Schuchert; LeVene: O. C. Marsh, p. 49.

> [...] I wish to consult you in regard to a question which I must soon decide, and on which you can advise me better than anyone else. I refer to my course of scientific study; and as I remember that you once spoke to me on the subject, I shall venture to ask your opinion on one or two points in regard to it. I intended to have done this before I left New Haven, but I was undecided whether to go into the army, or abroad until a few days before I sailed, and as you were then out of town most of the time, I had no good opportunity of doing so.
>
> Hitherto, as you know, I have devoted a good deal of time to Natural Science; but my studies have been much more general than I intend them to be in future. I am now sufficiently familiar with the German language and with scientific matters in Germany to pursue with advantage some particular branch of study, and I wish to commence upon it as soon as possible, and to concentrate all my efforts upon it. I am, however, nearly equally interested in Chemical Geology, Mineralogy, and Palaeontology: and my choice of one of these will depend on the prospect of making the result of my studies available on my return to America. Chemical Geology or Mineralogy I should certainly prefer, as my previous studies have been much more in that direction, and my cabinet and library, on which I have spent and am spending a good deal of time and money, relate more especially to these departments.
>
> Supposing for example, that during my stay in Europe, I should study Chemical Geology, including some of its practical branches, and should enlarge my cabinet and library in that direction, would there probably be an opportunity of making such attainments useful on my return? Would it be more advisable for me to devote myself to *Palaeontology*, – making an especial study of some one of its branches; e.g., the *Vertebrates*?
>
> From your familiarity with science in America you can easily advise me what branch it would be most advantageous for me to pursue, and I shall be greatly obliged for your opinion. From this point of view it is very difficult to judge, especially as the war is changing affairs in America so rapidly.[313]

The letter shows that in May 1863 Marsh was still undecided which discipline he should pursue, somewhat opportunistically asking Silliman, Jr. which specialization would provide him with the best future perspectives. The letter also underlines Marsh's inner conflict whether to further his scientific career or to serve his country/home State in

313 Othniel Charles Marsh, Heidelberg to Benjamin Silliman, Jr., New Haven, CT, 10 May 1863, quoted after: Schuchert; LeVene: O. C. Marsh, pp. 49–50.

the Civil War – or at least shows that he wanted to create the impression of being a conflicted patriot to Silliman.

He further wrote:

> Prof. Dana's Geology is spoken of in the highest terms by the geologists here who have seen it. At a meeting of the Geological Soc. of Germany, which I attended in Berlin, Prof. Ehrenberg showed a copy he had just received and said it was the best book of the kind ever published. I heard Prof. Geinitz say the same, while showing his copy to the Nat. History Soc. of Dresden, and he added that the Author was the 'Humboldt of America.'...[314]

The latter part of the correspondence shows how much appreciation some US-American scientific publication already received in Europe. After consulting with Dana, Silliman replied with the anticipated career instructions:

> It now seems clear that you have to fit yourself by suitable studies for duty *here* in connection with the science of Geology and *Palaeontology*. It is Prof. Dana's view that you should devote yourself with zeal and your well-known perseverance to the subject of Palaeontology, and especially to the palaeontology of the older rocks. Palaeozoic rocks abound in the U.S. and demand by far more study than they have received. It requires you to amass as soon as you can a good collection of fossils in this department of European geology for comparison. Our collections are weak in European Palaeozoic except in the Permian of which Dr. Geinitz has sent Prof. D. a good series. In obtaining this special knowledge you will of course study General Geology, so as to be prepared to give instruction, if required, on this subject in the post-graduate courses. Things now tend strongly toward placing these studies in what may be called the university studies, Mineralogy being now studied almost solely in that way. There is now every reason why I should write you with entire frankness on matters which so deeply interest us both and which are especially interesting to you personally. I may say then that you have only to show your fitness for such a chair as I have indicated, involving perhaps the curatorship of the museums, and you will receive the appointment. The fund contemplated in your uncle's codicil would be insufficient for such an endowment in view of all else we must do with it. But I can not doubt if he has this view of the subject before his mind that he will authorize you to go on and fit yourself and to use any available means to amass the collections needed to give a first class effect to your department. This done I do not question he will see the desirableness of having the museums on pro-

314 Othniel Charles Marsh, Heidelberg to Benjamin Silliman, Jr., New Haven, CT, 10 May 1863.

> cess of construction so that he may enjoy the pleasure of seeing you installed on his lifetime in a manner so honorable with both. The most desirable thing, of course, would be that he should authorize this – endow your professorship and have the fund named on his will unimpaired to sustain all the departments contemplated on our plan, there being no other means so sustain them and nothing more useless than costly establishments without foundations. I am well aware how very much this plan exceeds the limits named. But I have faith to believe that your noble relative will rise to the level of the occasion, if the subject is properly presented to his consideration, as I am sure you know how to present it.[315]

Simply put, the career advice was to study up on geology and paleontology, to buy an extensive fossil collection with Peabody's money, and to convince the uncle, whom Marsh had met in the meantime at Homburg, and would soon meet again in Wiesbaden, to provide further funding for a museum and professorship for his nephew.

Dana had also sent a letter to Marsh, underlining Silliman's advice:

> Prof. S. has spoken of my advice to you. I would say that there is no department that affords an opening now, excepting Paleontology; and if you could prepare yourself for Paleozoic Pal. (as this is specially needed in America), I think there would be no difficulty as to your appointment to the place. I wd recommend your studying abroad for a couple of years. Prof. Roemer of Germany wd be a good teacher, I think, and at the same time to attend to general Zoology and especially to Invertebrates (Mollusks, etc.) and then, besides this, to purchase and collect specimens of European species largely, for in no other way exct by handling specimens and their labels can you familiarize yourself with characters and names of genera or species.[316]

Marsh took the advice of both professors to heart, for he remained in Europe for almost two more years, familiarized himself with the European specimens, learned about the invertebrates after his return to Berlin, and paid a visit to Professor Roemer in Breslau (see below). First, and maybe most importantly, he had to secure further funding from his uncle:

> My Dear Uncle:

---

315 Benjamin Silliman, Jr., New Haven, CT to Othniel Charles Marsh, Heidelberg, 15 June 1863, quoted after: Schuchert; LeVene: O. C. Marsh, pp. 51–52.

316 James Dana, New Haven, CT to Othniel Charles Marsh, Heidelberg, 16 June 1863, quoted after: Schuchert; LeVene: O. C. Marsh, p. 52.

One immediate result of your munificent donation to Yale has been to more than realize my highest hopes of obtaining an honorable position when my studies here are completed. Although there is at present no vacancy at Yale the Faculty propose to create a new Professorship of Geology and Paleaontology (the science of fossil remains) and give me the position as soon as I can fit myself for its duties. [...] There is no other in America which I shall prefer. I correspond to that held by the great Agassiz at Harvard, and in point of rank is the same as that of Prof. Silliman Senior, or Prof. Dana. Prof. S. Jr. is only a Professor in the Medical Department at Yale and before obtaining even this, he served an apprenticeship of 6 years in a western college as have several other Yale Profs. Aside from these considerations the position itself will on many accounts be a very agreeable one to me. I shall have entire control of the cabinets of the *Peabody Museums* on which as *Trustee* I shall feel a special interest, and my other duties will allow me time for study, and for publishing any original results I may be able to obtain. [...]

By this appointment I shall at once be placed on a level with men all of much greater age and experience than myself. [...]

I shall do everything in my power to prove myself worthy of the confidence reposed in me, but it seems also necessary to give me an equal chance with the other Profs and to make my labors effective that I should have a library and cabinet in a measure equal to those possessed by my colleagues. A library and cabinet is to a Prof. of Science exactly what capital is to a man in business; with the advantage that in the former case that no risk of loss is incurred. Such a library and cabinet as the position requires can only be obtained in Europe and while I am here I shall have opportunities for collecting them such as I may never again possess.

The amount necessary for this object would be 3 or 4 thousand dollars. There is not, I think, a Prof. in Yale who has not an equal amount thus invested while Profs. S. and D. and several others have each private libraries and cabinets of much greater value. If I do not have similar means of study and advancement at my command, as the other Profs. possess, the disadvantages which at first I must necessarily labor under, on account of my inexperience, will be much increased.[317]

317 Othniel Charles Marsh, Heidelberg to George Peabody, Invergarry [?], 12 July 1863, quoted after: Schuchert; LeVene: O. C. Marsh, pp. 53–54.

A golden opportunity presented itself to buy a position in the top tier of US-American science, skipping some steps on the career ladder. Peabody could be convinced that his nephew should get everything he needed to catch up to the older and more experienced professors:

> With regard to your request that, in the event of your obtaining the prominent position in the College you name, I would allow you 3,000 to 4,000$ to enable you to supply yourself with the necessary Library, I beg to state that if such a Professorship is promised, with a liberal salary, I will give you the sum of five hundred pounds (now equal to $3,500) for that purpose, you can therefore get all the information that may be required before my return to London in November, when I will arrange to place you in a situation to meet your views.[318]

In October of the same year Marsh (after moving back to Berlin) wrote to his uncle, telling him that he was confident that he would get the position at Yale and preparing the business-minded uncle that there would most likely be no return on his investment, for the "honor" of working at Yale would be much more valuable than money and the salary would most likely be a comparatively meager one:

> The salaries of the Profs. at Yale, and I believe of those at Harvard also, are small, the honor of the position, being considered such an equivalent, that the applicants for every vacancy are always numerous, and are frequently willing to make considerable pecuniary sacrifices. A Prof. now in the law department, and ex-governor, had formerly, I have heard, a law practice worth $6 or $8,000 per annum, but preferred a Professorship with a salary of $1,600. This is about the average of the salaries at Yale, the difference between them depending I believe on the amount of the original endowment or its subsequent investment. Although such remuneration is small there is no position in the world that I should prefer to a Professorship at Yale, as it is for life, and besides the honor it confers, it will afford me unsurpassed opportunities for carrying out scientific plans, which in a small college I should have to relinquish.[319]

After the summer term was finished in Heidelberg, Marsh travelled through Switzerland, hiking and studying glaciers. Then he returned to Berlin and focused his education on paleontology and geology, like Dana had suggested. He worked with

---

318 George Peabody, Invergarry [?] to Othniel Charles Marsh, Heidelberg, 22 August 1863, quoted after: Schuchert; LeVene: O. C. Marsh, p. 55.

319 Othniel Charles Marsh, Berlin to George Peabody, [location unknown], 12 October 1863, quoted after: Schuchert; LeVene: O. C. Marsh, pp. 55–56.

Heinrich Ernst Beyrich (1815–1896), who taught geology and paleontology, Wilhelm Carl Hartwig Peters (1815–1883), zoologist and at the time director of the Berlin zoo,[320] and Christian Gottfried Ehrenberg (1795–1876), who gained international fame for his studies in microbiology and micro geology.[321] LeVene and Schuchert judge that Ehrenberg's lectures must have disappointed Marsh because of how little time was devoted to prehistoric reptiles; instead they focused on invertebrates. This changed in the summer of 1864, when Ehrenberg taught micro geology, and to Marsh's pleasant surprise, who kept a detailed notebook in Berlin, most of the specimens of Ehrenberg's were of American origin. Peter's lectures, being focused mainly on the systematics of ancient life, conveyed a solid foundation for Marsh's career in paleontology.[322]

Beginning with Peters' lectures Marsh made special note of any fossils discussed in the lectures. Most of the lectures he took note on also included some references to how the various scientific disciplines had evolved to their current state. On the last pages of his notebook Marsh took notes on how fish were conserved and on how stuffed birds were exhibited at the Berlin museum, including a sketch of a specimen preserved in a jar of alcohol and a picture and note on how birds were displayed in a

320 In 1880 Peters received a copy of Marsh's "Odontornithes" book (the very lavishly and elaborately designed book would later become evidence for Marsh's alleged squandering of government funds in a congressional investigation, see chapter 6. 5.). Peters calls Marsh "dear friend and colleague" ("Lieber Freund und college"). He mentions that the book has extra value to him, in addition to its scientific content he treasures it because he had a personal stake in Marsh's discoveries, presumably because he was Marsh's former teacher ("[das Buch] hat für mich doppelten Werth. Einmal wegen der ungemein intressanten Gegenstandes und der wissenschaftlichen gründlichen Beobachtung derselben und dann wegen der Theilnahme, welche ich persönlich an allen ihren Forschungen genommen habe.") See: Carl Hartwig Peters, Berlin, to Othniel Charles Marsh, New Haven, CT, 12 October 1880, MS 343, Series I. Correspondence, Box 25, Folder 1056.

321 There is one especially striking example of Ehrenberg's excellent repute in the US. In 1868 Oliver Wendell Holmes (1809–1894) wrote a poem in honor of Ehrenberg's 50-year anniversary as professor. Ehrenberg is celebrated as one who had "taught the teachers of mankind," whose "fame has journeyed westering with the sun, Prairies and lone sierras know" his name. See: Oliver Wendell Holmes: To Christian Gottfried Ehrenberg, for his "Jubilaeum" at Berlin, November 5, 1868, Boston, MA 1868. The poem was "was written at the suggestion of Mr. George Bancroft, the historian." Bancroft was among the first US Americans visiting and becoming familiar with the German system of higher education and in that respect a forerunner of Marsh. See: Holmes, Oliver Wendell: To Christian Gottfried Ehrenberg, for his "Jubilaeum" at Berlin, November 5, 1868, in: Oliver Wendell Holmes: The Complete Poetical Works of Oliver Wendell Holmes, Cambridge Edition, Boston, MA c. 1895, p. 206.

322 Marsh's took notes on Peters' 105 lectures between October 29, 1863, and March 15, 1864. The notes are a mixture of English with some German technical terms sprinkled in. At first many German words appear in Marsh's notes, but these grow scarcer in later lectures, most likely because Marsh had learned the English translation of these words. The notes are mainly comprised of lists upon lists of genera and species and other taxonomic ranks, complete with physiological descriptions with some anecdotes about the animals mixed in, at times even noting if an animal was good to eat. There is a list of German anatomical terms and their English translations on the last page of Marsh's first Berlin notebook, indicating that he was still learning the language. See: MS 343, reel 24, frame 489.

life-like manner.[323] Similar notes can be found on the last pages of Marsh's next notebook, elaborating on how insects were preserved and exhibited at the Berlin museum. Here Marsh goes into great detail, taking notes on the exact measurements of the glass cases specimens were displayed in. He further describes how vertebrates were prepared for exhibition at the museum.[324] Undoubtedly Marsh was already planning his own museum in New Haven, drawing inspiration from the state-of-the-art exhibition, preservation, and preparation techniques of the German museums, first in Berlin and later in Breslau (see below).

The notes on the paleontology lectures (84 lectures beginning on November 5, 1863, and ending on March 15, 1864) Marsh attended in Berlin are much less fragmentary and seem clearer and more complete than those that he took on Peters' lectures on biology. Like the lectures on biology, those on paleontology also begin with a short historical overview of the genesis of this discipline, incorrectly attributing the invention of the term paleontology to Hermann von Meyer in 1832.[325] There are also comprehensive notes on Beyrich's lecture on "Geognosy" ("Geognosie"), meaning "knowledge of the earth," an antiquated German term, later completely replaced by the term geology. There are notes on 58 lectures, beginning on May 4th and ending on August 5th, 1864. Marsh scarcely uses any German words anymore, indicating he had a better grasp of the language now and knew most of the English translations for the German terms used in the lectures. Noteworthy for Marsh's later career is that the only dinosaurs covered by Beyrich were *iguanodon*[326] and *archaeopteryx*, which Marsh refers to as the "[o]nly Bird (from Solenhofen [sic!])."[327]

In Berlin Marsh also met E. D. Cope for the first time. LeVene and Schuchert describe their first meeting as follows:

> They had met in Berlin in Marsh's student days, and apparently their first impression of each other had not been especially favorable, although their respective accounts of the meeting, given later, were undoubtedly colored by what followed.[328]

Cope was on his "grand tour" through Europe, as Jane Pierce Davidson notes in her Cope biography. She also adds that this was a behavior quite "typical [for an] wealthy

323 MS 343, reel 24, frames 708–710.

324 MS 343, reel 24, frames 840–842.

325 MS 343, reel 24, frame 879.

326 MS 343, reel 25, frame 298.

327 MS 343, reel 25, frame 304.

328 Schuchert; LeVene: O. C. Marsh, p. 262.

young American."[329] Davidson further writes that both US Americans met in Berlin, but that there is little information about this initial meeting:

> It was also during this trip [to Berlin] that he [Cope] first met O. C. Marsh who was doing graduate studies in Berlin at that time. Cope was in Berlin between mid-October and late December, 1863. He does not mention having met Marsh in his surviving letters, but Marsh later remembered the meeting and discussed it in his articles in the *New York Herald* in 1892 [sic!] when the Cope-Marsh War broke into the popular press.[330]

Davidson later elaborates on this and writes:

> Cope and Marsh first met at some point between 1863 and 1864 when both men were in Berlin. Marsh would later be quoted by Hosea Ballou as having `had doubts of his [Cope's] sanity.` Marsh was further quoted to say that he and Cope had then been on good terms and had `retained friendly relations... during the next five years.`[331]

Indeed, Marsh recalled his first meeting with Cope in unfavorable terms in an article published by the New York Herald on January 19, 1890 (see chapter 6. 5.):

> My acquaintance with Professor Cope dates back twenty-five years, when I was a student in Germany at the University of Berlin. Professor Cope called upon me and with great frankness confided to me some of the many troubles

329 Davidson: The Bone Sharp, p. 20. It is also noteworthy that Davidson argues that the Civil War was a deciding factor in favor of Cope's trip to Europe: "The outbreak of the Civil War itself no doubt had much to do with Edward's first trip to Europe which he made in 1863–1864. It was Osborn's opinion that this trip was arranged by Alfred [Cope's father] to get his son conveniently away from the draft and also away from any temptation on Edward's part to get into the war. This last possibility is more likely than Edward's being called in the draft, as his father could have bought him out of actual military service." See Davidson: The Bone Sharp, p. 26. However, the trip it appears was the final motivating factor in Cope's decision to pursue a career in science, much to the dismay of his father: "If Alfred sent his son to Europe during the Civil War to keep him away from the war and possible service in a non-belligerent capacity, he could not have been entirely pleased with the effect that this trip had in enhancing Edward's desires to be a scientist". See: Davidson: The Bone Sharp, p. 28. Contrary to this it seems that Marsh had made up his mind not to join the fighting and instead perusing a career in science before he left for Europe (see above).

330 Davidson: The Bone Sharp, p. 28.

331 Davidson: The Bone Sharp, pp. 72–73.

> that even then beset him, my sympathy was aroused, and, although I had some doubts as to his sanity, I gave good advice and was willing to be his friend.[332]

Later in 1864 Marsh travelled through southern Germany and the Harz Mountains. Here he studied the geology of the land and met with various scientists of those regions. Marsh even discovered some fossils whose descriptions were published in the American Journal of Science and the Journal of the German Geological Society (Zeitschrift der Deutschen Geologischen Gesellschaft).[333] From there he went back to Switzerland and Tyrol.

In the summer of 1864 Marsh was quite certain he would get a professorship at Yale, but not immediately. Therefore, he decided to extend his stay in Europe for a few more months, provided that his uncle was willing to extend his generous funding:[334]

> If the long cherished object of my ambition [the professorship] were still distant and uncertain as it was when I came to Europe, I should not think of asking for an extension of your already unexampled generosity to me, but as so high a position is now perfectly certain and so soon to be attained I have thought it but right to tell you just how the case stands, and I shall most cheerfully comply with any decision you may think best to give.[335]

With the approval of his uncle, Marsh went to Breslau in October of the same year. Here he worked under Adolph Eduard Grube (1812–1880), professor for zoology, Johann Heinrich Robert Göppert (1800–1884), who taught botany and paleontology, and Ferdinand von Roemer (1818–1891), who was professor for geology, paleontology, and mineralogy. Dana had told Marsh to go to Roemer and learn from him and indeed Roemer had a special interest in American geology, having visited Texas in 1845–1847 and

332 William Hosea Ballou: Marsh Hurls Azoic Facts at Cope, in: The New York Herald, 19 January 1890, p. 11.

333 Othniel Charles Marsh: Notice of a New Fossil Annelid (Helminthodes Antiquus) from the Lithographic Slates of Solenhofen, in: The American Journal of Science, ser. 2, vol. 38, no. 114, (Nov. 1864), p. 415. Marsh’s publications and participations in and lectures at meetings of the German Geological Society are mentioned in the journal of said society, but no articles of Marsh’s were published in it during the years 1864–1865, contrary to the bibliographical information of Marsh that LeVene and Schuchert provide. See for example: N.N.: Zeitschrift der Deutschen Geologischen Gesellschaft, vol. 16, (1864), p. 363. Also see: N.N.: Zeitschrift der Deutschen Geologischen Gesellschaft, vol. 17, (1865), pp. 13, 267–269.

334 Schuchert; LeVene: O. C. Marsh, pp. 56–60.

335 Othniel Charles Marsh, Berlin to George Peabody, [location unknown], 13 June 1864, quoted after: Schuchert; LeVene: O. C. Marsh, pp. 53–54.

having published on the land and geology of Texas and Western Tennessee.[336] Roemer was also interested in US-American politics and the ongoing Civil War and opened a lecture in November 1864 with "Three cheers for Lincoln," after news of Lincoln's reelection had reached the professor.[337] Roemer considered Marsh as an unofficial US ambassador, a representative of his nation, which is an example for how Marsh conducted – willingly, or not – "academic diplomacy." It may be that at this time and due to Roemer's reports from the American West, Marsh first decided to focus his later fossil hunting on that region. LeVene and Schuchert cite the following passage from Marsh's notebook:[338]

> The most inviting field for Palaeontology in North America is the unsettled regions of the West. It is not worth while to spend time on the thickly inhabited regions. It is not worth while to spend time on fossils that are indistinct, or in fragments that do not admit of full determination. There is enough to do with the good ones.[339]

Marsh's notes on Roemer's 80 lectures (beginning on October 31, 1864, ending on March 17, 1865) are very coherent and complete, more so than any of his other lecture notes, indicating he further mastered the German language and developed a heightened interest in the subject matter of Roemer's lectures. His notes include lists of all the American fossils at the Breslau museum,[340] and a list of American fossils described in Europe.[341] Marsh also took very detailed notes on the collection and storage of fossils at the Breslau museum, even going into the exact measurements of the drawers the fossils were stored in.[342]

Marsh believed he was the first US-American student to visit the University of Breslau and therefore he may have enjoyed some special privileges there.[343] Still, he returned to Berlin, which he considered his "European home" according to LeVene and

336 See for example: Ferdinand von Roemer: Texas. Mit Besonderer Rücksicht auf Deutsche Auswanderung und die physischen Verhältnisse des Landes nach Eigener Beobachtung Geschildert, Bonn 1849. Also see: Ferdinand von Roemer: Die Kreidebildungen von Texas und ihre Organischen Einschlüsse, Bonn 1852. Also see: Ferdinand von Roemer: Die Silurische Fauna des Westlichen Tennessee. Eine Palaeontologische Monographie, Breslau 1860.

337 Schuchert; LeVene: O. C. Marsh, p. 62.

338 Schuchert; LeVene: O. C. Marsh, p. 61.

339 Notebook of O. C. Marsh, quoted after: Schuchert; LeVene: O. C. Marsh, pp. 61–62.

340 MS 343, reel 25, frames 631–633.

341 MS 343, reel 25, frames 634–657.

342 MS 343, reel 25, frame 603.

343 "It appears that I am the 1st American student that has studied at this University, & I suppose it is owed [?] to this fact that the Profs here granted me special privilege & shown me much attention."

Schuchert, in March of 1865. Here he had made many friends, which was to prove very beneficial to his later career, for many letters, journals, and fossils would be exchanged through the network which had been established during this time in Berlin.[344]

From Berlin he travelled to Paris and, later, to London, where he worked in the British museum and attended meetings of the Geological and Geographical Societies. In London he established a lifelong friendship with Henry Bolingbroke Woodward (1832–1921), keeper of the geological collection of the British Museum. Before returning to the United States, he visited his uncle in Scotland in the summer of 1865. The enthusiastic hopes he had expressed to his uncle in the letter of June 13, 1864, were somewhat disappointed on Marsh's return to New Haven. It turned out he would not receive a professorship immediately. He had to wait until July 1866 when the Sheffield School established the first chair for paleontology in the United States and named Marsh professor. The professorship though was without a salary, meaning that Marsh was still reliant on the allowance from his uncle. On the plus side, a professorship without salary meant no teaching obligations and Marsh could focus purely on his research and exploration.[345]

Grinnell assesses Marsh's appointment as follows:

> Equipped with the best preparation afforded by the institutions of this country and of Germany, and endowed with ability, energy and perseverance, he assumed the duties of a professorship apparently the first established in that science.[346]

Among the memorabilia cataloged in the Marsh Papers is an advertisement for a July 4th party to be co-hosted by Marsh in Heidelberg in 1863. US-Americans from all over Germany were invited to come to Heidelberg, chosen for its "central position" within Germany. But not all US-Americans were welcomed; only those "Americans, both ladies and gentlemen, who" were "heartily in favor of the maintenance and perpetuity of the 'Constitution and the Union', ignoring minor political opinions," were invited.

Othniel Charles Marsh, Breslau to George Peabody, [location unknown], 23 [?] November 1864, MS 343, reel 22, frames 511–512.

344 Schuchert; LeVene: O. C. Marsh, p. 62.

345 Schuchert; LeVene: O. C. Marsh, pp. 63–66.

346 Grinnell: Othniel Charles Marsh, pp. 290–91. Quote on page 291. Note that Grinnell errs, the world's first professorship for paleontology was established in Paris in 1853. Still, Marsh was the first professor for paleontology in the United States.

This demonstrates that Marsh was part of a greater network of US-citizens living in and traveling about Germany.[347]

Further evidence for Marsh's integration into German society is another keepsake, a songbook from Heidelberg containing various humorous songs,[348] some of which concerned with scientific topics: there is a song about a comet, one about granite, and even one about an *ichthyosaur* witnessing the extinction.[349]

3.3 Correspondences with Former Teachers: Marsh's Transatlantic Network

The Marsh papers preserve five letters that Grube wrote to Marsh. The first three were written in 1865, when Marsh still resided in Berlin. The letters are social calls, inviting Marsh. But Grube also wrote to his former pupil in 1867. This letter is three pages long and in very clean and legible handwriting (in contrast to the letters written in 1865). He had received a scientific paper from Marsh (who apparently also included a personal story as remembrance to his former teacher; it is also possible that he sent some sort of keepsake to Grube, who thanks him for the "Andenken.") and also asked Marsh for further scientific papers from the US. In return, Grube would send some scientific papers originating in Europe. He also inquired about Leidy's postal address, and finally gave the regards of his wife, with whom Marsh was acquainted, as well as Prof. Roemer's (see below).[350] In the final letter, dated 1873, he addressed Marsh as his colleague ("Sehr geehrter Herr College"); back in 1867 he had addressed him more formally as "highly esteemed Professor" ("Sehr geschätzter Herr Professor"). Again, the handwriting is neat. He informed Marsh that his son Oscar, whom Marsh had met at Grube's home in Breslau, was coming to New York. Oscar was going to live and work in the US, and Grube asked Marsh whether he could lend Oscar some sort of assistance, if not with money then at least with some advice for adapting to US-American society and business. Again, he gave Marsh an update on Roemer's whereabouts.[351]

347 MS 343, reel 25, frame 661.

348 MS 343, reel 25, frames 679–694.

349 "Es starb zu selbiger Stunde die ganze Saurierei; Sie kamen zu tief in die Kreide, da war's natürlich vorbei." See: MS 343, reel 25, frame 682.

350 Adolph Eduard Grube, Breslau, to Othniel Charles Marsh, New Haven, CT, 7 April 1867, MS 343, Series I. Correspondence, Box 13, Folder 550.

351 Adolph Eduard Grube, Breslau, to Othniel Charles Marsh, New Haven, CT, 21 April 1873, MS 343, Series I. Correspondence, Box 13, Folder 550.

In the Marsh Papers there are eight letters preserved that Roemer had written to Marsh between 1867 and 1883. The first letter is written in German and Roemer addressed Marsh as his “dear sir Marsh” (“Mein lieber Herr Marsh”).[352] He congratulated Marsh for being allotted a professorship and the gift of the Peabody Museum to New Haven, which would in time, he prophesized, lead Yale to the forefront of natural sciences in America (see below).[353] Beside a professional exchange about fossil sponges, Roemer informed Marsh of the good progress his own museum in Breslau was making, that the new rooms were very nice indeed, especially in contrast to the old ones where Marsh himself had “so diligently worked” in the past, and invited him to return to Breslau to visit the new Museum.[354] Roemer gave Marsh an update on his friend Grube, told him that Grube would conduct research on the west coast of France, and that this would likely be the last time for some years that he would get an opportunity to do so, because there would probably be a war between France and Germany in the near future. This prophecy would prove true in the next few years, as true as his predictions concerning the important role Yale would play in US-American natural sciences, though the reasons Roemer gave for this conflict, namely that France could not bear a “united and strong Germany” (“Deutschland einig und mächtig”) and would like to maintain her hegemonic position, is up for debate.[355]

The next letter was sent in November 1868. It is written mostly in English and Roemer addressed Marsh as “Dear Sir.” The first sentence is written in German and Roemer remarked that he did not hear from Marsh in a long while. The letter consists of an update on the affairs of Roemer’s, what he had worked on, where he had gone, etc. It ends with a postscript mentioning that Grube was also doing well.[356]

352 Ferdinand von Roemer, Breslau, to Othniel Charles Marsh, New Haven, CT, 26 July 1867, MS 343, Series I. Correspondence, Box 28, Folder 1174.

353 “Ich habe mich sehr gefreut zu erfahren[,] dass Die bereits eine Professur für Palaeontologie erhalten haben. […] Ebenso habe ich mit grosser Befriedigung von den vielen prächtigen und reich dotierter [illegible] gelesen mit welcher die Universität von New Haven bereichert worden ist. New Haven wird mit allen diesen Hülfsmitteln [sic!] grossen Haupt-[illegible] für wissenschaftliche und vornehmlich für naturwissenschaftliche Bestrebungen in Amerika werden.”

354 “Die Aufstellung meines neuen Museums ist bereits weit vorangeschritten. Ich hoffe es soll eine der besten Sammlungen in Deutschland werden. Die Räume sind sehr hübsch und wenn ich jetzt an die erbärmlichen dunklen Zimmer denke in welchen sich die Sammlung früher befunden hat und in welcher Sie so fleissig gewesen sind, so muss ich lachen über den Contrast. Sie müssen nothwendig wieder einmal hierher nach Breslau kommen und das Museum ansehen.”

355 “Prof. Grube geht in den nächsten Wochen an die Westküste von Frankreich um zoologische Untersuchungen zu machen. Vielleicht wird die [?] nächsten Jahre eine solche Reine nicht möglich sein, denn ich glaube wir werden einen Krieg mit Frankreich haben. Die Franzosen werden ihn vom Zaune brechen, denn sie können nicht ertragen, dass Deutschland einig und mächtig wird und die nicht mehr allein in Europa befehlen sollen.”

356 Ferdinand von Roemer, Breslau, to Othniel Charles Marsh, New Haven, CT, 23 November 1868, MS 343, Series I. Correspondence, Box 28, Folder 1174.

The next letter is dated November 12, 1870, and is composed in English; Roemer addressed Marsh as "My dear Sir."[357] Marsh had not forgotten his former teacher, who then thanked Marsh for the "several interesting publications" authored by Marsh and sent to Breslau. He stated that by coincidence they had both been working on fossil serpents and that he had sent Marsh his own publications on the matter. Another similarity was that both Marsh and Roemer were establishing their respective museums at that time and Roemer inquired about future opportunities to exchanges specimens, once Marsh had his museum up and running:

> I should like to know whether you are not inclined to make an exchange of fossils with our Museum. Undoubtedly your University Museum possesses a great many duplicates of American fossils which would be highly acceptable [?] to me, and on the other side our Museum could offer a good many things which you want. The exchange would be profitable to both.

He then broadly instructed Marsh on what he would like to have sent to him and on how best to send it. In return he would send his fossils with the help of the Smithsonian Institution: "I could make my carry through the mediation of the Smithsonian Institution." He also informed Marsh regarding the goings-on of the Franco-Prussian War, the outbreak of which he had predicted two years earlier: "The streets of Breslau are swarming with captive French officers- About six hundred of them are here. The Hôtel de Rome, which you probably still remember, are their headquarters." As always, he gave an update on Grube's situation. He then again invited Marsh to visit Breslau and promised his assistance to any friends or students of Marsh's who might be travelling to Europe and wrote that he would "be glad to receive them." The greeting line at the end of the letter is written in German, followed by a postscript inquiring about Prof. Dana's supposedly feeble health and conveying the best for the colleague.

The next surviving letter was written in September 1874 and is held completely in German. This time Roemer addressed Marsh as his "most venerated friend and colleague" ("Verehrtester Freund und College!").[358] Roemer thanked Marsh for the many scientific books and periodicals he had sent to Breslau.[359] He continued his praises by congratulating Marsh on his "many interesting discoveries" which had so "enriched" the field of paleontology, especially concerning fossilized vertebrata, which he had

357 Ferdinand von Roemer, Breslau, to Othniel Charles Marsh, New Haven, CT, 12 November 1870, MS 343, Series I. Correspondence, Box 28, Folder 1174.

358 Ferdinand von Roemer, Breslau, to Othniel Charles Marsh, New Haven, CT, 27 September 1874, MS 343, Series I. Correspondence, Box 28, Folder 1174.

359 "Sie haben mir in den letzten Jahren so häufig wissenschaftliche Bücher [?] und Zeitungen zugeschickt, dass ich mich endlich einmal ausdrücklich bei Ihnen dafür bedanken muss."

"enriched" with many "extremely strange and new specimen."[360] He had further seen a picture of the future Peabody Museum shown in a newspaper (that Marsh had sent him) and thought it promised to become a most "magnificent Building" ("prächtiges Gebäude") in which Marsh could present his vertebrata outstandingly well.[361] He told Marsh of his own University Museum, which was now housed in completely new rooms and would be near unrecognizable to Marsh, who only knew the old museum. He admitted that it did not house any such gigantic skeletons as Marsh had found in the American West, but he still invited Marsh to visit the museum in Breslau and maybe draw some inspiration for his museum in New Haven.[362] This highlights yet another way in which European knowledge could be transferred to the United States, this time not only concerning the scientific method, but public education as well. Roemer again asked whether any exchange could be arranged between them and whether Marsh would send him some duplicate specimens he might have; he also expresses interest in any trans-Mississippian vertebrate fossils.[363] He also mentioned Grube again and this time Göppert as well, remarking that Marsh will surely remember his former teacher, ("Meine Collegen Grube und Göppert, derer Sie sich auch wohl noch erinnern warden"). In a postscript he told Marsh that he would send him a picture of himself (which he did, it is still a part of the Charles Schuchert Papers at Yale)[364] and asked for a photo of Marsh in return.

360 "[…] zugleich muss ich sie wegen der vielen und interessanten Entdeckungen, mit welchen Sie die Palaeontologie bereichert haben, beglückwünschen. Sie haben unsere Kenntniss der Fossilien Wirbelthiere mit einer ganzen Reihe von höchst merkwürdigen neuen Formen bereichert."

361 "Mit vielem Interesse habe ich auch in der zuletzt geschickten Zeitungs[illegible] die Abbildung des in New Haven neu zu errichtenden Museumsgebäudes gesehen. Nach der Abbildung und Beschreibung muss es ein prächtiges Gebäude werden. Darin werden Sie dann Ihre neuentdeckten Wirbelthiere vortrefflich ausstellen können und werden dann erst recht Ihre Freude an denselben haben."

362 "Seit dem Jahre 1866 ist auch unser Universitäts-Museum in schönen neuen Räumen untergebracht und Sie würden dasselbe in seiner jetzigen Gestalt wohl kaum wieder erkennen. Freilich [?] solche grosse Skelette wie Sie dieselben aus dem Westen holen sind nicht darin, sondern Alles ist kleiner und beschaulicher. Aber trotzdem würden Sie Manches darin finden, was Sie Interessieren würde. Besonders aber Bilde ich mir auf die Art der Ausstellung und Anordnung etwas ein. In dieser Beziehung kann sich kein Mineralogisches Museum in Deutschland mit dem meinigen messen. Ich würde sehr erfreut sein, wenn Sie vor der Einrichtung Ihres Museums nach Europa kommen und sich dann auch mein Museum ansehen wollten. Die eine oder andere Einrichtung meines Museums würden Sie, wie ich mir schmeichele, vielleicht praktisch genug finden und dieselben auch in New Haven einzuführen."

363 "Ich bin fortwährend eifrig bemüht die Sammlungen des Museums zu vermehren. Kann man von Ihnen nichts im Tausch erhalten? Sie haben gewiss Mancherlei als duplett abzugeben. Alles Palaeontologische aus den jenseits des Mississippi gelegenen Gegenden würde mir [illegible] interessant sein."

364 In the Charles Schuchert Papers collection of the Sterling Library at Yale there are two photos of Roemer (one dated 1865, the other 1874), which presumably had been found in Marsh's possessions and were used by Schuchert, when he wrote his Marsh biography. See: Charles Schuchert Papers (MS 435). Manuscripts and Archives, Yale University Library.

A rather short letter from 1878 follows, in which Roemer congratulated Marsh (whom he again addressed as "most venerated friend and colleague") for being appointed president of the National Academy of Science,[365] and asked him to initiate another trade.[366]

On October 10, 1880, Roemer sent another letter to Marsh, addressing him as "my dear Mr. Marsh" ("Mein lieber Herr Marsh!").[367] He thanked Marsh for his copy of the "Odontornithes," which had arrived in Breslau some days before. He told Marsh how delighted he was that Marsh had thought of him and how proud he was that he, as a teacher, perhaps contributed a little to Marsh's scientific education.[368] He had heard from visitors of the "greatness" of the Peabody Museum and its "paleontological treasures."[369]

The penultimate letter sent by Roemer to Marsh opens less enthusiastically. It was written in 1882 and is held in English. Roemer addressed Marsh as "My dear Professor Marsh," and opened by informing him that he had "lost a year of [his] life to illness."[370] He had suffered from "an inflammation of the lungs," and had only just recovered after he went on a cure to the Mediterranean. He could only now thank Marsh for sending him a "box with the very valuable casts of Your wonderful cretaceous bird." He regretted that now he was too old to visit the Peabody Museum and marveled at Marsh's discoveries there.

The last letter of Roemer's to Marsh is dated February 4, 1883. It is written in German and for a last time Marsh is addressed as "most venerated friend" ("Geehrtester Freund!"). He thanked Marsh for sending him a cast of *rhamphorhynchus* (a small pterosaur) and ensured him that he would gladly show this specimen to visitors of the museum and remember his famous pupil.[371]

---

365 Note that he was acting president of said society until 1879, when William Barton Rogers (1804–1882) was elected president. Marsh held the regular presidency of the NAS from 1883 to 1895. See: Charles Schuchert: Biographical Memoir of Othniel Charles Marsh, 1831–1899, Washington, DC 1938.

366 Ferdinand von Roemer, Breslau, to Othniel Charles Marsh, New Haven, CT, 19 December [?] 1878, MS 343, Series I. Correspondence, Box 28, Folder 1174.

367 Ferdinand von Roemer, Breslau, to Othniel Charles Marsh, New Haven, CT, 10 [?] October 1880, MS 343, Series I. Correspondence, Box 28, Folder 1174.

368 "Ich freue mich, dass Sie sich wieder freundlich erinnertn [?] und bin stolz darauf Sie einst [illegible] meinen Schuler gehabt zu haben, obgleich ich auf ihre wissenschaftliche Entwicklung erheblich eingewirkt zu haben kaum beanspruchen kann."

369 "Von Besuchern Ihres Museums habe die Grossartigkeit desselben und den Reichtum der darin aufbewahrten palaeontologischen Schätze allgemein [?] rühmen hören."

370 Ferdinand von Roemer, Breslau, to Othniel Charles Marsh, New Haven, CT, 17 May 1882, MS 343, Series I. Correspondence, Box 28, Folder 1174.

371 "Ich werde ihn alljährlich einmal meinen Zuhörern vorzeigen und dabei meines berühmten Schülers als Schenkgebers [?] gedenken."

Henry Woodward of the British Museum also stayed in contact with Marsh and wrote him various letters (thirty-one in total), most of them concerning the exchange of scientific papers, fossils, or casts. Like Roemer did several times, Woodward wrote Marsh specifically about his museum at New Haven, regretting that he would not be able to see it with his own eyes. He wrote: "If our Government were [sic!] more enlightened, they would send me to America on purpose to see your Museum."[372] The situation did not improve much, for in February 1890 Woodward ended a letter with:

> As the time of my trip to America approaches so possibility of coming diminishes & I fear, by August, it will vanish altogether, as [?] I shall be left to weep alone – such is the sad fate of
>
> Yours very sincerely
> Henry Woodward[373]

More than one year later (in June 1891) the lack of opportunity to visit New Haven still troubled Woodward's mind: "I fear I shall not be able to get as far as New Haven this year, but I may come later on! later on! when the powers that be favor my paying a visit to the States."[374]

It is likely that at this point in time the Yale museum was the forerunner in modern museum exhibitions. While inspired by the practices of the European museums, the US-American museums had overtaken the European ones and the knowledge transfer had changed directions; state-of-the-art US-American exhibitions now inspired the European museums.

Woodward praised the quality of the fossils of the American west. One time he joked that Marsh was "defending the locality [of the fossils] with Apaches!"[375] Underneath these lines there is a sketch of a man in stereotypical Native-American attire, attacking or scalping another man who bears a geological hammer in his hand. A caption underneath the picture reads: "fancy Cope! Or any other man!" (see figure 1).

372 Henry Bolingbroke Woodward, London, to Othniel Charles Marsh, New Haven, CT, 20 July 1889, MS 343, Series I. Correspondence, Box 36, Folder 1547.

373 Henry Bolingbroke Woodward, London, to Othniel Charles Marsh, New Haven, CT, 02 February 1890, MS 343, Series I. Correspondence, Box 36, Folder 1547.

374 Henry Bolingbroke Woodward, London, to Othniel Charles Marsh, New Haven, CT, 27 June 1891, MS 343, Series I. Correspondence, Box 36, Folder 1547.

375 Henry Bolingbroke Woodward, London, to Othniel Charles Marsh, New Haven, CT, 21 March 1894, MS 343, Series I. Correspondence, Box 36, Folder 1548.

Figure 1: Sketch by Henry Woodward, in: Henry Bolingbroke Woodward, London, to Othniel Charles Marsh, New Haven, CT, 21 March 1894, MS 343, Series I. Correspondence, Box 36, Folder 1548.

However, the correspondence with Woodward was not of a strictly professional nature. Even though he mostly addressed Marsh as "Dear Prof. Marsh," or "My dear Marsh," Woodward's letters to Marsh are often quite friendly and personal. Grinnell writes that Marsh and Woodward had a "warm friendship that ensued lasting throughout life."[376] Woodward often regards the best wishes of his wife and daughters and sometimes of personal friends of Marsh to the American. In a letter written on December 20, 1889, Woodward included a picture of a Christmas-card, wishing him a "very merry Xmas & a bright, happy New Year from all of the Woodwardian circle."[377]

Woodward later asked Marsh, jokingly (?), if he would get married:

> You say in the last letter: `This is probably the last one [a restoration and accompanying plate Marsh end to Woodward] I shall undertake for some time,` does this mean you are going to be married? Or that you are off again to the Rockies? Or that you are coming to Europe to see your friends? Do not keep us in the dark – we are burning to know. Especially the ladies who send their kind regards with my own to Prof O.C.M.[378]

In 1872 the German politician Theodor von Bunsen (1832–1892, not related to the aforementioned Professor Wilhelm Bunsen) visited the United States. He wrote a let-

376 Grinnell: Othniel Charles Marsh, p. 291.

377 Henry Bolingbroke Woodward, London, to Othniel Charles Marsh, New Haven, CT, 20 December 1889, MS 343, Series I. Correspondence, Box 36, Folder 1547.

378 Henry Bolingbroke Woodward, London, to Othniel Charles Marsh, New Haven, CT, 04 June 1894, MS 343, Series I. Correspondence, Box 36, Folder 1548.

ter to Marsh apologizing that he had not managed to come to New Haven to see the professor. He had "caught such a cold in Washington" that he had lost his voice.[379] This indicates that von Bunsen already knew Marsh, maybe they had met in Berlin. This further demonstrates how well Marsh was connected, not only within the scientific but also within the political high society, and not only in the US, but also abroad. Bunsen promised to return to America and told Marsh: "I look with great pleasure to being introduced by you to your friends, fossil or alive." When Bunsen returned to the US in 1874, he told Marsh in another letter that he would take him up on his kind offer to show the "palaiontological [sic!] trophies" of his expeditions to the West.[380] Evidently this time they met at New Haven, for in another letter Bunsen thanks his host.[381] In 1883 Georg von Bunsen, Theodor's brother, also a politician and member of the Reichstag, would visit the United States and attend the opening of the Northern Pacific Railroad. Theodor added that Georg might be visiting Marsh for he had "begged him to run over" to see Marsh, Yale College, and the paleontological collection.[382]

Georg von Bunsen paid Marsh a visit in New Haven, as attested by a calling card dated October 22, 1883. He informed Marsh that he would be coming to New Haven and bring his daughter.[383] On December 29 of the same year Bunsen wrote a letter,[384] thanking Marsh for his warm reception and the gift of Marsh's book on the *odontornithes*. He also wrote that Marsh did not accept the theory of evolution, a statement that will be discussed further in chapter 7. 2. of this thesis. He added that he saw again "how all Science is one and her Methods give power & command in all realms of human knowledge,"[385] a testament to how Bunsen and others saw an all-encompassing truth in science and how knowledge and discoveries as well as self-affirmation were circulated within the scientific networks. Finally, he invited Marsh to Berlin: "Do come and let me enjoy a little of your company. We have a room always ready for your reception, & a very warm welcome." The last part of the letter is comprised of a

379 Theodor von Bunsen, New York, to Othniel Charles Marsh, New Haven, CT, 25 January 1872, MS 343, Series I. Correspondence, Box 5, Folder 185.

380 Theodor von Bunsen, Newport, RI, to Othniel Charles Marsh, New Haven, CT, 14 October 1874, MS 343, Series I. Correspondence, Box 5, Folder 185.

381 Theodor von Bunsen, Washington, DC, to Othniel Charles Marsh, New Haven, CT, 18 October 1874, MS 343, Series I. Correspondence, Box 5, Folder 185.

382 Theodor von Bunsen, Washington, DC, to Othniel Charles Marsh, New Haven, CT, 09 August 1883, MS 343, Series I. Correspondence, Box 5, Folder 185.

383 Georg von Bunsen, New York, to Othniel Charles Marsh, New Haven, CT, 22 October 1893, MS 343, Series I. Correspondence, Box 5, Folder 184.

384 Georg von Bunsen, Berlin [?], to Othniel Charles Marsh, New Haven, CT, 29 December 1893, MS 343, Series I. Correspondence, Box 5, Folder 184.

385 Bunsen to Marsh, 29 December 1893.

rather lengthy rant on the rising tensions between nations and the intensification of nationalistic sentiment, which would lead to World War I and inner-societal quarrels:

> You will find the outward signs of prosperity, i.e. the love of spending & of aesthetic surroundings, greatly increased since you visited Europe last. And that in spite of interthreatening [sic!] each country, and Society in general, so to say at every turn. Great activity also prevails everywhere in scientific & historical research. It would be difficult to discover traces of actual decadence anywhere. Yet an observation often forces itself upon the outsider [?], viz. that joyousness is wanting which, perhaps, belongs more to such epochs of mankind as are allowed the luxury of drawing conclusions rather than the labour of specializing. Or is, perhaps, that lack of joyousness, perceptible everywhere, simply an outflow from the consciousness that Society at large & in all countries is attacked by, & nearly helpless in face of, howling masses as destructive as they are incapable, & that this threat reaches even the adyta of Science?
>
> In these countries the strange phenomenon can be noted that everything that succeeds (I do not speak of surface successes, as f.i. [sic!] the best rifle, or additions to the army & c) turn out to be the advantage of Socialism. Look at the 'Living Wage' in England, or the Progressive income Tax in Prussia.
>
> 'In Socialism ruere omnes' Tacitus would exclaim[.]

The Von Bunsen-Marsh correspondence is yet another example for how "academic diplomacy" worked within the networks of individuals, in this case one of them also being an official representative of his nation.

3.4 Money Matters: The Significance of the Peabody Patronage

LeVene and Schuchert write about the importance of the funding provided by George Peabody:

> It is already evident, from the foregoing chapters, that the financial background provided by his uncle, George Peabody, was an important factor in the career of Professor Marsh. If Marsh's mental equipment and his determined

> will were the two major elements on which his success was built, a third was certainly the money placed on his hands by the Peabody fortune.[386]

Afterwards they give a biographical overview of Peabody's life and detail some of the philanthropic contributions the businessman provided to US-American science and education. The assumption that Peabody's financial support was but one of three main reasons why Marsh had a very successful career and positively dominated the field of US paleontology in the last quarter of the nineteenth century is an understatement. The money of the businessman and banker was the prerequisite for his nephew's career, without it one can hardly imagine that Marsh would have received any higher education and certainly no college degree. There would have been no trip to Europe, which proved so important to Marsh's scientific education and his decision to focus on paleontology. And last but certainly not least he would hardly ever have obtained a professorship at Yale, and without the Peabody Trust and the erection of a Museum he would not have found himself in the luxurious position to commit himself solely to research, not having to worry about finances until the 1890s.[387]

Science as an integral part of the Yale curriculum was being established in the 1850s and 1860s and can be seen as part of an effort to modernize US-American higher education (see chapter 8. 3.). The Peabody-money that benefitted the Sheffield School directly (as did the Morrill Act of 1863) and allowed for the construction of the Peabody Museum is of much importance in this matter. Plans for a Museum that should stand for at least a "thousand years," to quote a letter Prof. Dana sent to Marsh in 1863,[388] began to take shape in the same year. Marsh was optimistic that the scientific collections at New Haven would, should, and could catch up to and surpass those displayed at Berlin:

> Will the Museum, as at present designed, be large enough for the requirements of the future? It would certainly not be large enough for the present Berlin collections and why should not those of New Haven soon be as extensive? [...] I am sure Yale has a glorious future before her.[389]

In October 1866 the fund for the erection and maintenance of the museum was established with an amount of $150,000. Silliman, Sr., Dana, and Marsh were on the Board of Trustees for life. Despite the seemingly enormous volume of the fund, it was necessary to let the money accumulate to finance the entirety of the ambitious proj-

386 Schuchert; LeVene: O. C. Marsh, p. 68.

387 Schuchert; LeVene: O. C. Marsh, pp. 68–93.

388 See: Schuchert; LeVene: O. C. Marsh, p. 81.

389 Schuchert; LeVene: O. C. Marsh, pp. 83–84.

ect. That meant the completion of the Museum would be postponed until 1876, which also meant that George Peabody would not see the fruits of his donations as he died in 1869 at the age of seventy-four. In the end, the construction of the Museum cost $175,000. The completed building encompassed roughly 3,158 square meters (34,000 square feet), housing laboratories, collections, and a lecture hall. Room enough to accommodate an extensive mineral collection (in 1876 the largest in America), a broad collection of zoological material, geological specimens, and a rather small paleontological collection. The collection contained mostly invertebrates, very few and fragmentary specimens of large vertebrates, and none of the spectacular dinosaur fossils that would elevate US-American paleontology to world fame. Marsh immediately set out to remedy the sorry state of the paleontological collection. At the end of the century over half of the museum space would be occupied by vertebrate fossils, yet only one room was dedicated to their exhibition, while most remained reserved for study. This only began to change in the 1890s, when Osborn's American Museum of Natural History began exhibiting lifelike restorations of prehistoric life to the general public. The dominating sentiment still being that serious paleontological work did not encompass public restorations, that those were rather art installations than scientific ones. The museum was torn down in 1917 to make room for another building (the Harkness Memorial Quadrangle Dormitory) and a new and even more spacious museum was built. It opened in 1925, this time with a great hall specifically dedicated to the dinosaur skeletons that Marsh acquired. Still, Marsh profited in one more way from the Peabody fortune: he inherited $100,000 from his uncle, $20,000 of which was to be kept invested.[390] LeVene and Schuchert end the chapter about Peabody with the following statement:

> The scale on which Marsh lived, and the money that he spent on his collections over a period of thirty-two years (1867–99), show that his income from the Peabody estate must have been very large. [...] it should be said here that the amount of money that came to Yale University directly from George Peabody and indirectly from him through his bequests to Professor Marsh was but little short of half a million dollars.[391]

---

390 Schuchert; LeVene: O. C. Marsh, pp. 84–93.

391 Schuchert; LeVene: O. C. Marsh, pp. 92–93.

3.5 Conclusion

In the end, perhaps Marsh’s attitude towards German science and higher education and the reason why he undertook the journey in the first place is best surmised by Marsh himself in a letter written to his father:

> I like the Germans very much and have the greatest respect for their intellectual qualities, which are surpassed by those of no other nation. They are a Nation of Scholars in fact, and the opportunities for study here are unequaled in the world.[392]

In a short biographical note Charles Emerson Beecher (1856–1904), who succeeded Marsh as curator of the Geological Collections at the Peabody Museum, wrote about Marsh’s European experiences that he “visited the most important localities in Europe, and obtained extensive collections.”[393] Beecher also lists the numerous international honors that were bestowed upon Marsh. Though he was honored for his scientific achievements in his later life, the foundation for said achievements was laid in no small part during the years 1863–1865, and in European institutions of higher learning. Furthermore, they underline how international, how global science was, or at least how scientists imagined themselves:

> The world was not slow to recognize his contributions to knowledge, for during his lifetime he received a large number of tangible evidences of distinguished consideration in the way of academic and scientific honors, medals, and membership in learned societies. In 1886, he received the degree of Doctor of Laws from Harvard University, and in the same year the honorary degree of Doctor of Philosophy from the University of Heidelberg. [...] In 1877, he was the recipient of the first Bigsby Medal awarded by the Geological Society of London, in recognition of his important labors on the Vertebrate Paleontology of the western territories of the United States. In 1898, the highly valued Cuvier Prize was given him by the French Academy, as one of the most able continuators of the science of which Cuvier had laid the foundations. Prominent among the various societies of which he was a member may be mentioned:

392 Othniel Charles Marsh, Berlin to Caleb Marsh, Lockport, NY, 18 August 1863, MS 343, reel 20, frames 752–755. Quote on frame 753.

393 Charles Emerson Beecher: Othniel Charles Marsh, in: The American Journal of Science and Arts, ser. 4, vol. 7, no. 42 (Jun. 1899), pp. 403–428. Quote on page 408.

> The National Academy of Sciences; Institute of France; Royal Academy of Sciences, Brussels; Royal Bavarian Academy of Sciences, Munich; Royal Academy of Science, Bologna; Royal Danish Academy of Sciences, Copenhagen; Royal Irish Academy; Geological Society of London; Geological Society of Germany; American Philosophical Society; Academy of Natural Sciences, Philadelphia; Zoological Society of London; Société Impériale des Naturalistes, Moscow; Geological Society of America, etc., etc.[394]

All international prestige bestowed upon Marsh and all the educational opportunities he had would have been unthinkable without the financial support George Peabody provided his nephew with. Marsh knew how much he was indebted to his rich relative, and that he had to prove to him that his investments would be worthwhile by attaining the highest prestige and honor the nineteenth-century Republic of Letters had to offer. In 1864, before his career had really taken off, Marsh therefore ensured his uncle that "I should do for science as much as you have done for your fellow men."[395]

The patronage of Peabody enabled Marsh to accept an unpaid position at Yale and in fact to build his own museum, creating his own position and job description, skipping a few steps on the career ladder, as detailed above. It is noteworthy that E. D. Cope, Marsh's later colleague and bitter rival, was in a similar position:

> There can be no doubt that the Cope family used their influence in both Philadelphia and in the Society of Friends to get Edward this position at Haverford College after his return from Europe. This was certainly an unusual practice among wealthy families at the time. O. C. Marsh benefited from the same type of family influence in obtaining his position and his museum at Yale. [...] Cope's obtaining position at Haverford under such circumstances may not be unusual, but it certainly fitted his pattern of reliance on the benefits which his family's wealth and position could provide him.[396]

This chapter has detailed how Marsh came to be the first professor for paleontology in the United States, in no small part thanks to the immense privilege of having a rich, philanthropic relative. It also shed some light on Marsh's early education and the factors that led to his decision to focus on a career in paleontology. It elaborated upon the most formative phase in his education in Europe, and especially in German academia, which influenced Marsh's opinions on how science was to be conducted in the context

394 Beecher: Othniel Charles Marsh, pp. 406–407.

395 Othniel Charles Marsh, Breslau to George Peabody, [location unknown], 21 October 1864, MS 343, reel 22, frames 505–506. Quote on page 506.

396 Davidson: The Bone Sharp, p. 32.

of higher education. His trip to Germany contributed greatly to the establishment of his professional network, the prerequisite for attaining the world fame of US science, and of paleontology in particular. In the context of this network, Marsh also gained access to his German assistants, who would then play an important part in Marsh's fall from grace during the "Bone Wars."

4

"Not with Pistols or Fists, but in Print" – The "Bone Wars"

The term “Bone Wars” refers to the “feud” between Marsh and Cope, the two most prolific US-American paleontologists of the nineteenth century. This scientific quarrel provides a vivid example for how interpersonal relationships affect the conduct of science and shall be outlined in all briefness in this chapter as to gain an understanding of the influence it had on the development of US-American paleontology. Because the “Bone Wars” are very well documented, this thesis can only expand on this documentation through the analysis of the correspondence of Marsh's assistants with Osborn (see chapters 6. 2. 3. and 6. 2. 4.). Cope and Marsh attacked each other over the scientific validity of their discoveries on the pages of scientific journals. They blamed each other regarding their professional conduct towards their colleagues and employees, most evidently on the pages of the “New York Herald” in 1890 (see chapter 6. 5.). They tried to hire professional bone hunters away from each other, accused each other of espionage in museums and in the field, and they raced for the discovery and description of paleontological specimens. This race did not only bring about many striking new discoveries, but the hate and urgency with which it was conducted lead to numerous errors in the descriptions.

Marsh, for example, published a reconstruction of *brontosaurus* – a very impressive and to this day popular sauropod – in 1883. However, no *brontosaurus* skulls had ever been found and thus Marsh decided to give his reconstruction a speculative head resembling that of a *camarasaurus*. The skull of this smaller sauropod had been found at another quarry nearby. Paleontologists later realized that *brontosaurus*, which was by then called *apatosaurus*, was more closely related to another sauropod – *diplodocus* – and not as closely related to *camarasaurus* as Marsh had suspected. Some argue that Marsh could have noted this himself, if he had conducted his studies more thoroughly, but he did not, due to his dispute with Cope and the haste in which he raced for the scientific description of as many specimens as possible. Parsons, who describes the whole *brontosaurus-camarasaurus* confusion in more detail than this thesis can, agrees with two paleontologists – Berman and McIntosh – and writes that “Marsh was not incompetent,” and that “his mistakes were due to his feud with Cope, which was at its height at the time of the *Brontosaurus* and *Camarasaurus* discoveries.” He furthermore agrees that the “feud led to sloppiness and poor judgement as descriptions were rushed into print on the basis of inadequate portions of the type skeletons.”[397]

The first subchapter will give a very brief overview of the “Bone Wars.”[398] It will also cite one example of how the “Bone Wars” were conducted in the fossil hunting

397 Parsons: Drawing Out Leviathan, pp. 2–21. Quotes on page 13.

398 Note that the content of the subchapter is in large parts a translated reproduction of a chapter of the author's Magister thesis: Philipp Wendler: Die “Knochenkriege”, die U.S.-amerikanische Presse und atlantische Wissenschaften in der zweiten Hälfte des 19. Jahrhunderts, unpubl. Magister thesis, Hamburg 2011, pp. 81–88.

grounds of Bridger Basin, Wyoming. It also exemplifies how the narrative of the noticeable scientific feud was fused with the "frontier myth."

As stated in the introduction, engaging with the history of a science is always to some extent biographical in nature. In addition, Mitchell reveals how the personalities of Cope and Marsh inspired their legend and are also part of the reason this chapter in US-American paleontology is still remembered vividly:

> Cope and Marsh are legendary figures in the history of the quest for American dinosaurs. They were pioneers in the 'bone rush' that accompanied the gold rush in the western United Stated after the Civil War. They have also come to epitomize contrasting styles of the scientist as a cultural figure. Marsh was a plodding, careful scholar, a skillful administrator, and a master of public relations who parleyed his Uncle Peabody's fortune into the first professorship of paleontology at Yale. Marsh may well have been the most famous scientist in America in the late nineteenth century, and he further inflated his reputation with inflated stories about his frontier heroism and his friendship with the Indians. Cope, by contrast, was a brilliant, moody prodigy who made hundreds of original discoveries and exhausted his family's modest fortunes in his insatiable quest for fossils. Cope and Marsh's 'fossil feud' was waged over priorities in naming, describing, and classifying new species, and over the bones themselves, which often became as hotly contested as mining claims. Marsh ultimately got the upper hand with his superior financial and institutional support.[399]

Due to this personal and biographic component of the analysis of the "Bone Wars," Cope and Marsh will be characterized in the words of their own contemporaries in the subchapters two and three. The fourth subchapter will describe how the increasing belligerence, and the emotional and financial investments of Marsh and Cope ousted the – up until then – most productive and famous US paleontologist, Joseph Leidy.

4.1 A Brief Summary of the "Bone Wars"

Some controversies and long-lasting personal feuds start out as scientific disputes which then spiral out of control. In this case, the argument concerned the head of the *elasmosaurus*. Marsh's personality and his tendencies for grandeur and secrecy fueled the feud. So did his behavior towards his assistants.

399 Mitchell: The Last Dinosaur Book, p. 29.

Most scholars agree that the "Bone Wars" started around 1868, when Cope acquired the fossilized bones of an enigmatic reptile. He studied and later described the reptile. In the corresponding presentation at the Academy of Natural Sciences, Cope identified the new discovery as an aquatic reptile, dubbed *elasmosaurus* ("thin-plate reptile"). He described it as being similar to *plesiosaurus*, the major difference being that "[t]he general form was different from Plesiosaurus in the enormous length of the tail, and the relatively shorter cervical region."[400] But Cope had made a grave error in his observation of the skeleton and he had put the head of the animal on its tail, not its neck. In 1869 Cope published an illustration of the wrong-headed *elsamosaurus* in the "American Naturalist."[401] When Marsh saw Cope's reconstruction, he immediately noticed the mistake his younger colleague had made. When Leidy later inspected the skeleton, he concurred with Marsh's assessment. Cope was mortified. He even sought to buy all the copies of the journal containing his incorrect reconstruction to prevent their circulation, but his efforts were not entirely successful. In 1870 Cope published a corrected description of *elasmosaurus*, but the damage was done, and the friendship between Cope and Marsh had ended for good. Before this incident they were quite cordial toward each other. They had met in Berlin (see chapter 3. 2.), had exchanged some letters afterwards, and even went fossil hunting together in Haddonfield, New Jersey.[402]

Soon after the *elasmosauraus* falling-out both Cope and Marsh began a race for the fossil-rich western regions of North America. Here lay a treasure-trove of undreamt-of magnitude. The expeditions and their adventurous exploits were covered by the press, the discoveries were published as soon as possible – one might say hastily – in the scientific journals.

In 1872 many vertebrate fossils were found at Bridger Basin, Wyoming, turning the location into a battleground in the "Bone Wars." Cope and his team arrived at the Bridger Basin in mid-July 1872. Cope intruded on Marsh's hunting grounds, trying to bribe some of his rival's paid bone collectors, but retreated to the Washakie Basin once Marsh arrived.

---

400 Edward Drinker Cope: Remarks on a New Large Enaliosaur, in: Proceedings of the Academy of Natural Sciences of Philadelphia 20 (Mar. 1868): 92–93. Quote on page 92.

401 Edward Drinker Cope: The Fossil Reptiles of New Jersey (Continued), in: The American Naturalist, vol 3, no 2 (May 1869), pp. 84–91.

402 Jane Pierce Davidson: Bonehead Mistakes: The Background in Scientific Literature and Illustrations for Edward Drinker Cope's First Restoration of Elasmosaurus Platyurus, in: Proceedings of the Academy of Natural Sciences of Philadelphia, vol. 152 (2002), pp. 215–240. Also see: Davidson: Bone Sharp, pp. 35, 71–73. Davidson elaborates on Cope's mistake and provides several examples of how marine reptiles were misrepresented in artistic reconstructions throughout the nineteenth century: Jane Pierce Davidson: Misunderstood Marine Reptiles. Late Nineteenth-Century Artistic Reconstructions of Prehistoric Marine Life, in: Transactions of the Kansas Academy of Science, vol. 118, no. 1–2 (Apr. 2015), pp. 53–67, DOI:10.1660/062.118.0107.

> Marsh regarded as his own any fossil field that he discovered, and he had no intention of having it invaded by a rival [...]. The years that followed saw a fight for supremacy in the western fossil fields that had many comic aspects, although it led to much bitterness.[403]

Though Cope's expedition was smaller and lacked the military backup, Cope managed to find various fossils, describing them then and there and sending his descriptions back east via letter or telegram. "Cope's publications from the field in the late summer of 1872 began to fall like bombshells into Marsh's literary garden, since both were describing and naming fossils collected in the same general area."[404]

The law of priority dictated that a newly discovered species would be named by the first person who managed to publish the discovery in an established scientific journal or a monograph. In the case of Cope and Marsh this led to a race to publication, entailing many errors, oversights, and great chaos in the scientific nomenclature, with some species being named twice or even more often. It took the next generation of paleontologists decades to bring order to this chaotic taxidermy. The ensuing battle over the nomenclature was fought in print, filling many pages of scientific magazines. This went on until Cope collapsed from overwork and an infection in October and returned home a few weeks later.[405] Both scientists now sought to employ scouts and fossil hunters, tried to entice them away from their respective rival, or even employed them to spy on the enemy:[406]

> Information about new fields was sought by both from every possible quarter, collectors were lured away from one 'bone sharp' only to reappear in the next year on the payroll of the other, great precautions were taken to keep the excavations secret, and there were as many false trails and 'salted' clues as one might find in a lusty melodrama.[407]

This established a pattern of behavior that would continue until Cope's death in 1897. The conduct of field paleontology had changed: it had become a very hostile working-environment with fierce competition. At this time Leidy quit fossil collecting as "he was disgusted by the turn events were taking"[408] (see below). A letter sent by Cope to Marsh in 1873 suggests that their friendship might not have been irreparably dam-

---

403 Schuchert; LeVene: O. C. Marsh, p. 262.

404 Lanahm: The Bone Hunters, pp. 112–13.

405 Lanahm: The Bone Hunters, pp. 110–124.

406 Davidson: Bone Sharp, p. 73–74, also see: Schuchert; LeVene: O. C. Marsh, p. 178.

407 Schuchert, LeVene: O. C. Marsh, p. 262.

408 Buffetaut: Vertebrate Palaeontology, p. 131.

aged, since Cope informed Marsh that he sent him a small fossil from Kansas that had ended up with him.[409] But in his reply to the letter Marsh clearly states that their friendship had ended when Cope had lured away one of Marsh's fossil collectors the previous year. Besides, Marsh had expected more fossils from Kansas than Cope had sent him and was now suspicious whether Cope was holding back the other fossils:

> The information I received on this subject made me very angry, and had it come at the time I was so mad at you for getting away Smith (whom I had given valuable notes about localities etc.). I should have 'gone for you,' not with pistols or fists, but in print. I came very near publishing this with some of your other transgressions [...] but my better judgement prevailed. I was never so angry in my life.[410]

As mentioned above, the need for speed in describing specimens led to many errors. Cope even began describing his findings in the field and sending his descriptions back east via telegrams, leading to even more mistakes when, for example, the complicated names of some specimens were misunderstood and Cope's *loxolophodon* became *lefalophodon*, a nonsensical name. Between 1877 and 1879 in the fossil fields of Wyoming when a quarry was abandoned fossils that were deemed to be of poor quality were destroyed on Marsh's orders, lest they should fall into the hands of his hated rival. Quarries were under armed guard to deter possible thieves and spies. It is astonishing that no expedition members were seriously harmed during the "Bone Wars."[411]

In 1878 Marsh became the vice-president of the National Academy of Sciences (NAS). In the same year congress had begun to restructure the surveys of the western territories. Marsh organized the restructuring of the geological survey, his work was ratified by the NAS almost unanimously with just one diverging vote, because Cope could not bring himself to agree with Marsh even once. When Marsh had sought to join the NAS in 1874, Cope had been the only member to vote against his admission. Now, with the establishment of the United States Geological Survey (USGS), Cope lost out because J.W. Powell appointed Marsh to be the vertebrate paleontologist of the survey in 1882, giving him access to vast federal resources. This position would soon bring some trouble for Marsh when it became harder to distinguish between the fossils he had acquired with his private funds and those for which the government had paid. The new administrative obligations also occupied much of Marsh's time now, making it harder to publish original scientific work and therefore necessary to rely on

409 Reingold: Science in Nineteenth-Century America, p. 241.

410 Othniel Charles Marsh, New Haven, CT to Edward Drinker Cope, Haddonfield, NJ, 27 January 1873, quoted after: Reingold: Science in Nineteenth Century-America, p. 242.

411 Davidson: Bone Sharp, pp. 75, 84–85.

the work of his assistants, which in turn led to accusations of intellectual theft in 1890 (see chapter 6. 5.).[412]

Beginning in 1873, the "Bone Wars" were fought out in a series of articles in the "American Naturalist". Cope published a description, and Marsh published an article pointing out Cope's mistakes. In the next issue Cope had corrected said mistakes, but had missed to acknowledge Marsh's input, which in turn was a great source of irritation for Marsh: "Prof. Cope has at last adopted nearly all my views [...] as well as most of my corrections of his errors, although without giving credit in either case."[413] This went on for some months until the editors of the "Naturalist" published the following statement:

> We regret that Professors Marsh and Cope have considered it necessary to carry their controversy to the extent that they have. Wishing to maintain the perfect independence of the NATURALIST in all matters involving scientific criticism, we have allowed both parties to have their full say, but feeling, that now the controversy between the authors in question has become a personal one and that the NATURALIST is not called upon to devote further space to its consideration, the continuance of the subject will be allowed only in the form of an appendix at the expense of the author.[414]

Indeed, Cope and Marsh continued their battle via privately financed articles, and if anything, the tone of the dispute grew even harsher, as evidenced by a nine-page attachment to the "Naturalist," written and paid for by Marsh. He continued to list all of Cope's supposed inaccuracies, and even called one a "stupid blunder."[415] Davidson attests that

> Marsh got nasty [...] it was Marsh's oldest critical refrain; Cope was a worthless paleontologist, a sloppy and ill-educated man, much in over his head. The Loxolophodon telegram, the source of their controversy in the beginning, Marsh labeled as 'merely an unintelligible telegram of no scientific value whatever...' [...] Finally in his summary, Marsh returned to the 'blunders' Cope made concerning the Dinocerata which Marsh said 'are without parallel in the annals of science.'[416]

---

412 Schuchert; LeVene: O. C. Marsh, pp. 262–272.

413 Quoted after Davidson: Bone Sharp, p. 79.

414 N.N.: Notes, in: The American Naturalist, vol. 7, no. 6 (1873), p. 384.

415 Davidson: Bone Sharp, p. 83.

416 Davidson: Bone Sharp, p. 83.

Cope responded in the next issue but stated that such attacks would not “render further discussion of the trivial matters upon which we disagree necessary.” And thusly ended the war in the “Naturalist.”[417] In 1890 the dispute reached a wider, non-scientific audience, as discussed in chapter 6. 5.

4.2 “Slow to Forgive” – Characterizations of O. C. Marsh

Because Marsh’s scientific career is discussed in chapters 3, 5, 6 and 7, this subchapter will skip a general sketch of Marsh’s career and will instead focus on a characterization through his contemporaries and through later scholars of the history of US-American paleontology.

In an obituary Beecher wrote about Marsh he states:

> Among the leading men of science in America, Professor Marsh was unquestionably one of the best known, and had one of the strongest personalities. The world-wide reputation he enjoyed, however, is not altogether attributable to the particular department of research in which he stood without a peer, for, added to his attainments in Vertebrate Paleontology, he possessed an unusual number of mental qualifications in other lines, as well as marked personal characteristics which made him known and felt where his science could never reach.[418]

He adds:

> Another element in his success was seen in the improvement he made in the methods of collecting, preserving, and developing vertebrate fossils, so that even forms long known only from fragmentary remains were represented in his collections by almost complete specimens, presenting nearly the same degree of novelty shown in forms actually new. […] The first Mosasaur was obtained in Holland previous to 1785. It remained imperfectly known for nearly a century, when Marsh, by his contributions to its anatomy, made possible a clear understanding of its structure and affinities. In the same way it could be shown that to many old descriptions of genera and species based upon single

417 Edward Drinker Cope: On Professor Marsh’s Criticisms, in: The American Naturalist, vol. 7, no. 7 (Jul.1873), Appendix.

418 Beecher: Othniel Charles Marsh, p. 403.

> teeth, he was enabled to add a knowledge of the remainder of the animal. Not only did he thus contribute the missing information in regard to many previously described forms, but he brought out a host of entirely new types and made his science one of the most complete exponents of the doctrine of evolution.[419]

As to Marsh's character, he assesses:

> In making an estimate of his character, it must not be forgotten that he developed wholly without the influence of family and home ties, which in most men profoundly mark their mature life. Self-reliance is probably the strongest trait fostered by the absence of immediate family connections. This, Marsh possessed to an extraordinary degree, and it naturally led to a self-centering of his life and ambitions. Out of it came, also, an absence of the complete exchange of confidence which normally exists between intimate friends. Even where perfect confidence existed, he seldom revealed more about any particular matter than seemed to him necessary or than the circumstances really demanded. As a friend, he was kind, loyal, and generous. As a patron of science, he has seldom been equaled. Honest work in any department appealed to him strongly, and he was ever ready with aid and counsel, even at the expense of a personal sacrifice. His disposition was a most happy one, and he was always keenly appreciative of the humorous and ludicrous and fond of relating amusing experiences and anecdotes. The sunny side of his character was nearly always uppermost, and the consideration of subjects of the greatest gravity was enlivened by constant sparkles of wit from his exhaustless store. He was normally restive under restraint, and met all opposition with power and fearlessness. Having practically created the modern science of Vertebrate Paleontology in America, he resented any encroachment upon the particular fields of research in which he was engaged. This attitude frequently developed feelings of hostility in other investigators, and often alienated him from co-workers in his department of science. Nevertheless, he labored faithfully for the truth as revealed in his work, and was ready to change opinions and published statements whenever facts seemed to warrant it.[420]

Beecher also said that Marsh's vertebrate fossil collection "was pronounced by Huxley, who examined it with care in 1876, to be surpassed by no other in the world; and Dar-

419 Beecher: Othniel Charles Marsh, p. 404.

420 Beecher: Othniel Charles Marsh, pp. 405–406.

win, in 1878, expressed a strong desire to visit America for the sole purpose of seeing it."[421]

This is high praise, but at least the declaration that Marsh had "practically created the modern science of Vertebrate Paleontology in America" is a cornerstone of this thesis and the reason why Marsh and his scientific network are the focus of this study.

As for a contemporary German perspective, Geinitz reported Marsh's death in the "Leopoldina", putting things into an international context right from the beginning by writing that Marsh's death was a "severe blow to science, not just in America, from which's soil his great discoveries were extracted, but it would also affect international science of all nations."[422] He later adds that Marsh had visited the museum in Dresden several times, and had even donated several of the exhibits which were employed in the education of future naturalists. Geinitz himself had received all of Marsh's publications and had also published excerpts and abstracts of Marsh's work; he was confident that many German scientists were familiar with Marsh's texts.[423]

In 1931 Henry W. Farnam characterized Marsh as being a bit selfish because he essentially was a self-centered person, who seldom, if ever, had to consider the concerns of his fellow human beings:

> Some of these oddities were, I think, the result of never having been obliged to consult the wishes and convenience of other people in his domestic arrange-

421 Beecher: Othniel Charles Marsh, p. 413.

422 "Der am 18. März nach kurzer Krankheit plötzlich erfolgte Tod von O. C. Marsh ist ein harter Schlag für die Wissenschaft, nicht nur in Amerika, dessen Boden seine grossen Entdeckungen entnommen sind, sondern es wird dadurch die internationale Wissenschaft in allen Ländern unmittelbar betroffen, und manches Jahr vergehen, bevor die vielen kostbaren im Peabody Museum zusammengehäuften Schätze in dem Sinne des voranstürmenden Entdeckers weiter gesichtet und präpariert werden können." Hanns Bruno Geinitz: Othniel Charles Marsh. Professor der Paläontologie an Yale Univeristät, New-Haven, Conn. und Landes-Paläontolog für Wirbelthiere in den Vereinigten Staaten Nordamerikas, in: Leopoldina, vol. 35, no.7 (Jul. 1899), pp. 122–124. Quote on page 122.

423 "Professor Marsh hat unser Königl. Mineralogisch-Geologisches Museum in Dresden wiederholt besucht und beschenkt mit werthvollen Gaben der Abbildungen und Abgüsse verschiedener Unica seiner Riesenthiere, unter denen nur hervorgehoben sein mögen: Abbildungen und Abgüsse von Hesperornis, Abguss des interessanten Eosurus-Wirbels aus Steinkohle von Nova Scotia, Abguss der vollkommensten Trilobiten und seiner Vervielfältigungen des Rhamphorhynehus von Eichstädt, den ich selbst für ihn seiner Zeit angekauft habe, nachdem dieser kostbare Fund weder in München noch in Dresden ein Unterkommen gefunden hatte. Für Marsh war ein Preis nie zu theuer. Ich habe lange Zeit den Vorzug gehabt, als alter Freund der Yale University und in meiner früheren Stellung als langjähriger Redacteur des neuen Jahrbuchs für Mineralogie, Geologie und Paläontologie die meisten Druckschriften von O. C. Marsh, die von 1861 an bis 1899 in dem hochschätzbaren American Journal of Science, New-Haven, erschienen sind, persönlich erhalten zu haben und darüber in den mir zugänglichen Blättern berichtet, so dass ich wohl annehmen darf. dass die wichtigen Marsh'sehen Arbeiten wenigstens zum grossen Theile den Männern der Wissenschaft auch in Deutschland bekannt sind." Geinitz: Othniel Charles Marsh, p. 122

> ments as he would have had to do had he been married. They do not indicate any lack of public spirit or consideration for others.[424]

In 1940 George F. Eaton, the self-proclaimed last surviving student of Marsh's, had read Schuchert and LeVene's book and felt compelled to defend his former teacher. He suggests that Marsh possessed "really charming personal traits – a gracious gentleness and thoughtfulness for the welfare of others, rarely found in men of vigorous, aggressive character, and of all men I have known he was one of the most appreciative of kind, friendly courtesies."[425]

He continues his defense by writing:

> But these lovable qualities were never displayed to persons whom he did not trust. Slow to forgive acts of treachery and hostility, he was yet able to forgive and forget past injuries when convinced that his former enemy had buried the hatchet, as in the case of Dr. Jacob L. Wortman who came voluntarily to Marsh and confessed his error in having, for some years, been active in the Cope-Osborn camp.[426]

In 1910 George Bird Grinnell produced a Marsh biography as part of a book called "Leading American Men of Science." He begins by underlining Marsh's enormous significance to US-American paleontology, and calls him one of the best scientists of the USA.[427] Grinnell does not mention the "Bone Wars," or any rivalry with Cope, though in a private letter written in 1919 he describes Cope as "Marsh's hated rival."[428] In the same letter he then describes Marsh as follows:

> Many of his ways of life were distinctly individual, and the people were disposed to laugh at his unusual ways rather than to observe the sterling qualities which lay beneath them. Marsh was a peculiar man and did not often show his real self to those with whom he casually came in contact. His great enthusiasm for the study of these fossils and his constant thought about them made it often hard to learn his views about things in general. Moreover, for many

424 Henry W. Farnam, New Haven, CT to Ernest Howe, New Haven, CT, 6 May 1931, MS 343, reel 26, frame 469.

425 George F. Eaton, New Haven, CT to Donald Adams, New York 17 July 1940, MS 343, reel 26, frame 559.

426 George F. Eaton, New Haven, CT to Donald Adams, New York 17 July 1940, MS 343, reel 26, frames 559–560.

427 Grinnell: Othniel Charles Marsh, p. 283.

428 George Bird Grinnell, New York to Ernest Howe, New Haven, CT 19 February 1929, MS 343, reel 26, frame 305.

> years, he had been much alone. He had few or no near relatives and was really attached to very few people. Hence, in great measure his thoughts were about himself and he had become somewhat selfish. Yet where his own interests were not involved, he was most kindhearted, and was often ready to take great trouble to be helpful to others.[429]

Samuel Wendell Williston worked for Marsh as a fossil hunter for eleven years. When Williston quit in 1885, he was extremely dissatisfied with his employer and accused him of willfully wasting the time of his assistants. Later Williston apparently switched sides and wrote some letters concerning Marsh to Cope. The latter used Williston's statements (it seems Williston thought them private and personal at the time of their writing, and definitely not for publication) in his newspaper offensive of 1890 (see chapter 6. 5.). Williston then stated his letters were written some years prior and assured Marsh that he "refused to have anything to do with the subject [...] The whole subject no longer concerns me, and is distasteful [...] I have no personal grievances."[430]

Pauly notes that Marsh was indeed a particularity in the landscape of higher education during the early second half of the century:

> The most famous life scientist at Yale was the United States' only 'professor for paleontology,' Othniel C. Marsh. Marsh received no significant support from the university. His uncle, London-based merchant banker George Peabody, supplied both his salary and the funds to build the Peabody Museum of Natural History. The federal government, in the persons of Marsh's longtime professional allies Braid and Powell, provided the bulk of operating funds and boxed specimens that made New Haven the chief center for study of American dinosaurs.[431]

About Marsh and his relation to his coworkers and colleagues Merrill writes:

> It was through him, ably assisted by Hatcher, more than any other man, that was brought about the enormous improvement in the manner of collecting and preparing fossils above referred to. 'He not only had the means and the inclination, but entered every field of acquisition with the dominating ambition to

429 George Bird Grinnell to Ernest Howe, 19 February 1929.

430 Quoted after: Wallace: The Bonehunter's Revenge, p. 225. For a complete biography of Williston see: Elizabeth Noble Shor: Fossils and Flies. The Life of a Complete Scientist. Samuel Wendell Williston (1851–1918), Norman, OK 1971. Especially pp. 71–111, describing his field work for Marsh, and pp. 117–123, depicting his involvement in the newspaper affair of 1890.

431 Pauly: Biologists and the Promise of American Life, p. 114.

> obtain everything there was in it, and leave not a single scrap behind.' This, and a natural disposition to rent the intrusion of others into a field which he felt he had created, to a considerable extent alienated him from coworkers in his particular department.[432]

This is a most interesting observation. Marsh, the first professor for paleontology in the US, who had invested a considerable portion of his own (i.e., his uncle's) money into the attainment of specimens indeed might have thought that the whole field of vertebrate paleontology belonged to him by the rights of acquisition, that it was his prerogative to describe any new fossil. Marsh may have concluded that if all of society was molded by capitalism, science was as well. With his uncle's financial support, it became possible for Marsh to purchase his prime position in paleontology by obtaining the most interesting fossils for himself, and in some cases locking them away for years. Did he then presume that any work drawn up by the men he hired and paid (often with government money) was his to publish? Marsh's scientific conduct and his treatment of his assistants evokes questions of this nature.

4.3 "Enjoying the Fight for its Own Sake" – Characterizations of E. D. Cope

Since the focus of this thesis lies on the analysis of Marsh's professional network and his working relations with his assistants, the assessment of Cope's character will take up less room within this chapter.

Even as a child Cope showed great interest in nature and kept a diary during a trip to Boston in 1847, in which he recorded (and illustrated) his field observations. Still Cope never received a formal higher education, but he spent a year studying under Leidy. In 1859 he joined a group of young scientists who worked for the Smithsonian; here he published his first scientific papers, and many more followed, mainly concerning reptiles, fish, and snakes.[433] In 1863 he travelled to Europe, visited museums and fossil collections in Great Britain, France, the Netherlands, Austria, and Prussia. After he returned to the United States in 1864, he became a professor at Haverford College, Pennsylvania, where he stayed until 1867 when he quit his position to travel and study the West. The number of Cope's publications rose rapidly, and he began to focus on vertebrate paleontology. In 1872 he joined Hayden on his survey, and many of the fos-

432 Merrill: The First One Hundred Years of American Geology, pp. 529–530.

433 Benjamin, Marcus: Edward Drinker Cope. Paleontologist, in: Jordan, David Starr (ed.): Leading American Men of Science, New York 1910 (Repr. Ed. 1973), pp. 313–340, see pp. 315–319.

sils Cope studied came from out West.[434] Osborn met Cope in Philadelphia in 1877 and sided with the Quaker in his battle with Marsh. Osborn had previously met Marsh in New Haven, and had studied some of Marsh's fossil collections at the museum. Marsh is said to have limited Osborn's access to less important specimens and even to have silently followed Osborn through the museum to keep an eye on him. In 1885 Osborn aided Cope in inciting a rebellion among Marsh's assistants (see chapters 6. 2. 3. and 6. 2. 4). In later years Osborn even helped Cope with his financial troubles, giving Cope $ 2,500 to help his friend and former mentor, who felt isolated and needed the professional and personal support.[435] Osborn even stated that Cope enjoyed the feud, and links this, in a typical Osbornian (i.e., racist) move to Cope's heritage:[436]

> Meanwhile, Cope's correspondence with Osborn reveals the inside details of this great warfare, and Cope's thoroughly humorous Celtic attitude towards it, namely: that he was thoroughly enjoying the fight for its own sake and enjoyed not only giving hard blows, but the indirect consequences of the onset.[437]

Regal even describes Cope and Osborn's relationship as follows:

> Like characters in a Greek drama, Cope was the wise old sage who pointed the young hero in the right direction, imparted secret knowledge to him, and gave him the weapons needed to defeat his foes. Once a master himself, Osborn returned to save the dying king [...] Osborn's biography, *Cope: Master Naturalist (1931)*, is in large part an attempt to resurrect Cope's memory, paint him as a victim in the bone wars with O. C. Marsh, and generally show Cope to have been a man of good humor and affection, not the bitter, angry eccentric most thought him.[438]

---

434 Benjamin: Edward Drinker Cope, pp. 319–331.

435 Regal, Brian: Henry Fairfield Osborn. Race, and the Search for the Origins of Man, Aldershot 2002, pp. 53–77.

436 Note that Osborn harbored some strong opinions about "race", the assumed "dilution" of the pure "Nordic race" with some of the ones Osborn found to be less desirable. Osborn considered it necessary to prevent excessive immigration of south European and Asian phenotypes; to help the "multiracial" children in New York to improve themselves to "fulfill their potential;" and to preserve the natural order of "races," classes and sexes against the erosive forces of the civil rights and women's rights movements. "Preservation" was generally an important concept of Osborn's great project – he advocated not only for the conservation of nature and animal species, but also for the preservation of, above all, the "Nordic race." See: Sommer: History Within, p. 26. For Osborn's convictions on race (and especially Cope's influence on that matter) see Regal: Henry Fairfield Osborn, pp 102–135.

437 Osborn: Cope, p. 408.

438 Regal: Henry Fairfield Osborn, pp. 75–76.

A short Cope biography was published in "Leading American Men of Science" in 1910. According to Marcus Benjamin, the author, there had never been a more genius geologist than Edward Drinker Cope. Edward was born to the prestigious Cope family of Philadelphia; the family name can be occasionally found in the annals of the City of Brotherly Love. When Edward was born in 1840 his family had accumulated a handsome fortune, thanks to their involvement in the parcel-business (just another way in which the establishment of the communications infrastructure would greatly further science, in this case in an admittedly roundabout way).[439] Benjamin notes some of Cope's scientific achievements: he received an honorary doctorate from the University of Heidelberg, and the Bigsby Medal in London. In 1866 he became a member of the American Philosophical Society, and he joined the NAS in 1872. In 1864 he was elected a corresponding member of the Zoological Society of London, in 1881 a member of the Geological Society of London, and in 1878 a member of the Geological Society of France.[440]

In 1871 British paleontologist Harry Seeley (1839–1909), an expert on dinosaurs, wrote to Cope and thanked him for a synopsis on the extinct American reptiles and frogs ("batrachian"); he also tells him that: "Clearly you for the first time enable Europeans to understand American fossils."[441] This praise seems somewhat hyperbolic, presumably to get on Cope's good side, for he then asks Cope to send him more insight into his work, but it still underlines the distinguished position Cope, Marsh, and only a few more US paleontologists had within the transatlantic paleontological network.

Url Lanham adds a religious dimension to Cope's relationship with science:

> Even for several years after his return to the United States Cope, while on the surface an active and brilliant professional scientist, was in private life a religious fanatic, embarrassing even his devout Quaker compatriots by his outpourings of religious fervor.[442]

Finally, Davidson stresses that Cope's artistic talent contributed in a significant way to his scientific publications: "An amazingly large amount of paleontological art by Edward Drinker Cope is extant, and his impact as an artist and a designer of books was significant."[443] At this point it is noteworthy that in 1899 Frank Bond sketched

---

439 Benjamin: Edward Drinker Cope, pp. 313–315.

440 Benjamin: Edward Drinker Cope, p. 337.

441 Harry Govier Seeley, Cambridge to Edward Drinker Cope, Philadelphia, 17 February 1871, Haverford College Quaker & Special Collections Edward Drinker Cope papers, HC.MC-956.

442 Lanahm: The Bone Hunters, p. 68.

443 Jane Pierce Davidson: A History of Paleontology Illustration, Bloomington, IN 2008, p. 80. Davidson describes Cope's paleographic art on pages 79–83 of the same book.

stegosaurus in a completely wrong way, placing its tail spikes on its body and interpreting the upright plates on the beast’s back as body-fitting armor plates, similar to the exoskeleton of a beetle. Adding insult to injury he posed Cope as the supposed discoverer next to the inaccurate monstrosity, while Marsh published the first description of *stegosaurus* (see figure 2).[444]

Figure 2: Restorations of ***Stegosaurus ungulates***, in: Gilmore, Charles Whitney: Osteology of the armored Dinosauria in the United States National Museum, with special reference to the genus Stegosaurus, in: Smithsonian Institution United Stated National Museum Bulletin, no. 89 (1914), plate 33.

When comparing both scientists, Marsh, the older of the two, appears as a thorough, careful scientist who dedicated his life to his work. While described as charming and socially apt, Marsh remained a bachelor all his life and devoted himself to his scientific research, the responsibilities that came with it, but also enjoyed the privileges surrounding it. He was described as strategic, even plotting at times. Marsh was well connected with the social elite and a skillful political negotiator, which granted him government connections and funding.

Cope was more impulsive and irritable than Marsh. He was born into a Quaker dynasty, very religious, and had a family of his own. As opposed to Marsh, Cope was mostly self-educated and had enjoyed a less formal education. Though privileged, he had fewer financial resources and government funding than Marsh and was less skillful at strategically forging social connections. Cope was a quick worker and being the younger and less formally educated scholar he had more to prove than Marsh and

444 Davidson: A History of Paleontology Illustration, pp. 141–143.

therefore published more frequently. But Cope also committed more scientific errors in his work, perhaps since he had fewer employees to support his research.

As can be seen from the descriptions of their contemporaries, Marsh and Cope shared certain characteristics, which contributed to their later feud. Both were financially privileged and studied with the leading paleontologists of their time. Their backgrounds likely shaped their outlook and understanding of the world: both Marsh and Cope were rather self-opinionated, aggravating their scientific (and sometimes personal) arguments. They also shared dogged and vindictive character traits, which prolonged their various quarrels. Their scientific conduct, the way they treated their employees, and their role within the scientific community were shaped by their distinctive characteristics and biographic circumstances.

4.4 "Tackling 'Scylla and Charybdis'" – Leidy Quits the Race

Before Marsh and Cope became the foremost authorities on US-American vertebrate paleontology Joseph Leidy described most American vertebrate fossils, including the first dinosaurs discovered by Euro-Americans (see chapter 2. 6.). Geologist John Evans was the first scientist to study the Badlands in 1849, prompted by random fossil finds through travelers who took them as souvenirs. In 1850 T. A. Culbertson was sent to the Badlands by the Smithsonian Institution to collect fossils there; the collected fossils were sent to Leidy. Together with Ferdinand Vandiveer Hayden, Evans returned to the Badlands in 1853; this expedition was financed by Hall. The presence of two German collectors annoyed Leidy and perhaps even offended his sense of patriotism. Hayden later continued collecting in 1854, when doing so became more dangerous due to the war with various Native American nations and the defeat at Fort Laramie. Gouverneur Kemble Warren, who would also become a famous traveler of the American West, came as an army-engineer to the Badlands. Both undertook various surveys of the region in the 1850s; the imminent danger of a confrontation with the Native Americans made the expeditions more exciting and heroic in the public eye.[445] By the 1850s Leidy had specialized in fossil bones, and had written a number of short scientific papers on that subject.[446] Before Cope and Marsh cornered the market and made professional bone hunting profitable, most vertebrate fossils had been sent to Leidy, the only US-American full-time paleontologist, for free.

---

445 Lanahm: The Bone Hunters, pp. 32–35.

446 Lanahm: The Bone Hunters, p. 19.

Cope had been a student of Leidy's, but in the early 1870s their relationship had soured. Still, Leidy's ties to Marsh and Cope remained cordial even after the "Bone Wars" broke out over Cope's "boneheaded mistake" and even besides the fact that both professors failed to mention Leidy's discoveries in their papers, which intellectual honesty would have demanded.[447] According to Warren, Cope had become a pariah due to his quarrelsome nature that led him to commit some shameful deeds within the scientific community. Therefore, he had immense trouble finding work within the scientific establishment and at times even infuriated the pacifistic Leidy.[448]

Thomson argues that there must have been a dispute between Cope and Leidy before 1869, when Leidy pointed out Cope's mistake concerning the *elasmosaurus* head in the "American Journal of Science". This seems rather likely due to Leidy's usually pacifistic and non-confrontational behavior. Thomson suggests the two Philadelphians had their falling-out because Leidy had excluded Cope's *laelaps* from the planned Paleozoic Museum in New York.[449]

In the early 1870s Leidy had to opt out of the business of vertebrate paleontology, which had indeed become commercialized:

> the important fossils coming to light were purchased by Cope and Marsh, Leaving Leidy with empty hands. Remaining in the field of paleontology would have meant a ferocious battle with two superb, ambitious, aggressive, paleontologists who, because of their wealth, preempted the field.[450]

Url Lanham quotes Leidy saying to a "British colleague" in 1870:

> Formerly every fossil one found in States came to me, for nobody else cared to study such things, but Professors Marsh and Cope, with long purses, offer money for what used to come to me for nothing, and in that respect I cannot compete with them.[451]

Warren describes an episode of the "Bone Wars" which may have been the reason why Leidy finally decided to quit vertebrate paleontology:

---

447 Warren: Joseph Leidy, p. 188.

448 Warren: Joseph Leidy, p. 210.

449 Keith Stewart Thomson: The Legacy of the Mastodon: The Golden Age of Fossils in America, New Haven, CT 2008, pp. 163–164.

450 Warren: Joseph Leidy, p. 151.

451 Lanham: The Bone Hunters, p. 18.

> A particularly venomous battle occurred between Cope and Marsh at Fort Bridger during Leidy's visit in 1872, though the particulars of the dispute and Leidy's part in it are not known. Perhaps outright theft was involved, or even fisticuffs, but Leidy remained silent on the subject [...] Leidy, who abhorred witnessing improper behavior in others, recoiled from the unseemly contest, dubbed the 'bone wars' by a gleeful public, and he withdrew from paleontology to pursue his old interests in parasitology and protozoology. Leidy, a proper Victorian gentleman who carefully guarded his reputation and honor, insisted in a world of peace, order, and bonhomie.[452]

Leidy did not only lack funding, but also had a great distaste for squabbling and fighting: "A love of peace was one of the essential characteristics of his nature [...] He disliked controversy exceedingly [...]"[453] Maybe it is due to his polite and humble nature that Leidy is far less well-known today than Cope and Marsh, beside his great scientific contributions.[454]

Cope and Marsh had almost completely cornered the US-American fossil market by 1880, thanks to their private fortunes and government support, and "tackling 'Scylla and Charybdis' (Marsh and Cope) at this point [1880] was a suicidal effort."[455] Reingold also attests that Cope and Marsh were "robber barons trying to corner the old-bone market."[456] Still, Cope and Marsh owed a lot to Leidy, for he had introduced them to Haddonfield and the Bridger Basin, from which they then blocked him.[457]

4.5 Conclusion

Why do the "Bone Wars" matter? As stated in the introduction, the "Bone Wars" fought between Marsh and Cope embody the most popular episode in the history of US-American paleontology. An abundance of books and articles, as well as novels (traditional and graphic), children's books, documentaries, and even a card game covering the "Bone Wars" have been published. This study does not focus primarily on the

452 Warren: Joseph Leidy, p. 187.

453 Osborn: Biographical Memoir of Joseph Leidy, p. 349. This passage also tells a lot about Osborn's personal political and (pseudo)scientific opinions: "Descent from patriotic German-American stock enables us to understand the sources of Leidy's fine moral qualities."

454 Warren: Joseph Leidy, pp. 245–252.

455 Regal: Henry Fairfield Osborn, p. 46.

456 Reingold: Science in Nineteenth Century-America, p. 238.

457 Warren: Joseph Leidy, p. 186.

conflict between Cope and Marsh, but the "Bone Wars" nevertheless cast their shadow over the entire history of US-American paleontology during the nineteenth century. The conflict influenced Marsh's professional network, his scientific career, and the scientific conduct of paleontology in the United States in several ways. Even though Thomson states that scientists like Marsh and Osborn "nurtured the young tradition of scientifically oriented scholarship in the United States," and that "[r]ather than talk about an academic ideal, they lived one,"[458] they too lived in a social environment and their interactions were not limited to the conduct of "pure science." Instead, they also harbored personal aversions towards each other, and a dispute that had started as scientific quickly became personal. On the other hand, the "Bone Wars" influenced paleontological discourse, which was chiefly conducted on the pages of scientific publications. The competition between Cope and Marsh also led to an unprecedented increase in scientific descriptions, especially of dinosaur skeletons.

> Leidy, Cope, and Marsh were great contributors to paleontology. Prior to their work only 98 genera and species of North American fossil vertebrates were known; to this number they added 2193 genera and species. Of this total Leidy contributed 375, Marsh 536, and Cope 1282.[459]

It is noteworthy that the dinosaurs discovered by Cope, Marsh, and their contemporaries became themselves frequently used metaphors for conflict and confrontation within science.[460]

Since their conclusion, the "Bone Wars" have repeatedly been discussed by various natural scientists, who assess the impact they had on paleontology as a discipline. Grinnell, for example, judges the "Bone Wars" as follows:

> In some respect this rivalry was unfortunate. It led to hasty examinations of the collections and sometimes this haste caused grave errors. [...] this rivalry was greatly to the advantage of those men employed by the paleantologists in collecting fossils for them, for the wages paid these collectors were high and sometimes bonuses were given for special discoveries. All this was a long time ago and has been forgotten by most people, but the few who remember the

---

458 Laurence R. Veysey: The Emergence of the American University, Chicago 1965, p. 153.

459 Reingold: Science in Nineteenth-Century America, p. 239.

460 Helen Haste: Dinosaur as Metaphor, in: William A. S. Sarjeant (ed.): Vertebrate Fossils and the Evolution of Scientific Concepts. Writings in Tribute to Beverly Halstead, by Some of His Many Friends, Amsterdam 1995, pp. 359–380.

> written and printed combats of those days still look back on them with hearty amusement.[461]

He adds: "It is entertaining enough to look back for a generation and to remember how vital at the time seemed the subjects over which we fought and, when we look at them today, to see how unimportant they appear."[462]

On the other hand, one of the most prominent paleontologists of the late twentieth century, Jack Horner, writes that the immense popularity of dinosaur paleontology caused it to be viewed as of no theoretical, scientific value. The standing of dinosaur paleontology within the scientific community suffered accordingly. The "Bone Wars" reinforced this prejudice as the public mudslinging was below most serious scientists and detracted from the truly extraordinary discoveries unearthed by the quarreling bone hunters. Increasingly, paleobiology (the broader discipline dinosaur paleontology is a part of) acquired the reputation of a respectable science with a theoretical foundation. While some of the early paleontologists were mere collectors, eccentrics prone to showmanship, others studied dinosaurs within a theoretical framework and sought to define their place within their ecological environments (it can be argued that Cope and Marsh were amalgamations of both stereotypes). Horner further elaborates that dinosaurs were somewhat disregarded by paleontologists around the middle of the twentieth century, who would rather study mammals. This trend changed slowly during the 1960s when forward-thinking scientists like John Ostrom (1928–2005) and Robert Bakker (*1945) heralded the start of a revolution of dinosaur paleontology. The bird-like attributes of dinosaurs were (re-)discovered, recently discovered nests allowed insights into the behavior of the animals. Ever more detailed anatomical studies, now including microscopic bone structures and the study of blood cells, shed new light on the placement of dinosaurs in the evolutionary process. Horner ends his essay describing his vision for the future of paleontology as a truly interdisciplinary science and suggesting that soon the fruits of collaboration would produce a dinosaur grown from a chicken egg.[463]

461 George Bird Grinnell, New York to Ernest Howe, New Haven, CT 19 February 1929, MS 343, reel 26, frame 305.

462 George Bird Grinnell, New York to Ernest Howe, New Haven, CT 19 February 1929, MS 343, reel 26, frame 306.

463 John R. Horner: Dinosaurs at the Table, in: David Sepkoski; Michael Ruse (eds.): The Paleobiological Revolution. Essays on the Growth of Modern Paleontology, Chicago 2009, pp. 111–121, DOI: 10.7208/chicago/9780226748597.001.0001.

5

Fossil-Hunting in the “Wild West”

"The relative completeness of Hadrosaurus, as well as the abundance and excellent preservation of other fossils collected subsequently in the US West, conveyed a competitive advantage to American paleontologists over their European rivals, whose fossils were often fragmentary and fewer in number."[464]

The modern conceptions of "science" and "America" were invented in sixteenth- and seventeenth-century Europe, respectively. While the conception of science has been the topic of chapter 2, this chapter will examine how the (English) colonies in North America, and later the United States, were conceptualized as a homogenous Nation-state, how the natural land of North America became the basis for nineteenth-century US-American identity, and how conceptions of masculinity, class, and race contributed to this process.[465] All this happened during a time when enormous swaths of land on the North-American continent were claimed by Euro-Americans, and the subjugation of the supposedly wild continent was often equated to the "conquest" of nature by science: "[...] the Western conception of America, as a singular integrated place inflected with the historic tropes of naturalness and conquest, parallels the imagining of scientific progress as a cumulative mastery of nature."[466]

But why link the history of US-American paleontology to the history of Euro-American expansionism? Were the centers of scholarship, study, and research not located in the east? As previous chapters have hinted at, the most spectacular American fossils were found in the west, not the east of the continent. Furthermore, the genesis of US-American paleontology in the nineteenth century is inevitably linked to Marsh and Cope, the true innovators, and pioneers of this discipline in the United States. The story of Cope and Marsh is in turn inextricably linked to their rivalry, the "Bone Wars." The introduction of this thesis lists numerous examples of pop cultural adaptations of the "Bone Wars." To elaborate on one of those: before Michael Crichton wrote Jurassic Park – arguably the most influential modern "dinosaur novel" – he had written the script for a story which was published posthumously under the title "Dragon Teeth" in 2017.[467] It tells the story of a Yale Student who becomes involved in the "Bone Wars." However, the rivalry between Cope and Marsh is only the means of getting the story started. The bulk of the story is centered on typical dime-novel mo-

464 Paul D. Brinkman: Paleontology, in: Montgomery, Georgia M., Largent, Mark A. (eds.): A Companion to the History of American Science, Chichester 2016, pp. 227–240, DOI:10.1002/9781119072218. Quote on pages 231–232.

465 For an excellent elaboration on the mechanisms of US-territorial expansion and empire building, with a special focus on "racial" relations, see: Paul Frymer: Building an American Empire. The Era of Territorial and Political Expansion, Princeton, NJ 2017.

466 Adam R. Shapiro: Science Education, in: Georgia M. Montgomery; Mark A. Largent (eds.): A Companion to the History of American Science, Chichester 2016, pp. 320–332, DOI:10.1002/9781119072218. Quote on page 320.

467 Crichton: Dragon Teeth.

tives, such as hostile encounters with Native Americans and the exploits of gunslingers like Wyatt Earp. Crichton had started writing the novel in his early literary career in 1974 and never published it. The story is a little rough around the edges and reproduces all the stereotypes usually found in Western novels and movies. Arguably, the prevalence of these stereotypes is why the "Bone Wars" are still remembered today: the conflict became interwoven with the most American of all American Stories – the territorial expansion of the United States, or to invoke the spirit of Theodore Roosevelt (1858–1919): "The winning of the West."

The first part of this chapter will focus on how the American West was "discovered" by US Americans and how the US government was involved in, profited from, and furthered the exploration. It will also tell of the scientists, among them paleontologists, who were tasked with the exploration.

The second subchapter will detail how the exploration became the foundation for a significant part of US-American identity, a national mythology, so to say. It will also examine how conceptions of masculinity and class affected the mythmaking-process, and how these conceptions were in turn shaped by the "frontier experience." This chapter will also deal with the emergence of US pop culture, the prerequisite for the success of the "frontier myth." Because "Buffalo Bill" Cody was one of the most prolific and popular propagators of the myth (both domestically and internationally), close attention will be paid to his career and practices.

How paleontology and especially the "Bone Wars" between Cope and Marsh became intertwined with "frontier mythology" will be explained in the third part of this chapter. This subchapter will mainly focus on primary sources, such as Marsh's unpublished autobiography and his correspondence.

The last subchapter will detail the involvement of Native Americans in the "Bone Wars," the hunt for fossils, and how Marsh became a political activist for the rights of Red Cloud's Oglala people.

All in all, this chapter will explore how science – paleontology in particular – became part of the national identity of the United States and how its propagation was shaped by US-American nationalism. The so-called "frontier experience" is an integral part of US-American identity, nationalism, and also the development of an US-specific paleontology.

5.1 Early Western Exploration

William H. Goetzmann begins his study of the exploration and the "winning" of the American West by quoting Marsh's unpublished autobiography (see below), in which Marsh recounts how he discovered the fossil remains of various species in a pile of earth near a railroad station in Nebraska in 1868.[468] Marsh certainly knew that the region "promised rich rewards for the enthusiastic explorer in this new field," and that his "own life work seemed laid out" before him. Because the exploration of the West was, for the most part, sponsored by the US government, it must be studied in the context of the various scientific and government institutions, which sprang up during the nineteenth century. To this end, Goetzmann divides the exploration into three chronological periods. The first one begins with the voyage of Lewis and Clark and ends roughly in 1845. It is a period of international competition for the West. The second period he describes as an era of statement and investment against the backdrop of *Manifest Destiny*. The last major period lasts from 1860 to 1900 and was a time of surveying and scientific study.[469]

A utopian idea of "the West," as a place for expansion and freedom had existed in the imaginations of North Americans for a long time. That is why the Royal Proclamation of 1763, which forbade settlement in the lands west of the Appalachian Mountains, fueled the discontentment of British colonists and thus greatly exhilarated the desires for American independence, becoming one of the main contributing factors to sever ties with Great Britain. A desire for westward expansion was therefore baked into the self-imagination of the United States from the very beginning. After Independence was won, when Lewis and Clark explored a seemingly "unclaimed"[470] land with an abundance of natural resources – which also turned out to be the geographical key to the vast East Asian markets – they stirred the desires of many an empire to claim this "wilderness." Lewis and Clark were both soldiers, employed by the government. President Jefferson held a special interest in the exploration of the territories which he had acquired for the US in the Louisiana Purchase in 1803. Beside this patriotic and statesmanlike motivation, he had vested interest in scientific discovery: as a member of the American Philosophical Society, he arranged the support of this organization for the early exploration of the West (for Jefferson's personal interest in the scientific exploration of the West, see chapter 2. 6.).

Davidson writes about the importance of the Lewis and Clark Expedition for the development of US paleontology:

> The import for paleontology of the expedition lies not so much in what fossil the collected, but rather that the president and the army had included collect-

468 William H. Goetzmann: Exploration and Empire. The Explorer and the Scientist in the Winning of the American West, New York 1966, pp. ix–xi.

469 Goetzmann: Exploration and Empire, p. xiii.

470 Long before the popular expedition, the American northwest had been explored and claimed by the Spanish, who were the first Europeans to set foot on the land. Of course, long before that, Native Americans had been the first humans to discover the region and settle there.

> ing fossils in their concept for this trek. For all practical purposes, [...] this was the first geological survey sponsored by US government. Many more nineteenth-century geological surveys would follow this pattern of congressional support and military execution. [...] It would be fairly safe to say that Jefferson established the connection between the US government, its military, and geology, and paleontology. Since a large number of very significant paleontological discoveries were made, or at least published under the aegis of the federal surveys and the US Geological Survey, one can say that President Jefferson is responsible for many of the most important contributions not only to American paleontology, but to the development of the science in the nineteenth century as well. In addition, the Lewis and Clark Expedition presented a model for a number of early nineteenth-century state surveys, which would begin as early as 1831 with surveys established in Massachusetts and Tennessee.[471]

The first period of western exploration was characterized by the rivalry between the US, Great Britain, and Spain (later Mexico), all nations expanding into the Northwest, competing for the economic exploitation of the era, primarily for the lucrative fur trade. John Jacob Astor`s (1763–1848) economic interests were an all-important driving force behind the United States' expansion into the region. Astor, who was a German immigrant, was a part of the German-American network that played an important role in the development of science in the United States. After the fur-trader, the soldier and the scientist arrived in the disputed western regions of the Louisiana Purchase. While the fur traders, or "Mountain Men," were immortalized almost immediately as all-American heroes, taming the "wild frontier" for the "civilization" that followed them, the settlers were the deciding factor in the battle for imperialist supremacy in the American West. By their numerical superiority US-American settlers "won" the disputed Oregon territory from Britain (c. 1848), and began settling in California and Texas, where they incited the Texas Revolution (1835–1836) and the Bear Flag Revolt (1846). The Treaty of Guadalupe Hidalgo (1848) and the Gadsden Purchase (1854) underlined the United States' official acquisition of the West. The exploits of explorers such as John C. Fremont (1813–1890), which were widely publicized in books, complete with illustrations of the grandiose landscape and maps of the regions – until then unknown to the US-Americans – were part of the de facto acquisition of the land by the citizens of the US and an ever-increasing influx of immigrants.[472]

---

471 Davidson: Patrons of Paleontology, pp. 44–45.

472 Goetzmann: Exploration and Empire, pp 3–180, 240–264. For more information on the early exploration of the American West see: A. Hunter Dupree: Science and the Federal Government. A History of Policies and Activities, Baltimore, MD 1957, pp. 91–119.

The exploration of the West also offered new horizons to romantic artists and scientists. Recently discovered plants were the basis for Thomas Nuttall's (1786–1859) Botanic Garden in Harvard College. Nuttall's publications were well received in Europe and "despite his frontier orientation, Nuttall became a man of renown on both sides of the Atlantic."[473] Still, at this time the West was merely a source for raw data and US-American scientists had, for the most part, not acquired world renown. Goetzmann concludes that

> Thus their [the American scientists'] labors resulted in many novelties but few new theories, though the eccentric Constantine Rafinesque, prowling the backwoods of Kentucky, clearly foreshadowed Darwin's hypothesis. Rather, as Americans were to do for many decades, they provided the background research, the collection, classification, and description of plants and animals and minerals that were to make the sweeping generalizations of midcentury comprehensible to the scientific world.[474]

The desire for the establishment of a transcontinental railroad was of prime importance to the exploration of the West. The government sent explorers and scientists to further this undertaking in the name of the nation's frequently evoked *Manifest Destiny*, though the scientists often had other intentions than the government, the railroad executives, or the populous in general:

> Whereas the pioneers and settlers looked upon the West as a place to live, and the prospector saw it as a place to exploit, to the government scientist the West was a vast natural laboratory – a bonanza of exotic specimens and wonders of nature whose meaning and interconnectedness it had been the job of the scientists to describe since the eighteenth century.[475]

Later these differences, namely the attempt of pure, non-utilitarian science, would pit Marsh and the USGS as a whole against the Senate in the 1890s, and would lead to a cut in funding, immensely affecting the conduct of paleontology in the United States (see chapter 6. 5.).

In North America fossil collecting (conducted in the name of science by Euro-Americans) had been associated with the "frontier" since its inception in the eighteenth century. In the middle of the nineteenth century, Ferdinand Vandiveer Hayden (1829–1887) would become the first of many US citizens to investigate the fossils of the

---

473 Goetzmann: Exploration and Empire, p. 182.

474 Goetzmann: Exploration and Empire, p. 184.

475 Goetzmann: Exploration and Empire, p. 303.

West. He had some paleontological education and was sent on his first fossil-hunting expedition by the paleontologist James Hall (1811–1898) in 1853. Further expeditions in 1854, 1855, and 1856 ensured that in 1856 "Hayden knew more about the paleontology of the northern plains than any man alive."[476] The fossils that were unearthed by the expeditions were sent back east to Fielding Bradford Meek (1817–1876) of the Smithsonian Institution in Washington, D.C., and to Leidy in Philadelphia. The main goal of the pre-Civil War exploration of the West was not to find fossils but to map out the topography of the region for far more utilitarian purposes.[477] Still, the importance of this early period of collecting should not be underestimated in its impact on the genesis of the US-American Geo Sciences: "Under the influence of Hayden, Meek, and Leidy, and stimulated by the great Western explorations, the science of paleontology in America began to have worldwide consequences."[478]

After the Civil War, the soldier explorer was replaced by the civilian scientist, who had most likely spent some time at a European university and was a specialist in his field of choice. He was in all probability a member of the East Coast elite with close ties to the centers of education, like Harvard or Yale. The era of the great government surveys began with the 1860s. The first of these was the California Survey, conducted between 1860 and 1874 under the leadership of newly appointed state geologist Josiah D. Whitney (1819–1896), a Yale graduate. Clarence King (1842–1901), another Yale alumnus, who later became the first director of the USGS, earned his spurs with the California Survey. The Yosemite Grant of 1864 was a direct result of the survey: land (which would later become the Yosemite National Park) was set aside in an effort to preserve the wild nature of the West. The establishment of such preservation measures constitutes another aspect of the conception of the mythical West and will be discussed in more detail later in this chapter.[479] King and the other surveyors had to defy the elements and ran into some violent encounters with the Native Americans of the region. This made their undertaking all the more fascinating to the general public and lent an adventurous and heroic touch to the scientific exploration of the West. Marsh's abandoned autobiography and his "Wigwam" full of his Western parapherna-

---

476 Goetzmann: Exploration and Empire, p. 310. For more information about Hayden's explorations see: Dupree: Science and the Federal Government, pp. 198–199.

477 Goetzmann: Exploration and Empire, pp. 309–321.

478 Goetzmann: Exploration and Empire, p. 320.

479 The "wild" nature had to be conserved and with it important aspects of the "American character" that had been formed through the interaction with and subsequent "conquest" of said nature. See David M. Wrobel: The End of American Exceptionalism. Frontier-Anxiety from the Old West to the New Deal, Lawrence, KS 1993, pp. 44–47.

lia and mementoes show that he himself fostered the adventurous tradition (see part three of this chapter).[480]

After the Civil War had concluded, the army was once again employed in the interest of land-hungry settlers, surging West to fulfill the *Manifest Destiny* of their (adopted) country and their own individual *American Dream*. The army was tasked with the protection of the settlers against the dispossessed Native Americans, who were fighting the invaders. Army and settlers needed reconnaissance of their region of operations, thus the exploration of the territories became a secondary objective for the army. In 1867 King was employed by the army and put in charge of the scientific exploration of the 40th parallel – from the California border to the Great Plains. A similar survey was undertaken by army lieutenant George Montague Wheeler (1842–1905), who began his work in 1871. In 1867 John Wesley Powell (1834–1902), who had lost most of his right arm at the battle of Shiloh and would later become the second director of the USGS, led his first expedition into the Rocky Mountains. In the wake of the army explorations another future National Park, the Yellowstone area, was rediscovered. Yellowstone further inspired awe and appreciation for the grand natural lands of the West, in the US and in visitors like Zittel (see chapter 6. 1.). In accordance with public demand, congress sent Hayden to explore Yellowstone in 1871, leading to the establishment of the National Park in 1872. In 1874 the famous 7th Calvary under the command of the even more famous George Armstrong Custer accompanied a scientific staff (including the paleontologist George Bird Grinnell of Yale) to the Black Hills, where, it was rumored, gold was to be found.[481] The gold rush that followed stirred bloody conflict with the Lakota, who inhabited the area, for the Black Hills were part of their reservation.[482]

Clarence King's exploration of the Fortieth Parallel (1867–1873), too, was a military affair. He had received his commission directly from Secretary of War Edwin McMasters Stanton (1814–1869) and was attached to the Chief of Army Engineers Andrew Atkinson Humphreys (1810–1883). The expedition depended upon army outposts and supplies. King enjoyed enormous freedom of choice anyway, he wrote up his own orders and employed his own staff, which consisted entirely of civilians. King's geologists were embodiments of a new generation of American scientists, open-minded and educated in Germany. His expedition into Nevada, Utah, Colorado, and south-

480 Goetzmann: Exploration and Empire, pp. 355–389. On the expeditions of King also see: Dupree: Science and the Federal Government, pp. 195–196. Davidson adds further detail on the early surveys of the North American West, see: Davidson: Patrons of Paleontology, pp. 50–58, 73–77.

481 In a series of letters Grinnell wrote to Marsh, his employer at the time, he wrote of the great support Custer and his men had lent to the fossil-collecting efforts. See: MS 343, Series I. Correspondence, Box 13, Folder 548.

482 Goetzmann: Exploration and Empire, pp. 390–424. For more information on Wheeler's expedition see: Dupree: Science and the Federal Government, pp. 196–197.

ern Wyoming was in part conducted along the local railroad lines and accompanied by a military escort (both very common modi operandi in later expeditions). King immortalized his scientific efforts in "Mountaineering in the Sierra Nevada," published in 1872. The book tells a semi-fictionalized, exotic adventure story, and it links the expedition to the popular image of the "frontier" (also a feature of most subsequent scientific expeditions west of the Mississippi). In 1873 King quit his expedition due to chronic illness and returned to the East Coast, where his rock-specimens were inspected with the newest microscopes, imported from Germany, and in accordance with Ferdinand Zirkel's theories,[483] called microscopic petrography.[484]

Wheeler was, as mentioned above, also an army explorer and, as Goetzmann argues, the last of his kind. His career really began in 1871, when he started his first expedition into Nevada and the Arizona Territory. Along with a cavalry escort, Wheeler was accompanied by Frederick Wadsworth Loring (1848–1871), a reporter from Boston, who documented the expedition and greatly contributed to Wheelers renown. The expedition ran into trouble almost from the get-go, and had a very hard time exploring Death Valley. Next the expedition turned to the Colorado River and travelled upstream. The journey was filled with hardship and documented photographically, adding to the mystification of the Western land. Wheeler also produced a map of the river. Shortly after the party had split up in November, some members, including Loring, took a stagecoach to California. It was ambushed by a band of Native Americans, who killed all but two of the passengers. The event became known as the "Wickenburg massacre" (the stagecoach had just left Wickenburg in the Arizona Territory) and contributed to the dangerous and adventurous legacy of Western exploration, which should forever accompany and motivate the scientific ventures.[485]

In 1874 conflict with Hayden arose. Hayden accused Wheeler and the army, vehemently and publicly, of superfluously exploring territory that had already been "discovered" and explored before by civilian scientists like himself. This effort to carve out a monopoly for federally funded exploration Hayden found not only to be immoral, but also a waste of government money. Other prominent scientists, such as Dana and Marsh, now turned against Wheeler. But Wheeler managed to weather the storm, thanks to the support of President Grant (1822–1885). This is a fine example showing how the infighting for resources and prestige influenced scientific conduct in the nineteenth century. In addition, the 1870s were the last decade of military exploration in the US. Though some of the later expeditions were accompanied by army

483 Zirkel was the doctoral adviser of Otto Meyer, who later worked for Marsh (see chapter 6. 4.).

484 Goetzmann: Exploration and Empire, pp. 430–460.

485 Goetzmann: Exploration and Empire, pp. 467–477.

detachments, the westward exploration bow rested firmly in the hands of civilians.[486] In 1869 the Department of the Interior had formed the Geological Survey of the Territories, which competed with the army surveys and was from its inception onwards controlled by civilian scientists. In 1853 Hayden, who was working for the survey, explored the Dakota Badlands. The paleontologist Bradford Meek accompanied him to study the abundant fossils deposits of the region. Thanks to this excursion and others that followed to the West, as well as a very detailed map of the Yellowstone and the Upper-Missouri, which Hayden drafted with the help of Meek, they rose to some prominence. Most of the collected fossil specimens were sent to Leidy in Philadelphia. Hayden served as a surgeon in the Civil War, which interrupted his scientific career. After the war Hayden successfully (and with the help of O. C. Marsh amongst others) secured money from the State of Nebraska for a survey of his own in 1867. His survey was renewed and imbued with $ 5,000 in additional funds in the following year. In 1869 the survey was continued, was appropriated $ 10,000 and was officially named the Geological Survey of the Territories. The appropriation for the survey was more than doubled in 1870, when it received $ 25,000.[487] The survey provided paleontologists back east with ample material for their scientific work. Cope, for example, popularized some of the fossils which had been provided by the survey in various descriptions, some of them illustrated. Cope generally profited from the survey, owing much of his early scientific reputation to the fossils that it yielded. Together with paintings and photographs of the grandiose natural scenery of the West, the descriptions inspired the collective imagination of the nation and contributed to the rise of popular culture and education.[488]

In 1877 Hayden's survey clashed with Powell's survey. Powell employed the assistance of Carl Schurz, then Secretary of the Interior, and the Survey of the Territories was later absorbed by the USGS. Hayden nonetheless continued to work as an employee of the USGS, until he retired for health reasons in 1883.[489] In summary, it can be stated that Hayden's collaboration helped Cope's career as much as Powell's partnership helped Marsh with his respective occupation.

Goetzmann describes John Wesley Powell as the "greatest explorer-hero since the days of Frémont;" and as "a casually educated, self-made scientist with a driving ambition." He was an "outstanding representative of political men who came to the fore in the late nineteenth century."[490] This "breed of men" are unambiguous examples for

486 Goetzmann: Exploration and Empire, pp. 478–488. For more information on the transition from military to civilian exploration see: Dupree: Science and the Federal Government, pp. 184–194.

487 Goetzmann: Exploration and Empire, pp. 489–499.

488 Goetzmann: Exploration and Empire, pp. 510–511, 514.

489 Goetzmann: Exploration and Empire, pp. 526–527.

490 Goetzmann: Exploration and Empire, p. 530.

the "soft," the societal factors that shaped science in the nineteenth century, and still today. These are examples for how the pursuit of personal glory and standing mixed with politics – and not the pursuit of "pure" academic knowledge – are key factors in the conduct of science. Goetzmann associates Powell with the emerging business tycoons, the captains of industry, some of whom, like Andrew Carnegie, had their hand in paleontology, or had relatives who were paleontologists themselves, like in the case of Osborn. But back to Powell: he stemmed from a family of staunch abolitionists and joined the Union Army at the start of the Civil War in May 1861. During his service he met and befriended Ulysses S. Grant, a friendship that really paid off once the popular General became President of the United States. In 1862 Powell lost most of his right arm fighting at the "Hornet's Nest" during the battle of Shiloh. He nonetheless remained in the army, rose to the rank of major and was appointed brevet lieutenant colonel. He stayed with Grant (and Sherman) and collected fossils in the trenches of Vicksburg. His contacts and the familiarity with bureaucracy would serve Powell well in his later career. After the war Powell became a professor of natural history at Illinois Wesleyan University and put an emphasis on the more practical aspects of science and field trips. He then became a professor at the Illinois State Normal University and began to lobby for the establishment of a museum for natural history (he became the first curator of the museum, newly founded in Bloomington, Illinois) and funds for the exploration of the West. Powell started on his first expedition (which was backed by U. S. Grant, who was still the Commanding General of the United States Army) in June 1867. Other expeditions followed, including the exploration of the Colorado River and the Grand Canyon in 1869, when Powell and his company became the first Euro-Americans to do so successfully. This feat made the one-armed scientist an instant hero when he returned to the East. The fact that three party-members had left the expedition and were most likely killed by Piute contributed to the public attention paid to this now tragic and adventurous undertaking. Beside his scientific interest, Powell began scrupulously researching the ethology of some of the Native American tribes of the Southwest (the Ute and the Paiutes for example). He became fluent in various Native-American languages and dialects of the region,[491] and in 1873 he began to expose several serious problems and mistreatments experienced by the Natives he studied. Powell began lobbying on their behalf (all the while Powell still held the belief that the Native Americans had to be "civilized / Americanized"), much like Marsh would do in 1875 for Red Cloud (1822–1909) and the Oglala.[492]

---

491 He later authored a monograph on that subject, see: John Wesley Powell: Introduction to the Study of Indian Languages, with Words, Phrases, and Sentences to be Collected, Washington, DC 1877.

492 Goetzmann: Exploration and Empire, pp. 551–576. Also see: Dupree: Science and the Federal Government, pp. 199–235.

In the late 1870s serious efforts were made to tighten and consolidate the various government surveys of the West; the result was a vicious behind-the-scenes struggle for control over the institutions. In 1878 the National Academy of Science was tasked by Congress with managing the unification. In May of the same year Joseph Henry (1797–1878), the president of the NAS, died and Marsh became his successor as acting president of the NAS. Marsh was undoubtedly a friend and ally of Powell and King, and an adversary to Hayden, whose paleontologist of choice had been Cope. Furthermore, Marsh and the other members of the NAS-committee, among them Agassiz and Dana, were proponents of the civilian-led exploration of the West, and Hayden clearly a champion of the army-way of doing exploration. Inspired by the predictable recommendations of the committee, Congress created the USGS in March 1879. After a brief struggle for control, King was chosen to be the first president of the survey. The mission statement of all subsequent expeditions had been changed from discovery to the assessment of natural resources. King's reign, however, was short; in 1881, roughly one year after he had accepted his commission, he resigned from the presidency to tend to his private business ventures, and Powell succeeded him. With the help of Marsh Powell tightened his grip on the institution and managed to increase its funding. His downfall began in 1890, when members of Congress, mostly from western states, began to argue against Powell's unchecked control of the USGS, using Marsh's elaborate (and seemingly thriftless) publications as evidence of corruption and the misappropriation of public funds by the USGS, resulting in Powell's abdication (see chapter 6. 5.).[493]

5.2 Preliminary Conclusion

In the first half of the nineteenth century the "discovery" of the trans-Mississippi West was furthered by political ambitions. The US government sought territorial expansion into the West and saw US ambitions contested by the national interests of Great Britain and Mexico. Consequently, most early explorers were soldiers first and scientists second. These army-explorers paved the way for settlers and later railroad companies. This national quest for new territory was conventionalized into the divinely ordained *Manifest Destiny* of the young nation. In the 1850s the army-explorers and – even more

493 For further details on Powell's life and career, see: Mary, C. Rabbitt: John Wesley Powell: Pioneer Statesman of Federal Science, in: The Colorado River Region and John Wesley Powell: A Collection of Papers Honoring Powell on the 100th Anniversary of His Exploration of the Colorado River, 1869–1969 (Geological Survey Professional Paper 669), Washington, DC 1969, pp. 1–21. For more information on the founding of the USGS, Marsh's appointment, and his budget see: Dupree: Science and the Federal Government, pp. 208–235.

importantly – the settlers "won" the West for the United States through negotiations and violence, mainly directed against Native Americans and Mexicans. The Civil War had transformed the Union, had centralized power, and grew Washington's populace and bureaucracy immensely, "giving the federal government a more active role in developing the country."[494] Thus, after the Civil War the West was surveyed for the acquisition and exploitation of natural resources. These surveys were usually funded with government money but led by civilian scientists. They provided even more raw material for paleontologists and were the prerequisite for the rise of US paleontology to the top position it held globally at end of the century. All the while, stories of heroic exploration and the "frontier life" in general seeped into the national consciousness and an emerging popular culture transformed the "frontier experience" into a cornerstone of US identity.

5.3 The "Frontier Myth" and US Popular Culture

"The story of the West is our Trojan War, our Volsunga Saga, our Arthurian cycle, our Song of Roland."[495]

According to Frederick Jackson Turner's "frontier theory," the "frontier experience" gave birth to authentic and genuine US-American identity:

> In the settlement of America we have to observe how European life entered the continent, and how America modified and developed that life and reacted to Europe. Our early history is the study of European germs developing in an American environment. Too exclusive attention has been paid by institutional students to the Germanic origins, too little to the American factors. The frontier is the line of most rapid and effective Americanization.[496]

Turner's thesis was contested from its inception and hardly any twenty-first-century historian would agree with Turner wholeheartedly. Nonetheless, Turner's theory is a product of its time, it expressed the significance of the "frontier" in a nutshell and shaped all public discourse thereafter:

---

494 Pauly: Biologists and the Promise of American Life, p. 47.

495 Thomas King Whipple: Study out the Land. Essays by T.K. Whipple, Berkeley, CA 1943, p. 59.

496 Frederick Jackson Turner: The Significance of the Frontier in American History, in: Annual Report of the American Historical Association, 1894, pp. 119–227. Quote on page 201.

> Turner summoned the frontier from the dim academic backcountry, but in popular culture the frontier already stood squarely in the foreground. Turner did not have to tell Americans about the frontier; he could mobilize images they already knew. [...] Americans had recognized for generations the cultural utility of the frontier in their politics, folklore, music, literature, art, and speech. All Turner had to do was to tell the Americans about the SIGNIFICANCE of this familiar frontier.

Patricia Nelson Limerick goes one step further in criticizing Turner, she mocks the term "frontier" by calling it the "f-word," and concisely describes why it is a difficult one:

> The term 'frontier' blurs the fact of conquest and throws a veil over the similarities between the story of American westward expansion and the planetary story of the expansion of European empires. Whatever meanings historians give the term, in popular culture it carries a persistently happy affect, a tone of adventure, heroism, and even fun very much in contrast with the tough, complicated, and sometimes bloody and brutal realities of conquest. Under these conditions, the word 'frontier' uses historians before historians can use it.[497]

In his monumental "The Fatal Environment," Richard Slotkin depicts how the rise of popular culture in the United States was essential to the formation of the "frontier myth." He chronicles the evolving nature of the myth in great detail and with an emphasis on fictional literature, while always keeping the economic and social history, race relations, and class struggle in mind.[498] Slotkin suggests that the "frontier myth" came into being through the fundamental changes happening in all European and North American societies, caused by the rapid modernizations the nineteenth century brought:

> The Myth of the Frontier is the American version of the larger myth-ideological system generated by the social conflicts that attended the 'modernization' of the Western nation, the emergence of capitalist economies and nation-states. The major cultural tasks of this ideology were to rationalize and justify the

---

497 Patricia Nelson Limerick: The Frontier in the Twentieth Century, in: James R. Grossman: The Frontier in American Culture. An Exhibition at the Newberry Library, August 26, 1994 – January 7, 1995, Berkeley, CA 1994, pp. 67–102. Quote on page 75.

498 Richard Slotkin: The Fatal Environment. The Myth of the Frontier in the Age of Industrialization 1800–1890, New York 1985. Slotkin continues his examination of the "frontier myth" during the late nineteenth and the twentieth century in "Gunfighter Nation." See: Richard Slotkin: Gunfighter Nation. The Myth of the Frontier in Twentieth-Century America, New York 1992.

> departures from tradition that necessarily accompanied these developments. Progress itself was to be asserted as a positive good against aristocratic and peasant traditions that emphasized stasis and permanence in productive techniques and social relations. The styles, interests and values of the new classes of entrepreneurs were to be defended against those of old aristocracies and the peasantry. Progress itself was to be interpreted in economic terms – an increase on wealth of productive capacity, of levels of consumption from year to year and decade to decade. Individualistic assertiveness and achievement were to be justified as values in themselves, and reconciled with the traditional claims of corporate solidarity and deference. Social bonds were to be redefined, with free contract replacing customary fealties, and social standing varying according to achievement as well as birth.[499]

Only after the Civil War local identities merged under the umbrella of a national identity, islands of production and industry were connected rapidly, and the whole domestic market of the US changed. As the agrarian economic basis of the nation evolved and commerce and industry became more prevalent, class struggle intensified and became more evident.

An important part of the genesis of pop-culture (and Slotkin's thesis) is technical innovation, which allowed the spreading of ideas to a rapidly growing audience. Pop-culture in return shaped the national identity of the USA:

> The history of the development of the forms and institutions of commercial or mass popular culture is directly related to the development of a political ideology of American nationality and to the creation of nationwide networks of production distribution. The basic structure of this commercialized national culture were [sic!] developed between the Revolution and the Civil War with the emergence of national parties and the development of a nationwide trade on books, magazines, and newspapers utilizing an ever-expanding transportation network. Between the Civil War and the Great War the nascent 'culture industries' took advantage of new technologies to meet demands of an ever-growing and increasingly polyglot culture with varied and complex needs and tastes.[500]

John M. Coward also covers how the emergence and unprecedented growth of mass media (newspapers in this case) helped to form a national identity and create a popular culture. Coward elaborates on the emergence of the newspaper popular culture

499 Slotkin: The Fatal Environment, p. 33.

500 Slotkin: Gunfighter Nation, pp. 9–10.

during the course of the nineteenth century, and on how the newspapers shaped the self-perception of the nation (it was widely accepted that *Manifest Destiny* intended for the nation to rule the continent). The depiction of Native Americans (and, one can argue, paleontology as well, though it is not part of Coward's study) was shaped and determined by the prolific newspaper industry. The media's depiction of Native Americans as ruthless warriors (some brave, some devious), as enemies of "civilization," or as "noble savages" still resonates today.[501]

Another aspect characterizing the changing self-image of the young nation was hunting and its implications for masculinity. Hunting game is linked closely to the emergence of the "frontier myth" and US-American nationalism. Daniel Justin Herman showcases a very comprehensive history of hunting and the US-American self-imagination.[502] He states that:

> Americans were products of the frontier; it was natural that the backwoods hunter had emerged as American hero [...] abundant game and Americans' skill in harvesting it seemingly had made possible the spirit of independence itself.[503]

The connection between hunting, the "frontier," and patriotism was further strengthened in the nineteenth century, when real life "frontiersmen" like Daniel Boone, Davy Crocket, Kit Carson, and the fictional Natty Bumppo of James Fenimore Cooper's Leatherstocking Tales, captured the imagination of the nation.[504] Romanticized retellings of true events and completely fictional adventure stories inspired a new wave of self-promoting adventurers like William "Buffalo Bill" Cody and Theodore Roosevelt. The stories also motivated more scientifically inclined explorers, like John C. Fremont and George Bird Grinnell, who Herman calls "hunter-naturalists." He directly associates the boom of natural history during the nineteenth century with the enthusiasm for hunting:

> In an era of science, however, mastery of the land involved more than killing wild animals. Any man could shoot a deer. Hunting might make white Americans equal to American Indians but not superior. What was needed was a courage greater and a knowledge beyond those of Indians. Thus while Americans

---

501 John M. Coward: The Newspaper Indian. Native American Identity in the Press, 1820–90, Urbana, IL 1999, pp. 13–20.

502 Herman: Hunting and the American Imagination.

503 Herman: Hunting and the American Imagination, p. 1.

504 For a detailed description of the Leatherstocking tales and its influence on the "frontier myth," see: Slotkin: The Fatal Environment, pp. 81–106.

> demonstrated their courage through ritualized hunting excursions into the wilderness, they demonstrated their knowledge through a scientific discourse called natural history.[505]

Herman then further elaborates:

> To know natural history was to claim dominion over the earth; that is what made natural history so attractive to hunters. The hunter with his gun and the naturalist with his pen were Janus faces of the same man; often they were the same man. Indeed, the so-called Great Reconnaissance of nineteenth-century naturalists might be called the Great Hunt. Hunters and naturalists together entered nature to take command of it.[506]

Later, Grinnell argued passionately for the establishment of reservations to conserve the "pure" American nature and its original inhabitants, including various Native American populations.[507]

Through hunting one could experience the true nature of the land, meaning the hunter formed a close bond with the land, making him a better patriot and, in a sense, a more genuine US American. The glorification of hunting and the "frontier" was accelerated by the rapid development of the east-coast metropolises through immigration, industrialization, the rapid growth of the market, and the transportation revolution. All of which seemed suspiciously European and conveyed the impression of a corruption of civilization through luxury. It was believed that civilization and the US-American ventures could only experience regeneration through contact with the raw nature of the land, strengthening democratic values such as self-reliance and individualism. Moreover, in the minds of many contemporaries hunting was "manly;" the act of hunting and killing, and the exposure to nature strengthened one's "manliness," for it taught martial skills (mainly proficiency with firearms) and physical fitness. This trend intensified as the nineteenth century progressed and it is fair to say that hunting became a substitute for the war experience, as other perceived proving grounds of manliness, like the battlefields of the Civil War, faded into memory.

The legend of the "US-American hunter" originated in the early nineteenth century. It was nurtured by authors who produced a plethora of romantic and exciting liter-

505 Herman: Hunting and the American Imagination, p. 7. Grinnell was the editor of the sporting periodical "Forest and Stream," the magazine also printed frequent reports on the geological surveys of Wheeler and Hayden (see below). Herman: Hunting and the American Imagination, pp. 161–162.

506 Herman: Hunting and the American Imagination, p. 8.

507 Brian W. Dippie: The Vanishing American. White Attitudes and U.S. Indian Policy, Middletown, CT 1982, pp. 223–228.

ature, and was adapted by countless men who identified with their supposedly superhuman, patriotic, and pioneering ancestry. The enthusiasm for hunting was mainly a social construct, not entirely based on the factual past:

> Yet between 1820 and the end of the nineteenth century (or earlier, if we count the literature of Daniel Boone), Americans began to recall their forbearer primarily as hunting people and secondarily as farming people. Like eighteenth-century Britons who traced English liberty to the Saxon past, writers and artists traced the American love of hunting – and liberty – to a quasi-mythical pioneer past. In doing so, they sought to give American men a mythic standard to live up to, yet the creation of the myth required the obfuscation of the source. In truth pioneers did not single-handedly sacralize hunting; it took educated men to do that.[508]

Add to this that it became even more important for scientists to imbue themselves with the appearance of the “frontier’s men.” Paleontologists and other “over-civilized” types were being perceived as weak and unmanly, as demonstrated in a Thomas Nast Cartoon in 1879.[509] The cartoon shows a scrawny, lanky man, wearing glasses and studying, amongst other subjects, paleontology (symbolized by a tome bearing the letters “Paleontology” on its back and a sketch of a *pterodactylus* skeleton on its front) in the foreground. A buff and supposedly rather unintelligent looking man is depicted in the background. A caption underneath the picture states: “Education. Is there no middle course?” (see figure 3). The message is clear: being well-educated bears the risk of being what is pictured here as an “unmanly, unfit specimen,” letting go of education completely bears the risk of becoming what Nast depicts here as “a primitive brute.” It is fair to say that Marsh and his students would be perceived as being in the first, the “over-civilized” camp, and therefore had to redouble their efforts to imbue themselves with the manly activities of the “frontier,” such as hunting.

---

508 Herman: Hunting and the American Imagination, p. 60.

509 Thomas Nast: Education. Is There No Middle Ground?, in: Harper’s Weekly, vol. 23, no. 1183 (30. Aug. 1879), p. 696.

Figure 3: Thomas Nast: Education. Is There No Middle Ground?, in: Harper's Weekly, vol. 23, no. 1183 (30. Aug. 1879), p. 696.

The most remarkable game present in the far West was the bison.[510] It became a symbol for the North-American continent itself, much like the Native American. The bison hunt and its near extinction through hunting is often seen as a metaphor for the "conquest" of the American West by Euro-Americans. Stories, retelling the exploits of courageous bison hunters, are therefore directly linked to the *Manifest Destiny* of the nation. Also, the vanishing of the bison was directly linked to the existential threat Native American nations of the prairies faced, for the bison herds were an important food source for these tribes, and their extinction in turn threatened Native American existence as well.

In conjunction with hunting, a growing tourism business contributed to, and drew from, the emerging "frontier myth." Tourism to the American West developed hand in hand with hunting and the railroad infrastructure. Fascination with untamed nature and the "vanishing" Native population inspired a generation of wealthy tourists.[511]

The emergence of pop culture through dime novels etc. even influenced how scientists experienced nature, proving once more that there is no such thing as "pure science:"

> Even so strict a scientist as John Wesley Powell operated under the romantic tradition, although he explored in the West and wrote about it after the Civil War, during the so-called Age of Realism in American literature. In much of his *Exploration of the Colorado River and Its Canyons,* Powell is the observant field sci-

---

510 "The buffalo inspired awe; it represented American nature at its most sublime." Herman: Hunting and the American Imagination, p. 201.

511 Earl Spencer Pomeroy: In Search of the Golden West. The Tourist in Western America, 1st Bison Book pr., Lincoln, NE 1990.

> entist or scientific explorer, telling us of river courses, canyon depths, and rock strata. Or he is the adventurer narrating the perils through which he and his followers have passed. But on occasion, in little set-pieces scattered throughout his book, he is the conscious describer of the beauty of a landscape.[512]

Bryant also argues that early conceptions of North American nature were established in US pop culture during the nineteenth century, that they became an indicator of Americanness and therefore influenced how contemporaries experienced "the frontier:"

> [...] that shift in taste back in the eighteenth century, from fearing wild nature as chaotic and threatening to enjoying as sublime, beautiful, or picturesque, came just in time to play a major role in the exploration and opening of the American West. The conventions were developed in time to glorify the typical Western landscape. Our sustained concept of the West as a heroic landscape and our cultural tradition of the sublime landscape grew up together. Without the Rockies and the Sierra Nevada, Bierstadt would have been limited to the Catskills. Without Yellowstone and the Grand Canyon, what would Thomas Moran have painted? The west has become established in our national consciousness as our frontier land, as that which makes America special, and the West is a sublime landscape. Thus America is, imaginatively, on a sublime scale – in many things beyond only landscape.[513]

David Wrobel argues that this nostalgia was in part due to a "frontier anxiety," which gained a foothold in many US-American minds, even before Turner's formulation of his famous "frontier thesis" in 1893. The anxiety came with the realization that the "frontier," the source of American values and an agrarian "Garden of Eden," was about to vanish. The colonization of the continent was about to be concluded. Not only would a societal safety valve vanish, but this also meant that generations of American men had to find another field in which to prove their manliness.[514] "The frontier was mas-

512 Paul Bryant: Nature Writing and the American Frontier, in: Paul Bryant et al. (eds.): The Frontier Experience and the American Dream. Essays on American Literature, College Station, TX 1989, pp. 205–216. Quote on pages 211–212.

513 Bryant: Nature Writing and the American Frontier, p. 214.

514 This corresponds with a crisis of masculinity that occurred at the close of the nineteenth century in the United States. Native-born "white" males rapidly lost their control over women, workers, and African Americans. The class conflict became very visible as the old elites were losing their monopoly on political and economic power. For a comprehensive introduction to this theme, see: Matthew G. Hannah: Governmentality and the Mastery of Territory in Nineteenth-Century America, Cambridge 2000, pp. 84–93.

culine; machines and cities were its antithesis. They emasculated men, robbed them of their true manhood."[515] So far

> The ritualized killing of wild animals incorporated a rich vocabulary of gestures, objects, sayings, clothing, and images, that when woven together, told a story about masculine triumph over nature. These frontier stories legitimated the power these men exercised – over women, workers, colonial subjects [...][516]

Westward expansion had been lending opportunities to men who had failed in the overcrowded cities of the east. Those failed individuals could start anew at the "frontier." Now it was suspected that social pressures would mount in the overpopulated cities. The inner societal struggle for resources, it was believed, had been halted thus far by the abundance of unclaimed land to the west. This cheap land was the source of the superiority of the United States over the old European countries. Also, it was this "free land" which transformed the European "proletarian hordes" into true US-citizens. If the "frontier" was to be closed, these troublemakers were to overwhelm US-society with their foreignness. This "sense of crisis" dominated US-political thought in the 1890s. Labor unrest and an economic depression caused growing awareness of the class struggle. Only the Spanish-American war and the "acquisition" of "new frontiers" in the Caribbean and the Philippines could recharge national confidence.[517] The same tactics that helped to "win the West" were employed in the Philippines to fight the Filipino resistance. Generals, who before were successful "Indian Fighters," unleashed the same genocidal violence against the Philippine natives. A long and brutal conflict ensured, which seems to foreshadow the conflict in the Vietnamese "Indian Country" some seventy years later.[518]

> Scholars of American masculinity have argued that a mythic West reassured and inspired white middle-class men shaken by economic change, women's rights movements, immigration, and labor unrest. The mechanization and impersonality of modern life distanced men from older norms of masculinity that emphasized the role of the family patriarch and Christian gentleman. Continuously in struggle for success, men sought to blunt the edge of compe-

515 Richard White: Frederick Jackson Turner and Buffalo Bill, in: James R. Grossman: The Frontier in American Culture. An Exhibition at the Newberry Library, August 26, 1994 – January 7, 1995, Berkeley, CA 1994, pp. 7–66. Quote on page 49.

516 Monica Rico: Nature's Noblemen. Transatlantic Masculinities and the Nineteenth-Century American West, New Haven, CT 2013, p. 4.

517 Wrobel: The End of American Exceptionalism, pp. 3–15, 29–41.

518 Walter L. Hixson: American Settler Colonialism. A History, New York 2013, pp. 167–184.

tition with male camaraderie. In compensation, they turned to western fantasies of adventure that reassured them about their virility.[519]

5.3.1 The Americanization of Nature

North America was at first perceived as something of a blank canvas in the minds of Europeans, a quasi-uninhabited continent onto which European societies could be transplanted. Names such as New England, Nieuw Nederland, and Nouvelle France bear witness to this style of thinking. At the same time America was conventionalized as a place without culture or civilization. Native American achievements could not be considered because that would have delegitimized the right of the colonial powers to the land, and later the idea of a *Manifest Destiny*. Thus, no ruins or other witnesses of original American cultures could be claimed to prove American greatness, but a grand landscape and nature, which a Christian god supposedly had created, could; nature itself was the best proof that God had selected the young nation for greatness. In a sense this constituted a continuation of Jefferson's conception of a natural religion (see chapter 2. 6.). In the first half of the nineteenth century the landscape of New England had dominated in romantic art depictions of North America. During and after the Civil War the scenery of the American West would take the place of New England's nature. The West could be claimed by North and South, it was much less political, a quasi-bipartisan conception of American nature that could stand as a symbol for the Nation as a whole. Thus, places of immense natural beauty had to be preserved in National Parks, such as Yosemite and Yellowstone.[520]

In his book "Wilderness and the American Mind" Roderick Frazer Nash describes in great detail the ambivalent relationship between the United States and the "Wilderness," and underlines his claims with many vivid examples.[521]

He reflects on the term "Wilderness," its etymology, and its ever-changing meaning. He works out the different view-points Europeans and (Euro-) Americans held with regards to the topic. At the beginning of the nineteenth century Europeans admired the wild, there was very little "wild nature" left in the Old World. But to US Americans the Wilderness was still a life-threating obstacle to be overcome in the name of civilization and Christianity: "Civilizing the New World meant enlightening darkness, ordering chaos, and changing evil into good. In the morality play of west-

519 Rico: Nature's Noblemen, p. 9.

520 Claudia Schnurmann: Frontiers, Landscape and Ideology in the 19th-century USA: the Yosemite, in: Steven G. Ellis; Raingard Eßer (eds.): Frontiers and the Writing of History, 1500–1850, Hannover 2006, pp. 275–299.

521 Roderick Frazer Nash: Wilderness and the American Mind, 5th ed., New Haven, CT 2014.

ward expansion, wilderness was the villain, and the pioneer, as hero, relished its destruction.”[522]

At first America was imagined as a new Garden of Eden, but most settlers were severely disappointed when their expectations clashed with seventeenth-century colonial realities. Because Native Americans were closely associated with the Wilderness from the beginning of European colonialism, they were seen as part of the same evil as the deadly nature. Certain seventeenth-century ideas survived into the nineteenth century and into a time in which it became the *Manifest Destiny*, ordained by God, to push West.

Then, during the course of the eighteenth century, nature and the “Wilderness” became romanticized. Now nature was associated with religion. US-American Romanticism originated in the cities, far away from the “frontier” and the perils of the “Wilderness:”

> In early nineteenth century, for the first time in American history, it was possible to live and even to travel widely without coming into contact with wild country. Increasingly people lived on established farms or in cities where they did not experience the hardships and fears of the wilderness. From the vantage point of comfortable farms, libraries, and city streets, wilderness assumed a far different character than from a pioneer’s clearing.[523]

With Romanticism came Primitivism, which implied that a lifestyle closer to nature would foster some virtues that were since lost to modern civilization. Instead of hating and fearing the “wild people,” one should learn from the “noble savage.”

After its independence, the young Nation needed to define itself, and that happened by playing up the contrast to the Old World:

> Creation of a distinctive culture was thought to be the mark of true nationhood. Americans sought something uniquely ‘American,’ yet valuable enough to transform embarrassed provincials into proud and confident citizens. Difficulties appeared at once. The nation’s short history, weak traditions, and minor literary and artistic achievements seemed negligible compared to those of Europe. But in at least one aspect Americans sensed that their country was different: wilderness had no counterpart in the Old World.[524]

---

522 Nash: Wilderness and the American Mind, p. 24.

523 Nash: Wilderness and the American Mind, p. 57.

524 Nash: Wilderness and the American Mind, p. 67.

Thus, the wilderness, or nature itself, became the singular feature of the United States. This unique characteristic was utilized to defend the young Republic against European snobbery, see for example Jefferson's "Notes on the State of Virginia," and the Buffon-debate that inspired it (see chapter 2. 6.). It was in this cultural environment that Cooper achieved his literary fame. The acquisition and exploration of the trans-Mississippi West furthered the appreciation for nature as a distinctly American phenomenon. Then, in the second half of the century the realization settled in that nature was endangered:

> Appreciation of wilderness led easily to sadness at its disappearance from the American scene. What to do beyond regretting, however, was a problem, especially in view of the strength of rationales for conquering wild country. But as the Romantic and nationalistic vindications of wilderness developed, a few Americans conceived of the possibility of its deliberate preservation.[525]

The establishment of National Parks for the preservation of the wilderness was soon to follow. But the relationship of US Americans to the wilderness remained ambivalent. Newfound appreciation for nature and the urge to preserve it in allotted niches did not mean that US Americans would stop the exploitation and destruction of nature and its inhabitants (including Native Americans) in other places.

5.3.2 Popularizing the West: Buffalo Bill

"Cody shared a destiny with the buffalo and Plains Indians; indeed, one cannot utter William F. Cody's immortal alliterated sobriquet without conjuring buffalo. Indians and buffalo are symbolically related both metaphorically and in reality. Buffalo Bill, Lakota Indians, and buffalo are forever linked in American legend and mythology."[526]

Robert W. Rydell and Rob Kroes argue that US-American mass culture, meaning culture being mass-produced as well as consumed by "the masses," was born on May 10, 1869, when the golden railroad spike fused the nation into a communicative unity through the transcontinental railway. Technical innovation – most of all the railroad as well as the telegraph – made the mass transportation of immigrants and swift exchange of goods possible, transformed the internal commercial market of the US as well as the marketplace of ideas, and led to the emergence of mass culture. Within these circumstances the circus spectacles of P. T. Barnum (though he began his career

525 Nash: Wilderness and the American Mind, p. 97.

526 Bobby Bridger: Buffalo Bill and Sitting Bull. Inventing the Wild West, Austin, TX 2002, p. 7.

in the 1840s) and Buffalo Bill Cody were born and grew up to be commercial successes. Mass culture now also "Americanized" the immigrant who arrived in ever increasing numbers as the nineteenth century wore on. In that regard mass culture took on the role the "frontier" had played according to Turner. So it seems natural to link the (mythical) "frontier" to US-American mass culture, and the "frontier" was indeed the most prevalent source for mass culture.[527]

After the Civil War, national unification made great strides. One of the tools for nationalization were world fairs, like the Centennial International Exposition of 1876. These expositions

> [...] acted as powerful nationalizing forces in American life. Just when American society was becoming increasingly turbulent, world fairs promised material progress well in the future and laid down a blueprint for a racially exclusive technological utopia.[528]

In addition, the fairs were the ideal stage for Cody and his own contribution to the "nation building" of the United States. These processes were also a reaction to the social upheavals of the second half of the century, which also saw other measures installed to ensure the continuance of US nationalism: "Flag-raising ceremonies, like Columbus Day itself, were very recent innovations in the nation's schools and, like world's fairs, were direct responses to the social and political struggles of the Gilded Age."[529]

Hixon argues along the same line for Cody's emerging popularity in the wake of the development of US pop culture:

> As Americans began to cultivate the mythology of the frontier, Indians participated in the 'Wild West' shows made famous by William 'Buffalo Bill' Cody. However hackneyed, the shows were wildly popular and many of the 'show Indians' enjoyed playing their roles as well as the travel and other opportunities.[530]

---

527 Robert W. Rydell; Rob Kroes: Buffalo Bill in Bologna. The Americanization of the World, 1869–1922, Chicago 2005, pp. 14–46.

528 Ryell; Kroes: Buffalo Bill in Bologna, p. 49. Note that Turner's all-influential "The Significance of the Frontier in American History" was first read to a group of historians at the world fair in Chicago in 1893. See: Ryell; Kroes: Buffalo Bill in Bologna, pp. 47–53, 73.

529 Ryell; Kroes: Buffalo Bill in Bologna, p. 56.

530 Hixson: American Settler Colonialism, p. 186.

"Cody produced a master narrative of the West as finished and culturally significant as Turner's own," judges Richard White.[531]

Inspired by Catlin's depictions of Native American life,[532] P. T. Barnum (mis-)represented Native Americans in his circus in the 1830s and 40s.[533] Cody, in contrast, really knew the West and its inhabitants, including some Native American cultures, and in his show he tried not just for spectacle, but also for realism in representing Native American culture:[534] "Cody realized that people – Indian and non-Indian – wanted desperately to hold on to some vestige of the untamed West which was departing North America."[535]

Rico elaborates further on the "edutainment" aspect of Cody's enterprise:

> Indeed, Cody avoided the word 'show' altogether, insisting that this was no mere circus act but an uplifting display of historical and ethnographic information. This didacticism was part of Cody's bid of middle-class respectability after nearly a decade spent in the world of melodrama, a world associated with rough working-class audiences.[536]

Another reason for Cody's popularity was that he was featured in many spectacular dime novels.[537] "For Cody western masculinity was a way of climbing up in the world. Under the flowing locks and broad-brimmed hat was the mind of a Gilded Age businessman with his eye out for new opportunities."[538]

Cody rose to popularity when journalist and publicist Edward Zane Carroll Judson Sr. (1821/23–1886) began writing about the soon-to-be-legendary exploits of Buffalo Bill after he had met Cody in 1868. He did so under the pseudonym Ned Buntline. The first of many dime novels about Buffalo Bill was published the next year, soon many other publications, including a theater production, followed. In 1872 Cody met Buntline again in New York and attended said theater production. He then began his

531 White: Frederick Jackson Turner and Buffalo Bill, p. 9.

532 His paintings as well as the fact that Catlin toured Europe with a troupe of Native Americans, contributing to his depictions via dance performances. See: Linda Scarangella McNenly: Native Performers in Wild West Shows. From Buffalo Bill to Euro Disney, Norman, OK 2012, p. 22. For more on the enactment of Native American culture before Buffalo Bill's Wild West see: Lester George Moses: Wild West Shows and the Image of American Indians. 1883 – 1933, Albuquerque, NM 1996, pp. 10–20.

533 Barnum was inspired by the success of Catlin's exhibit, see: James W. Cook: The Arts of Deception. Playing with Fraud in the Age of Barnum, Cambridge, MA 2001, p. 133.

534 Bridger: Buffalo Bill and Sitting Bull, pp. 291–292.

535 Bridger: Buffalo Bill and Sitting Bull, p. 224.

536 Rico: Nature's Noblemen, p. 139.

537 Rico: Nature's Noblemen, pp. 135–136.

538 Rico: Nature's Noblemen, p. 162.

cultural ascension in earnest when he realized how others made a fortune exploiting his name and decided to do the same.[539] With the help of Buntline he started his acting career, playing himself, but often returned to the "frontier." During the summer Cody guided hunting parties (comprised of the wealthy east-coast elite) in the West and then starred in theater productions about his own embellished deeds in the fall season. He soon parted ways with Buntline, who nonetheless continued writing about Buffalo Bill on his own accord.[540]

James Butler "Wild Bill" Hickok (1837–1876) joined his friend Cody (they had met in their youth, later they served together in the Civil War) in 1873, also portraying himself on stage. Wild Bill left only a few months later, he had lived up to his nickname, constantly clashing with his fellow actors, the audience, and east coast society in general.[541]

In the early 1880s Cody created his Wild West show, which was inspired by his earlier performances, always playing himself, relatively honest, never adopting a real stage persona. The show was a more honest representation of "frontier life" and Native American (mostly Lakota and Dakota) culture, but it was still an amalgamation of reality, drama, and romantic sentiment a la Cooper. From the very beginning Cody and Nate Salisbury (who organized the shows with Cody) planned to bring the Wild West to Europe (and her profitable markets). Sitting Bull (c. 1831–1890) joined the show for one season in June of 1885.[542] At the time he was a living legend himself, and although he was branded the "slayer of Custer,"[543] he used his fame to survive in a world dominated by mostly hostile Euro-Americans (much like Red Cloud, see below). Sitting Bull and the now world-famous Native American warriors lent the Wild West further credibility, here real Native Americans portrayed themselves.[544] Furthermore, their

539 Bridger: Buffalo Bill and Sitting Bull, pp. 178–180, 198–203.

540 Bridger: Buffalo Bill and Sitting Bull, pp. 209–219.

541 Bridger: Buffalo Bill and Sitting Bull, pp. 220–226.

542 Bridger describes the meeting of Cody and Sitting Bull as follows: "Mythological forces – subliminal and conscious – brought the two men together in order for Cody to present the 'vanishing' West to the world at the very moment it was passing. The moment had arrived for the birth of the Wild West." See Bridger: Buffalo Bill and Sitting Bull, p. 305.

543 For more on the vilification of Sitting Bull in the US newspapers see: Coward: The Newspaper Indian, pp. 159–195.

544 Sitting Bull had become a symbol for Custer's defeat and for the resistance to US-colonial expansion. His apprehend appreciation of the industrial achievements (a newspaper office, a shoe factory, etc.) he was shown in city of St. Paul, Minnesota, was interpreted as a sign that even the staunchest enemy of assimilation could be convinced by the marvels of modern society: "Sitting Bull had been considered a tough nut to crack in terms of his (un)willingness to be subjugated. His visits to St- Paul served as proud examples that even the most savage of them all could be convinced that civilization was best for the Indians." Nadja Martin-Catherin: The Making of 'Indians'. Sitting Bull, Native Agency, and American Culture, Trier 2015, pp. 72–79. Quote on page 73.

participation generated publicity. The Wild West became a great financial success domestically, drawing huge crowds. Touring England was next on the agenda.[545] "With their freedom, horses and guns taken from them, Buffalo Bill's Wild West was the last place on earth for the Lakota to be able to resurrect and enjoy any vestige of their former way of life."[546]

Then in 1891 the Wild West became associated with the then very scandalous Ghost Dance movement.[547] The growth of the Ghost Dance movement within Lakota society can be read as a reaction to the suppression of their culture by the US government, which tried to assimilate the Laktota reservations using bureaucracy, and harbored intentions that can be described as cultural genocide. Although heavily influenced by the Christian religion, the movement was often misconstrued as a rebel rousting, which had to be forcefully subdued, leading directly to the massacre at Wounded Knee in 1890:[548]

> It is paradoxical that a religion with strong Christian elements that preached peace and cooperation with the whites came to be perceived as a hostile movement in South Dakota. Unfortunately for the Lakota, the general perception outside the dance camps was that the dances constituted preparation for an armed uprising rather than the rites of a peaceful religious movement. There are many reasons why this was the general view, and other people were to some extent responsible for distorting the perception of the Lakota Ghost Dance into a hostile movement, to suit their needs.[549]

For his 1892 tour through Europe (see below), Cody tried to hire a group of imprisoned Ghost Dancers. The War Department sought to get rid of the perceived troublemakers and after some bureaucratic back and forth the group joined the Wild West and left for Europe. The Ghost Dance was never performed in the show.[550]

The Wild West promoted a warlike image of Lakota culture. This was, in the eyes of the government in Washington D.C., an outdated image that should be dropped in favor of "reformation" and "civilization" of the Lakota:[551]

545 Bridger: Buffalo Bill and Sitting Bull, pp. 302–323.

546 Bridger: Buffalo Bill and Sitting Bull, p. 329.

547 Bridger: Buffalo Bill and Sitting Bull, p. 396.

548 Sam A. Maddra: Hostiles? The Lakota Ghost Dance and Buffalo Bill's Wild West, Norman, OK 2006, pp. 14–62.

549 Maddra: Hostiles, p. 45.

550 Maddra: Hostiles, pp. 100–107.

551 Maddra: Hostiles, pp. 63–64.

> Cody's use of Indian performers, and more specifically the image he presented of them, clashed with the idealized image of the Indian championed by reformers. Cody presented, and indeed celebrated, the Indians as wild mounted warriors and hunters of a bygone age. The reformers 'wished to foster the ideal of Indians as tamed humans in a tamed land, who were embracing civilization through land allotment, education and industry.' Consequently, government officials became concerned about the negative effects on their assimilation programs of Indian participation in Wild West shows.[552]

The Office of Indian Affairs (OIA) patronized Native Americans by regulating their employment. Therefore, they had to be convinced that the Wild West and other shows did not endanger the Native performers and, more importantly, did not ruin all efforts of the OIA to "civilize" them. Nonetheless, some Native American performers saw their employment by Cody & Co. as a great opportunity, giving them the freedom of mobility they were denied at home. While traveling with the Wild West show they could see the world, live the "old ways," resist being assimilated, gain some agency in representing their own culture, and make money. The relationship between the Native performers and Wild West shows is very complex, because for all the opportunities this stage presented, Native American performers were still somewhat exploited in a system of unequal power, performing the history of *Manifest Destiny* and the superiority of imperialism to a mostly "white" audience:[553] "OIA restrictions for employment were in place to protect Native performers, but these restrictions also coincided with government assimilation policies."[554]

Cody was already famous in the States as an exemplary "frontier" and showman when he and his Wild West show traveled to London in 1887. Being the first Wild West show that performed internationally, Cody's interpretation of the North American West was the first live representation of that region and its inhabitants most Europeans were exposed to. While the entertainment aspect of the shows cannot be overstated, education was another declared goal of Cody's venture, making it what would nowadays be titled "edutainment". The show was a huge success in London with over one million visitors, including Queen Victoria.[555]

---

552 Maddra: Hostiles, p. 85.

553 McNenly: Native Performers in Wild West Shows, pp. 39–54; 70–99. For more on the problems reformers had with Wild West shows as opposition to their "civilizing" of the Native Americans see: Moses: Wild West Shows and the Image of American Indians, pp. 60–79.

554 McNenly: Native Performers in Wild West Shows, p. 39.

555 Rico: Nature's Noblemen, pp. 132–135.

5.3.2.1 *The Wild West Tours Europe*

Buffalo Bill's Wild West first crossed the Atlantic in 1887. Cody became somewhat of a cultural ambassador representing US mass culture (reduced to the "frontier experience") in Europe. The premiere of this first tour through England was on May 5 in London. On May 11 Queen Victoria attended the show. This was the first public appearance of the Queen since her husband had died in 1861. 1887 was also the golden jubilee of Victoria's coronation, which further imbued Cody's show with meaning as this political connection made him even more of an unofficial ambassador:

> That the Wild West also held enormous potential for domestic politics was equally clear, especially when Queen Victoria asked for another command performance of the Wild West show on the eve of her Jubilee Day festivities. For this occasion, the kings of Belgium, Greece, Saxony, and Denmark, as well as an assortment of Europe's princes and princesses, including the future Kaiser William II, joined England's royal family to take in the Wild West performance and show their subjects that they too could delight in ordinary pleasures.[556]

Rico shows another interesting application of Cody's performances in that Britons and US Americans found common ground in the idea that Euro-Americans were predestined by nature to rule:

> The fact that the show celebrated the American conquest of the West as a national achievement did not preclude seeing that conquest, or the show representing it, as a *racial* achievement in which the British people could share. Cody and his publicists emphasized the show's presence in Britain as a gesture of Anglo-American friendship and unity. In this, they capitalized on the immediate context of the Wild West's London run: the American Exhibition to which it was adjacent, and the national celebration of Queen Victoria's Diamond Jubilee, with which it coincided. The Wild West show drew upon this context to make a reading of the show as a shared narrative of racial triumph more plausible.[557]

Similarly to the world exhibitions, the history that was performed at the Wild West was also fueled by and depicted scientific racism, the supposed superiority of industrialization, progress, and colonialism, selling it as edutainment.[558]

---

556 Ryell; Kroes: Buffalo Bill in Bologna, p. 108.

557 Rico: Nature's Noblemen, pp. 133–134.

558 McNenly: Native Performers in Wild West Shows, pp. 22–23.

Two years later Buffalo Bill's Wild West attended the world exhibition in Paris and was part of the American exhibition there. He then toured Spain and in 1890 Italy. In Rome Pope Leo XIII (1810–1903) met with Cody and other members of the Wild West. Then the show moved on to Germany and ended in England, where Queen Victoria again attended.[559]

The Wild West show was not the first instance ordinary Europeans experienced a more or less fictionalized representation of "frontier life":

> Each European country had at the time its own specific history of fictionalizing the American West. For instance, among European countries, Germany offers the clearest case of longtime infatuation with the American Indian. This may have had to do with a romantic, if not nostalgic, affiliation with the peoples threatened by the onward march of civilization, an affiliation that had the marks of a projection of feelings of loss of cultural bearings prevalent in a Germany undergoing rapid modernization itself.[560]

Note that the (fictional) literary works of Karl May were one of the reasons the Wild West was most warmly received in Germany. May had very much popularized the American "frontier experience" in *fin de siècle* Germany. His depictions of Native American life shaped the image of indigenous North American cultures for generations, and May's vision is still very prevalent today.[561]

In Germany, the logistical aspects of the show were studied closely by the military. The efficiency and speed with which the show set and broke camp and was loaded onto railroad cars was an unparalleled logistical feat. Understanding and copying Cody's efficient techniques would greatly aid in potential future wars.[562]

---

559 Ryell; Kroes: Buffalo Bill in Bologna, pp. 105–111.

560 Ryell; Kroes: Buffalo Bill in Bologna, p. 112. During a stay in Munich, it was suggested that the Wild West was of "high scientific interest" ("von höherem wissenschaftlichen Interesse"). On the one hand, this distinction meant that Cody had to pay less commercial tax due to the educational value of the show; on the other hand, the display of Lakota culture was of real ethnological interest. In this regard the Wild West fits in perfectly with other ethnological expositions, or human zoos (Völkerschauen), which were very popular at the turn of the nineteenth century. See: Sibylle Spiegel: Buffalo Bill's Wild West in München. Eine Veranstaltung Von "Höherem Wissenschaftlichen Interesse", Gerolzhofen 2002, pp. 23–29.

561 Bridger: Buffalo Bill and Sitting Bull, p. 359.

562 Ryell; Kroes: Buffalo Bill in Bologna, pp. 114–115.

5.3.2.2 *Cody and Marsh*

In September 1871 Cody guided a high-profile group of bison hunters, comprising members of the east coast elite. Besides being an enjoyable diversion, this was seen as part of the war effort against the Lakota, Cheyenne and various other Native American nations in a war that was never officially declared, but still had all the lethal consequences wars usually have. The mission was to destroy an all-important source of sustenance for the Natives. Hunting expeditions like this were the reason Cody was very well connected to the wealthy and influential elite, which in turn brought many opportunities for his show in later years.[563] One member of said elite was O. C. Marsh, who met Cody in 1871. Cody described his first encounter with the professor in his autobiography:

> During the summer of 1871, Professor Marsh, of Yale College, came out to McPherson, with a large party of students to have a hunt and to look for fossils. Professor Marsh had heard of the big bone which had been found by the Pawnees in the Niobrara country, and he intended to look for that as well as other bones. He accordingly secured the services of Major Frank North and the Pawnees as an escort. I was also to accompany the bone-hunters, and would have done so had it not been for the fact that just at that time I was ordered out with a small scouting party to go after some Indians. The day before the Professor arrived at the fort, I had been out hunting on the north side of the North Platte River, near Pawnee Springs, with several companions, when we were suddenly attacked by Indians, who wounded one of our number, John Weister. We stood the Indians off for a little while, and Weister got even with them by killing one of their party. The Indians, however, outnumbered us, and at last we were forced to make a run for our lives. In this we succeeded, and reached the fort in safety. The General wanted to have the Indians pursued, and said he could not spare me to accompany Professor Marsh. However, I had the opportunity to make the acquaintance of the eminent Professor, whom I found to be not only a well-posted person but a very entertaining gentleman. He gave me a geological history of the country; told me in what section fossils were to be found; and otherwise entertained me with several scientific yarns, some of which seemed too complicated and too mysterious to be believed by an ordinary man like myself; but it was all clear to him. I rode out with him several miles, as he was starting on his bone-hunting expedition, and I greatly enjoyed the ride. His party had been provided with Government transportation and his students were all mounted on Government horses. As we rode along he

563 Rico: Nature's Noblemen, p. 135.

> delivered a scientific lecture, and he convinced me that he knew what he was talking about. I finally bade him good-bye, and returned to the post. While the fossil-hunters were out on their expedition, we had several lively little skirmishes with the Indians. After having been absent some little time Professor Marsh and his party came back with their wagons loaded down with all kinds of bones, and the Professor was in his glory. He had evidently struck a bone-yard, and ‘gad!’[a favorite expression of Marsh’s] wasn’t he happy! But they had failed to find the big bone which the Pawnees had unearthed the year before.[564]

This is a telling account of their meeting. Cody manages to portray himself as a heroic “Indian Fighter,” and jovially writes about the killing of at least one Native, asserting that he had some “lively little skirmishes” shortly thereafter. He also describes how the government supported the Yale expedition, and mentions that the initial report about the “big bones” came from Pawnee sources. Cody’s description shows how Native American intelligence, private scientific expertise, and government expense all contributed to US-American paleontology. In a later autobiography, published in 1920, Cody describes how the news of the “very large bones” had reached the camp and were later relayed to Marsh:

> While we were in the sandhills, scouting the Niobrara country, the Pawnee Indians brought into camp some very large bones, one of which the surgeon of the expedition pronounced to be the thigh bone of a human being. The Indians said the bones were those of a race of people who long ago had lived in that country. They said these people were three times the size of a man of the present day, that they were so swift and strong that they could run by the side of a buffalo, and, taking the animal in one arm, could tear off a leg and eat it as they ran.
>
> These giants, said the Indians, denied the existence of a Great Spirit. When they heard the thunder or saw the lightning, they laughed and declared that they were greater than either. This so displeased the Great Spirit that he caused a deluge. The water rose higher and higher till it drove these proud giants from the low grounds to the hills and thence to the mountains. At last even the mountaintops were submerged and the mammoth men were drowned.
>
> After the flood subsided, the Great Spirit came to the conclusion that he had made men too large and powerful. He therefore corrected his mistake by cre-

564 William Frederick Cody: The Life of Hon. William F. Cody Known as Buffalo Bill the Famous Hunter, Scout and Guide. An Autobiography, Hartford, CT 1879, pp. 278–280.

> ating a race of the size and strength of the men of the present day. This is the reason, the Indians told us, that the man of modern times is small and not like the giants of old. The story has been handed down among the Pawnees for generations, but what is its origin no man can say.[565]

A few pages later he retells his meeting with the professor:

> Shortly after my return I received orders instructing me to accompany Professor Marsh on a fossil-hunting expedition into the rough lands of the Big Horn Basin. The party was to consist of a number of scientists besides Professor Marsh, together with twenty-five students from Yale,[566] which institution was sending out the expedition.
>
> I was to get together thirty-five saddle-horses for the party. The quartermaster arranged for the transportation, pack mules, etc. But General Sheridan, under whose direction the scientists were proceeding, always believed in my ability to select good horses from a quartermaster's herd.
>
> In a few days Professor Marsh and his companions arrived. The Pawnee Scouts, then in camp, had a year before unearthed some immense fossil bones, so it was decided that Major North, with a few of these scouts, should also accompany the expedition. Professor Marsh had heard of this discovery, and was eager to find some of the same kind of fossils.
>
> Professor Marsh believed that the Basin would be among the last of the Western lands to be settled. The mountain wall which surrounded it would turn aside pioneers going to Montana or northern Oregon. These would head to the east of Big Horn Mountains, while those bound for Utah, Idaho, and California would go to the south side of the Wind River Mountains. He was confident, however, that some day the Basin would be settled and developed, and that in its fertile valleys would be found the most prosperous people in the world. It was there that my interest in the great possibilities of the West was aroused.[567]

---

565 Cody: An Autobiography of Buffalo Bill (Colonel W. F. Cody), pp. 196–197.

566 In fact, no other fully qualified scientists and only eleven students accompanied Marsh on this expedition, see below.

567 William Frederick Cody: An Autobiography of Buffalo Bill (Colonel W. F. Cody), New York 1920, pp. 209–210.

Again, Cody underlines that Pawnee had discovered the bones, this time in more detail, reproducing the myth that associates the Pawnee with the bones. Again, Cody reflects upon how the government supported the expedition, and this time he adds a utilitarian dimension to Marsh's scientific lectures: he describes how the professor aroused his interest in the "the great possibilities of the West." This shows how science and the utilization (or more critically: exploitation) of the West were interconnected.

Cody and Marsh stayed in contact. It seems whenever Cody visited New Haven he tried to meet up with Marsh. Four letters documenting Cody's efforts to meet Marsh in New Haven are preserved in the Marsh papers (MS 343).[568] He invited Marsh to one of his performances in New Haven on December 22, 1874.[569] It seems Marsh had lent Cody $ 80, which the latter repaid on the evening of the performance.[570] Marsh's autobiography shows how the professor latched his scientific expeditions onto the aura of heroism that surrounded Cody (see below). Marsh used Cody's reputation and the popularization of the "frontier myth" to link paleontology to the *Manifest Destiny* of the nation.

5.3.3 Americanizing Paleontology on the "Frontier"

Not just the railroads greatly furthered the conduct of US-American science, but innovations and improvements in communications, technology, and infrastructure were also essential in nurturing science in the US. The scientific discoveries of the West could reach the East Coast with its centers of scholarship within minutes via telegraph. Specimens and letters were distributed quickly and safely via railroad and an improved postal service. These technical novelties also contributed to the development of a popular culture, which in turn adopted many scientific discoveries. The dinosaur became part of the public imagination, which in turn helped to find entrepreneurs and philanthropists willing to fund the expensive fossil hunting expeditions. Besides, the improvements in infrastructure and the postal service greatly furthered the public lecture circuit, beginning in the 1830s.[571]

---

568 William Frederick Cody, Springfield, MA to Othniel Charles Marsh, New Haven, CT 23 February 1873; William Frederick Cody, Bridgeport, CT to Othniel Charles Marsh, New Haven, CT 26 November 1873; William Frederick Cody, Bridgeport, CT to Othniel Charles Marsh, New Haven, CT 26 November 1873; William Frederick Cody, New Haven, CT to Othniel Charles Marsh, New Haven, CT 22 December 1874 (I); William Frederick Cody, New Haven, CT Othniel Charles Marsh, New Haven, CT 22 December 1874 (II), MS 343, Series I. Correspondence, Box 7, Folder 264.

569 New Haven, CT to Othniel Charles Marsh, New Haven, CT 22 December 1874 (I).

570 New Haven, CT to Othniel Charles Marsh, New Haven, CT 22 December 1874 (II).

571 David Hochfelder: The Communications Revolution and Popular Culture, in: William L. Barney (ed.): A Malden, MA 2001, pp. 305–316, see p. 307.

The spectacular dinosaur fossils found in the West arguably made paleontology authentically US-American. The call to adventure was an integral part of late nineteenth-century paleontology in America: "Young men were thrilled with the prospect of a dangerous, romantic adventure, all for the sake of science! Paleontology took a special aura, a mystique, that may underline the fact that an inordinately high percentage of participants wrote memoirs and autobiographies."[572]

The link between exploration and paleontology provided an argument in favor of the utility of science. This aspect was very important in order to justify the public expenditures for science in general, and a factor in the establishment of US-American scientific institutions:

> Finally, for the practical minded, natural history was utilitarian in the fact it furthered discovery and exploitation of the country's natural resources for the benefit of the people and proved to be of considerable importance in such practical endeavor as agriculture and medicine.[573]

"Frontier paleontology" with its products, and first and foremost the awe-inspiring dinosaur skeletons, was destined to become part of the national pop culture:

> In a visual culture marked by the gigantism in capitalism, architecture, and technology, and by the spectacle of Barnum and Bailey's Circus and the Ringling Brothers and Barnum and Bailey's Greatest Show on Earth, museum exhibits had to become more dynamic and captivating.[574]

A few decades before there were American dinosaur skeletons to exhibit Barnum had bought Peals Museum, which associated paleontological exhibits, like the "*American mammoth*," with US national pride in the form of portraits of the Founding Fathers (see chapter 2. 6.).

Still, some scientists like Osborn and Marsh thought that serious scientific exhibitions should distance themselves from the sensationalist entertainment industry.[575]

---

572 Warren: Joseph Leidy, p. 183.

573 Alexandra Oleson: Introduction. To Build a New Intellectual Order, in: Sanborn C. Brown; Alexandra Oleson (eds.): The Pursuit of Knowledge on the Early American Republic. American Scientific and Learned Societies from Colonial Times to the civil War, Baltimore, MD 1976, pp. xv–xxv. Quote on page xvii.

574 Sommer: History Within, p. 67.

575 Later Marsh recalled an incident in which he had met Barnum on a train from New York, northward bound. He started a conversation with Barnum, and it turned out that the latter was very frustrated because "some little cuss" had bought some valuable objects from Mexico in New York. The showman had intended to buy said objects himself. It turned out that Marsh was the "little cuss" who bought

There are no photos that depict Marsh with his fossils. There are photos of Marsh and there are some photos of his fossils, but he never posed beside them as many of his colleagues did, like hunters do with their prey. Perhaps this was too profane for the Yale professor, who, for similar reasons, did not want to open the Peabody Museum to the public but instead intended to reserve it for purely scientific audiences.[576]

The exploration of the American West after the Civil War marks a fundamental change in the history of paleontology in the United States, and indeed the history of paleontology in general. The Badlands held a treasure-trove of fossils, spectacular regarding their completeness and sheer size. It was here that paleontology became truly "American."

The nostalgia for and the mystification of the "Wild West" were born out of the uncertainty and upheaval at the turn of the century (*fin de siècle*), which stretched into the 1920s. These sentiments were reinforced by the feeling that the heroic era of the "frontier" had ended in 1890. Gone with it were a sense of individualism and closeness to nature, especially in the ever-growing, highly industrialized cities of the Northeast. Frederic Remington literally drew this picture, while Theodore Roosevelt employed the nostalgia for "the Old West" in his political career.

Buffalo Bill's Wild West shows promoted a picture of the progress of civilization and justified the "savage war" in the name of imperialism. General Custer's defeat in 1876 was reenacted countless times by Cody's troupe.[577]

Monica Rico further elaborates on the connection between Cody's show and a nostalgic sentiment for the "frontier:"

> Just as the frontier seemed to be slipping away into the past, Buffalo Bill's Wild West, in its raucous glory, could be seen as evidence that the American frontier would never really close. By renewing themselves with the entertaining and educational day with Buffalo Bill, audiences were partaking in a ritualized reconquest of the frontier and thereby constructed 'an idealized national memory' of the West that defined them as an American public with a particular history.[578]

the objects, which he then told the – assumingly very perplexed – Barnum. See: Schuchert, LeVene: O. C. Marsh, p. 349. On May 3, 1880 Barnum send two tickets for his "greatest show on earth" to Marsh, the tickets are also part of the Othniel Charles Marsh papers – they seem to be unused. See: Phineas Taylor Barnum, New Haven CT to Othniel Charles Marsh, New Haven, CT May 3, 1880, MS 343, Series I. Correspondence, Box 29, Folder 1247.

576 Davidson: A History of Paleontology Illustration, p. 99.

577 Slotkin: Gunfighter Nation pp. 53–87. For more on Cody's career see chapter 5. 3. 2. and Moses: Wild West Shows and the Image of American Indians, pp. 21–251.

578 Rico: Nature's Noblemen, pp. 132–135.

Linda McNenly suggests that conceptions of colonialism and progress fueled the "Wild West Shows," which gained rapid popularity at the close of the century:

> The frontier was expanding, and the tension between settlers and Native people was growing. It was a time of unrest and confrontation competing with the desire to settle the West and fulfill manifest destiny. Wild West shows grew out of this context of colonialism and progress as well as this history of the cultural display of others.[579]

Buffalo Bill was a true master of self-promotion; he embodied the image of the Western adventurer and gentleman impeccably. Marsh wanted to nurture a similar image for himself. Publicizing his friendship with Cody was a means to this end.

The buffalo hunt was a regular part of most Wild West shows, further evidence of its importance to the "frontier myth" and the self-image men like Marsh tried to promote.

5.3.4 Preliminary Conclusion

The mystification of the natural land of North America accelerated during the course of the nineteenth century. Nature and the so-called "conquest" of the "Wilderness" were to provide a foundation for the national identity of the young Republic. Beginning in the 1840s, more and more people believed it was the divinely ordained *Manifest Destiny* of the United States to rule the entirety of North America and to bring "civilization" to the "Wilderness" and its inhabitants. Native Americans played an ambivalent role within the context of the "frontier myth:" on the one hand they were perceived as savage enemies standing in the way of progress, a threat to the Euro-American settlers which was to be eliminated, on the other hand they were sometimes perceived as part of original nature and had to be protected with it to preserve an important part of US national identity. The "frontier" was understood as "both the engine of progress and the domain of real men who dominated other men and nature."[580] True men of science thusly had to prove themselves on the "frontier".

The emergence of mass media and a national popular culture allowed for the conception of a *Manifest Destiny* and with it the mystification of "frontier life." When in the second half of the century race and class conflicts, accelerated by mass immigration and industrialization, became more prominent, the significance of the "frontier myth" and its potential for "Americanization" as well as conservative promises

579 McNenly: Native Performers in Wild West Shows, p. 23.

580 White: Frederick Jackson Turner and Buffalo Bill, p. 49.

of a simpler life became more apparent. In 1893 Turner finally provided a theoretical framework for these processes with his "frontier theory." All the while dime novels and Wild West Shows had firmly established the significance of the "frontier" in the public consciousness. "Buffalo Bill" Cody was the most prolific and popular perpetuator of the "frontier myth" within the United States and abroad. While the world was "Americanized" through mass culture, paleontology (as well as many other sciences) was "Americanized" as well. Many of the US-American paleontological discoveries in turn became part of mass culture.[581] Depictions of dinosaurs (often locked in mortal combat) seeped into the collective consciousness through magazines and dime-novels. On a more scientific level Hawkins' Crystal Palace Dinosaur reconstructions and Leidy's *hadrosaurus* skeleton had done their part in cementing the image of the terrible lizards in the heads of contemporaries.

5.4 Marsh in the "Wild West"

"Yet, in spite of the scoffers, it is clear that Marsh and his student collectors, on their memorable expeditions into the western fossil fields, launched American vertebrate paleontology into its heroic period."[582]

The "Bone Wars" provide a metanarrative inseparable from Cope's and Marsh's paleontological contributions, which affects how the protagonists are perceived to this day. There is a plethora of evidence that the "Bone Wars" had a significant impact on how scientific work was done by Marsh and his crews in the West. Henry W. Farnam partook in the Yale expedition of 1873, and remembered in 1931:

> We found it very difficult to get any information from Professor Marsh on what we were doing. I cannot recall that he ever gave us a cursory lecture on the geological formations on which we were working or the possible significance of what we were finding. If we asked him questions, he was very apt to give a few of his characteristic grunts and return a noncomittal [sic!] answer. He was essentially a collector and not a distributor. At that time his bête noire

581 Allen A. Debus explains how Dinosaurs and other extinct creatures became pop cultural icons and illustrates his point with various examples from antiquity to the twenty-first century. See: Allen A. Debus: Prehistoric Monsters. The Real and Imagined Creatures of the Past That We Love to Fear, Jefferson, NC 2010.

582 Lanahm: The Bone Hunters, p. 91.

> was Professor Cope of Philadelphia, and I always thought that he was afraid that if he told us anything it might possibly leak back to his antagonist.[583]

The most interesting, or at least most popular episodes of the "Bone Wars" were fought out in the American West, and not on the pages of scientific magazines or the New York Tribune (see chapter 6. 5.). The previous parts of this chapter outlined how scientific explorations became interwoven with the "frontier myth" and popular culture. The rest of this chapter will detail how paleontology and particularly the "Bone Wars" became part of that narrative.

5.4.1 The Yale Expedition of 1870

In 1898 Marsh started to write – but unfortunately never finished – his autobiography. The draft was written by typewriter and some corrections were made by hand. The typed outline of the autobiography is part of the Othniel Charles Marsh papers at Yale College (reel 26 of the microfiche).

The sketch is titled "FOSSIL HUNTING IN THE ROCKY MOUNTAINS: A Narrative of Thirty Years' Work, with Reminiscences of Friends and Foes." It is dedicated to the memory of Thomas Henry Huxley, whom Marsh calls "guide, philosopher and friend." The first chapter details the professor's first journey to the Rocky Mountains in 1868. He describes it as his "first visit to the far West, and all was new and strange."[584]

Marsh's first fossil-hunting grounds lay in the east – at Haddonfield, New Jersey. Here he joined Cope (whom he still considered a friend), who had worked at Haddonfield for some time. Another important hunting ground was Greenfield, Massachusetts. In August 1868, however, Marsh attended a meeting of the American Association for the Advancement of Science (AAAS) in Chicago (and promptly was elected general secretary of the Association). Attendees of the congress were given the opportunity to join an excursion out west. Marsh and the other scientists followed the Union Pacific railroad into the Wyoming territory. Later Marsh remembered this first excursion to the West in his autobiography. On the way back East, he found some (a "hatful") fossils in a mount of discarded earth at Antipope Station, Nebraska. This early success convinced Marsh that the vast western American lands held an abundance of fossils, ready to be plucked from the dry ground.[585]

At this point in their biography, Schuchert and LeVene draw a direct correlation to Marsh's experiences in Europe:

583 Henry W. Farnam, New Haven, CT to Ernest Howe, New Haven, CT, 6 May 1931, MS 343, reel 26, frames 462–463.

584 MS 343, reel 26, frame 246 autobiography, p. 1.

585 Schuchert; LeVene: O. C. Marsh, pp. 94–99.

> When Marsh reached home in late summer of 1868, he at once set about making plans for the systematic exploration of the western plains, convinced that in them lay riches comparable to the wealth of vertebrate remains on which he had been trained in Europe.[586]

Marsh returned to the West in June of 1870, this time accompanied by eleven students (all heavily armed against supposed Native American attacks) from Yale. Of those eleven students, only George Bird Grinnell would later pursue a career in paleontology and be employed as a professional fossil hunter by Marsh. Thanks to a letter from General Sherman, the Yale expedition was provided, amongst other supplies, with an armed escort and four guides. Two of the guides were Pawnee, another was "Buffalo Bill" Cody, in 1870 already a "frontier-celebrity." This marked the beginning of a decades-spanning friendship – or at least a mutually beneficiary relationship – between the professor and the soon-to-be showman. The expedition chose the Union Pacific railroad as the basis of their operation, searching north and south of the tracks. The railroads "opened" the West to exploration, expansion, and exploitation. Later, various members of the expedition remembered their journey as an adventure under the constant threat of Native American raids (which never came). After hunting for fossils in Nebraska and Colorado, the party crossed into Utah. After a few weeks vacationing at Salt Lake City, San Francisco, and Yosemite, the group reengaged collecting in Wyoming, and then – in November – in Kansas. After spending Thanksgiving on the prairie, Marsh engaged in a supposedly spectacular buffalo hunt, which he fondly recalls later in his autobiography. The expedition returned to New Haven on the eighteenth of December. It was a huge success, for the party delivered thirty-six boxes of fossils to the museum, and the exploits of the expedition were widely published in magazines and newspapers, greatly contributing to Marsh's celebrity status and opening a wide range of social contacts, which in turn opened possibilities to receive funding for further expeditions.[587]

The first pages of the Marsh autobiography seem sketchy and unordered, they recall some anecdotes, describe the landscape and invoke a genuine fascination with the fossilized enigmas the West might hold:

> I was eager to explore it at once, as I felt sure that entombed in the soft sandy clays that filled the present basin to the brim, there must be remains of many strange animals new to science, long waiting to be brought to light, and to tell the tale of their life history to him who could read it.[588]

---

586 Schuchert; LeVene: O. C. Marsh, pp. 99–100.

587 Schuchert; LeVene: O. C. Marsh, pp. 100–120. Also see: Goetzmann: Exploration and Empire, pp. 425–429, and Lanahm: The Bone Hunters, pp. 79–85.

588 MS 343, reel 26, frame 246 autobiography, p. 1.

Marsh underlines the dangerous nature of his adventure to the West: "It was a dangerous place just then for explorers, as the Sioux Indians were on the war path."[589] He goes on to describe an instance in which a fellow traveler from the east had been killed and scalped by the Native Americans, and then the encounter of a railroad conductor who had escaped the same fate by a narrow margin (though he had been shot with arrows and scalped, but survived the ordeal).[590] He adds one more "scalping story," told in the quite humorous tone of one Methodist preacher, who, instead of being scalped, was just deprived of his hair piece by a Cheyenne Brave.[591]

The first of the few fully written out chapters of the autobiography details Marsh's first buffalo hunt in Kansas in 1870. Here Marsh links his exploits in the West to the bison, an established symbol for the American "Wilderness" and therefore the true "frontier experience," or at least what was meant by that term in in the 1890s. Furthermore, he participated in the manliest of rituals – the hunt. He underlines his government backing and mentions a letter from General Sherman, which he describes as "the magic sesame that had assured me a cordial welcome and ample assistance at every army post."[592] He then references the luring danger in the region, mentioning that the Cheyenne and other "hostile Indians"[593] had claimed the region as their hunting ground and were raiding the settlers. Marsh then mentions his first encounter with "Buffalo Bill," who served him as a guide.[594] Thus Marsh links his legacy as a "frontiersman" and explorer to that of Cody, who was well established in these matters by 1898. Afterwards Marsh describes his hunt in rather exciting terms, bringing down three bison including the lead animal, the "master of the herd."[595]

In the third chapter Marsh describes the aforementioned Thanksgiving dinner in the prairie. Marsh references Cope's wrongheaded *elasmosaurus*.[596] Under the ever-present threat of hostile Cheyenne, his team then searched the Twin Buttes, where the *elasmosaurus* originated, for further fossils. He describes their fear of a raid during the night and that they "feared most of all the silent but deadly Indian arrows."[597]

---

589 MS 343, reel 26, frame 249 autobiography, p. 4.

590 MS 343, reel 26, frame 249 autobiography, p. 4.

591 MS 343, reel 26, frame 251 autobiography, p. 6.

592 MS 343, reel 26, frame 253 autobiography, chapter 2, p. 1. The three letters that demanded support for Marsh's expedition in 1870, 71, and 73 survive in the MS 343, see: MS 343, Series I. Correspondence, Box 29, Folder 1247.

593 MS 343, reel 26, frame 253, autobiography, chapter 2, p. 1.

594 MS 343, reel 26, frame 254, autobiography, chapter 2, p. 2.

595 MS 343, reel 26, frame 258, autobiography, chapter 2, p. 6.

596 MS 343, reel 26, frames 262–263, autobiography, chapter 3, pp. 2–3.

597 MS 343, reel 26, frame 264, autobiography, chapter 3, p. 4.

The fourth chapter is titled "My First Pterodactyl," and again it opens with an allusion to the looming threat: "We were encamped on the Smoky Hills River, and as the Indians were also there after buffalo and other victims, we had an escort of United States troops from Fort Wallace to protect us and help on our work."[598]

Here Marsh also mentions that the "signs of Indians were plenty," and that they were potentially "savage enemies." He writes that at one point his companions thought he might have been "picked up" by Native Americans when he came late to camp.[599]

George Brid Grinnell, at the time one of Marsh's students and a member of the first Yale-expedition in 1870, later wrote: "None of us knew or cared anything about the objects for which it was being undertaken. Vertebrate fossils meant nothing to us, but we all longed to get out into the uninhabited and the unknown, West, to shoot buffalo and to fight Indians." [600] He also describes Marsh as "very much of an outdoor man and a keen sportsman."[601]

5.4.2 Later Expeditions

The expedition of 1871 consisted of only ten students, who accompanied their professor out West. Grinnell was not part of that endeavor, but Oscar Harger (1843–1887) – soon to be a loyal assistant of Marsh's – was. The expedition started out at Fort Wallace, Kansas, on July 2. Once more accompanied by a military escort, the party soon found various saurian fossils, amongst them parts of a pterosaur, complementing a discovery the expedition had made the previous year in the vicinity. After a four-day rest in Denver, the expedition reached Fort Bridger in Wyoming on August 22. The desert plains yielded many a fossil and eleven full boxes were sent back east, stuffed with, among other specimens, the fossils of several extinct horses. After another rest at Salt Lake City, the expedition went on to Oregon, where they arrived in October. Here they collected (as always guarded by a military escort) another eleven boxes worth. The party stopped their collecting for the winter and returned to New Haven, where Marsh arrived in January 1872.[602]

The expedition of that year consisted of only four students plus Marsh (and of course various other guides, the military escorts, etc.). The expedition spent the first

598 MS 343, reel 26, frame 269, autobiography, chapter 4, p. 1.

599 MS 343, reel 26, frame 271, autobiography, chapter 4, p. 3.

600 George Bird Grinnell, New York to Ernest Howe, New Haven, CT 19 February 1929, MS 343, reel 26, frame 302.

601 George Bird Grinnell, New York to Ernest Howe, New Haven, CT 19 February 1929, MS 343, reel 26, frame 303.

602 Schuchert; LeVene: O. C. Marsh, pp. 120–126.

half of their time in Kansas (where an even more spectacular and dangerous buffalo hunt was undertaken) and the second in Wyoming:[603]

One chapter of the autobiography is titled "A Ride for Life in a Herd of Buffalo" and within the first paragraph "hostile Indians" are evoked.[604] Marsh mentions General Sherman, calling him his "faithful friend," and notes the kindness of the General in providing him with an armed escort, especially because "the frontier posts had none too many troops to keep the Indians in check."[605] Again it were the Cheyenne who "left several of their victims on the plains, pierced by arrows characteristic of that tribe."[606] The rest of the fossil hunting expedition was also conducted in "Indian land."[607] The group encountered a massive bison herd (at least 50,000 heads strong) and in allusion to his many "frontier" traits Marsh writes that "My hunting weapons consisted of a cavalry carbine and a pair of navy revolvers, not too many for an Indian country,"[608] again linking the hunt for bison to the hostile Native Americans, referencing two of the most prominent Western tropes. He then shot the bison in "the exact manner, my first guide, Buffalo Bill, had taught me long before."[609] A Photo of the 1872-expedition underlines the martial undertones of Marsh's "frontier-experience": the depicted men are armed and dressed as "frontiersmen," ready for adventure (see figure 4). He then found himself surrounded by the stampeding herd, and killed various other animals to escape the situation. In the end he shot another bison and cut steaks from it to provide dinner for his small party. He never mentioned what happened to the other killed or severely wounded animals. Unfortunately, the script for the planned autobiography ends here. But it is very telling that Marsh started his tale of 30 years of paleontology not with a retelling of his education or his prehistoric subject, but with tales from the West, alluding to the dangerous bison hunt, the lessons Buffalo Bill gave him, and the constant threat of hostile Native Americans. He thus imbued his tale with some of the major aspects of the mythical "frontier," long since established in US pop culture. This, no doubt, made the proposed autobiography much more financially viable and interesting to an unscientific audience and linked paleontology to the "frontier," the poster child of US-Americanness.

603 Schuchert; LeVene: O. C. Marsh, pp. 126–132.

604 MS 343, reel 26, frame 276, autobiography, chapter 5, p. 1.

605 MS 343, reel 26, frame 276, autobiography, chapter 5, p. 1.

606 MS 343, reel 26, frame 276, autobiography, chapter 5, p. 1.

607 MS 343, reel 26, frame 278, autobiography, chapter 5, p. 3.

608 MS 343, reel 26, frame 279, autobiography, chapter 5, p. 4.

609 MS 343, reel 26, frame 279, autobiography, chapter 5, p. 4.

Figure 4: The Yale College Scientific Expedition of 1872, https://findit.library.yale.edu/catalog/digcoll:4339968, as consulted online on January 13, 2022.

1873 saw another expedition, and this time Marsh was accompanied by thirteen students (as per usual armed), including Harger. The expedition and its cavalry-escort left Fort McPherson, Nebraska, on June 12. After the party had returned to Fort McPherson, it moved to Wyoming in July and, after ten more days of successful fossil-hunting, the expedition went on to Salt Lake City for a short rest, during which Marsh met with the Mormon leader Bingham Young, with whom he discussed the extinct horses, which were of special theological interest to the Mormons. After several weeks of collecting in the John Day Basin in Oregon, the group retired to San Francisco in October and then returned to New Haven via Kansas. The whole expedition had yielded forty-nine boxes of material to study in the winter, and in fact in many more winters to come.[610]

The 1874 expedition contributed like no other to Marsh's mainstream popularity, not (just) because of the spectacular fossil discoveries it yielded, but because of the professor's public advocacy on behalf of Red Cloud and the Oglala Dakota. In the summer Colonel Stanton informed Marsh that the region around the Red Cloud and Spot-

610 Schuchert; LeVene: O. C. Marsh, pp. 132–138.

ted Tail Agencies in Dakota were "very rich in fossils."[611] He added that furthermore "the remains of a Mastodon are reported near Red Cloud," and that he saw "many astonishing teeth and bones" there. In another region nearby, he writes, "the ground is reported 'heaped up' in many places with fossil remains." He furthermore promised a military escort for any fossil-hunting expedition in the future. Stanton had written a dispatch to General Ord – commander of the Department of the Platte – informing him of the fossils. Ord then also wrote to Marsh to invite him to come West, also promising a military escort.[612]

This time traveling without students from Yale, as this expedition was more of a private venture than official Yale-business, he arrived at the Red Cloud Agency in November. Accompanied by some hired help and a cavalry escort that included Stanton himself, Marsh ventured into the Black Hills. Since 1868 and the conclusion of Red Cloud's War, the territory north of the Platte River and east of the Big Horns was reservation land. The Lakota and Dakota were obligated by treaty to stay on this reservation; in turn they had been promised territorial sovereignty and subsidies from the US government in the form of food, blankets, and other materials, sorely needed to live under the less than ideal reservation conditions. As soon as 1871 the sovereignty of the reservation was broken, when it was decided to build a railroad through reservation land. Construction of the railroad was to be enforced by the US military, which erected a Fort at the base of the Black Hills. To make matters worse, reports of gold findings in the Black Hills soon flooded back east and the US government sent a military expedition under the command of General Custer to survey the situation.[613]

The foreign military expedition worried the inhabitants of the reservation. Another conflict with the US government also deteriorated the mood: Agent J. J. Saville wanted to count the inhabitants of the reservation, but was denied time and time again; now he threatened to withhold the contractually-promised government supplies, unless the Natives agreed to be counted. Red Cloud tried to fire back, which Schuchert and LeVene describe rather condescendingly:

> Red Cloud, attempting to regain prestige [he supposedly had lost by striking a deal with the invaders he had fought successfully] by showing his people that he was a bigger man than the agent, had been complaining to the soldiers at Fort Robinson that the rations, blankets, etc., given the Indians by Saville were of poor quality, to which charges the military men probably lent a willing ear.

611 Thaddeus H. Stanton, Cheyenne, Wyoming Territory to Othniel Charles Marsh, New Haven, CT 3 June 1874, MS 343, Series I. Correspondence, Box 30, Folder 1303.

612 Edward Otho Cresap Ord, St. Louis, MO to Othniel Charles Marsh, New Haven, CT 3 June 1874, MS 343, Series I. Correspondence, Box 30, Folder 1303.

613 Schuchert; LeVene: O. C. Marsh, pp. 139–142.

> Saville, on the other hand, had countered by writing to the Indian Bureau that the Army was interfering with his business. It was into this general and particular love feast that Professor Marsh unwittingly stepped that morning of November 4, 1874, when he arrived at the Agency – the very day on which the council of chiefs had decided against the census demanded by Saville.[614]

Though the fossils fields he wanted to explore were outside of reservation land, Marsh decided to seek the cooperation and approval of the council of chiefs, to avoid any trouble. The professor explained his intent and the chiefs could hardly believe that someone would go through so much trouble to unearth some bones, but finally agreed to support the expedition. In turn, Red Cloud employed Marsh to voice the grievances of the Dakota in Washington and to the US-American public. Nonetheless, the chiefs changed their minds and later denied their cooperation[615] until Marsh and the other US Americans arranged a big feast and regained the sympathy of the chiefs.[616] LeVene and Schuchert paint a quite dramatic picture of the departure of the expedition, reproducing the whiff of adventure that came with western exploration and (fossil-) hunting:

> Exasperated by these delays, Marsh made up his mind to give the Indians the slip, and gave orders to start shortly after midnight. The soldiers made their way with extreme caution through the Indian villages, which lay between the Agency and the only spot for fifteen miles where the White River could be crossed. The dogs barked furiously as they went past the lodges, but, contrary to the usual idea of Indian alertness, the warriors slept peacefully on and the stolen march was not discovered until daylight.[617]

They continue to describe how the expedition party narrowly escaped a big war band looking for the fossil hunters soon after they had arrived at their hunting ground: "the

614 Schuchert; LeVene: O. C. Marsh, p. 143.

615 LeVene and Schuchert describe the mood at the agency as quite hostile, underlining the inherent and imminent danger that supposedly accompanied any dealings with the West and its inhabitants. Marsh would have been proud: "The sight of the soldiers and the wagons, actually drawn up on marching order, excited the Indians greatly, and they gathered around in great numbers, armed quite as well as the small band of soldiers, with breechloading rifles and revolvers of the most recent pattern. [...] In this tense situation, when a single shot, or an order to advance, would probably have precipitated a massacre, there was nothing to but to direct the soldiers to go back to Fort Robinson." Schuchert; LeVene: O. C. Marsh, p. 144.

616 Schuchert; LeVene: O. C. Marsh, pp. 143–144.

617 Schuchert; LeVene: O. C. Marsh, pp. 144–145.

expedition left the region less than twenty-four hours before it was scoured by a large war party in search of the Big Bone Chief [Marsh]."[618]

The meeting with Red Cloud had far reaching consequences for Marsh and his fame in the US. The Red Cloud – Marsh relationship is further discussed in Chapter 5.5.3.

With the expedition of 1874 Marsh quit the field and focused on his scientific work. From now on he relied on an army of professional bone diggers, who supplied his museum and laboratory at Yale. Even if he himself seldomly traversed the "Wilderness" after 1874, his "frontier"-legacy was still very much a part of Marsh's image.

Marsh had a home built, following his specific instructions. Construction began in 1876 and the interior was finished in 1881. LeVene and Schuchert estimate that the luxurious house with its three stories, eighteen rooms, and one tower, set the professor back $ 30000, and that the furnishing must have cost a similar amount. Today the Marsh house is the home of the Yale School of Forestry and Environmental Studies, the surroundings are known as the Marsh Botanical Garden. From 1881 forward the house was the home of a very passionate collector, the collections included more than 1,300 rare orchids (in a greenhouse that was later added), vast amounts of oriental rugs, and most importantly, at least for this study, Marsh's Western memorabilia.[619] Marsh liked to display this collection, this witness to his heroism and the evidence for his adventures, in a room he called the "Wigwam." It was the reception hall of his home and thus all guests must have passed the collection at least once when they came to visit. LeVene and Schuchert describe the "Wigwam" in some detail. Since said description says a lot about how Marsh liked to present himself to his guests (among them many students from Yale), and it says a lot about how the aforementioned authors wanted Marsh to be remembered, it is copied here in its full length:

> Its most striking room was the high octagonal reception hall which he called his 'Wigwam'; he relates that when Red Cloud came into it, he looked up as if in search of the hole where the smoke went out. This hall had a bewildering amount of art objects: paintings, Japanese and Chinese cloisonné and bronzes and *kakemonos*. To the left of the entrance was a very large round oak table of special design covered with western memorabilia, from which Marsh loved to pick up the peace pipe that he and Red Cloud had smoked in 1874, the Mormon Bible that Brigham Young had given him, or some other interesting souvenir, and talk about it to his guests. Ernest Howe, who remembered the Professor 'not as a scientist or partisan, but as a rather pompous but kindly old gentleman who had hunted buffalo in the dim past,' says that in the 'nineties, during

618 Schuchert; LeVene: O. C. Marsh, p. 145.

619 Schuchert; LeVene: O. C. Marsh, pp. 347–348.

> Sundays of the winter term, it was Marsh's custom to entertain small groups of students at his home, where After [sic!] a luncheon planned to satisfy even the appetites of Yale undergraduates we were taken to the 'Wigwam,' a sort of trophy room filled with mementos and treasures from all over the world. Here a scalp or a pair of buckskin leggings, or a frontiersman's pistol would recall some incident of the west and Yale seniors became small boys again, listening to tales of Indian savagery, or of hairbreadth escapes from stampeding buffalo.[620]

5.4.3 Preliminary Conclusion

Nearly everyone who wrote about Marsh or Cope reproduced the sense of adventure in their writing, this is true for LeVene and Schuchert, and also for Marsh himself in his autobiography, as seen above. Lanham for example remarks: "The image of himself as an intrepid frontier outdoorsman lay very near his heart, for Marsh sedulously promoted this aspect of his life for the public."[621] To give another example: Goetzmann describes Marsh's first expedition 1870 as very colorful and exciting, noting that "even the Indians joined in the fun of digging up the prehistoric beasts."[622]

Of course, the expeditions to the West did provide US paleontology not only with anecdotes of "frontier" heroism, but also with the most valuable fossils. For example, Marsh found a completely new type of dinosaur in the American West, the horned *ceratopsidae*, the most prominent member of this family being the *triceratops*. He later called them "the strangest animals yet discovered in any part of the world."[623] Dodson describes in detail how John Bell Hatcher (1861–1904) started his career collecting bones for Marsh and later published his own scientific articles, continuing Marsh's work. Dodson delivers a comprehensive history of the reception of *triceratops*. At the time it seemed that all horned dinosaurs were exclusive to North America (and that was still true in 1996, when Dodson wrote his book, until 2010, when *sinoceratops* was discovered in China).[624] At the end of the nineteenth century it seemed that horned dinosaurs were an original and uniquely American addition to paleontology.[625]

620 Schuchert; LeVene: O. C. Marsh, p. 348.

621 Lanahm: The Bone Hunters, p. 86.

622 Goetzmann: Exploration and Empire, p. 426.

623 Othniel Charles Marsh, New Haven, CT to John Wesley Powell, Washington, DC 4 February 1893, MS 343, Series I. Correspondence, Box 26, Folder 1096.

624 Xing Xu et al.: First Ceratopsid Dinosaur from China and its Biogeographical Implications, in: Chinese Science Bulletin, vol. 55, no. 16 (jun. 2010), pp. 1631–1635, DOI:10.1007/s11434-009-3614-5.

625 Dodson: The Horned Dinosaurs, pp. 56–75.

Marsh would also acquire an extensive collection of fossil horses and bird with teeth, both providing solid evidence for Darwin's theory of evolution (see chapter 7. 2.). The evolution of the horse is especially relevant in this regard because the horse is, maybe after the bison, the animal most associated with the "frontier," be it the loyal companion of many a cowboy, the cavalry horse, or the mustang of a Native American warrior.

5.5 Bones of "Thunder Beasts" – Native American Contributions to Paleontology

As the nineteenth century progressed, the conception most Euro-Americans had of Native Americans changed. Before the Civil War, Native Americans were seen as a vanishing people doomed to extinction. These "Noble Savages" lived in a more natural state than the "civilized" Euro-Americans in their cities.[626] Their presumed way of life was perceived to be very romantic, their seemingly unavoidable extinction stirred nostalgic sentiment. Brian Dippie deliberates on how the trope of the "Vanishing Indian" was used by Euro-Americans to, in a way, absolve themselves from any responsibility for the genocide, and on how Native Americans were perceived to be doomed to lose the struggle for survival anyway because of their natural inferiority: "By removing the Indian's fate from mortal hands, it stultified normal, humane concern."[627] The concept of the "Vanishing Indian" thus became a self-fulfilling prophecy, and proof for the irresistible progress of human civilization.[628] A healthy balance between the Native American's "natural state" and European progress that manifested itself on the "frontier" was understood to be the spirit of American democracy (a concept that was a quasi-forerunner to Tuners "frontier thesis"). On the other hand, it seemed that the reverse applied to the Native Americans, as it was presumed that: "Coming in contact with civilization, Indians surrendered what was good in their racial character and absorbed what was bad in that of the whites."[629]

Kevin Hutchings explores this link, stating that when the theory of extinction gained a foothold in the minds of scholars, beginning in the late seventeenth and early

626 For more on the trope of the "Noble Savage" see: Dippie: The Vanishing American, pp. 18–25. David Hurst Thomas shows how the stereotype of the "Noble Savage" (who oftentimes had to be protected against his barbaric and cannibalistic brethren) was utilized by Europeans since Columbus, see: David Hurst Thomas: Skull Wars. Kennewick Man, Archaeology, and the Battle for Native American Identity, New York 2000, pp. 3–10.

627 Dippie: The Vanishing American, p. 122.

628 Dippie: The Vanishing American.

629 Dippie: The Vanishing American, p. 12.

eighteenth centuries, the "cultivation" of North America was understood as a feasible solution to the overpopulation contemporaries like Malthus foresaw. While being the salvation of the European style agrarian societies, it spelled doom for the seemingly "primitive" Native Americans. The "extinction" of the indigenous societies seemed an inevitability to many US-Americans, even in the first half of the nineteenth century. In this scenario Native Americans were lumped together with the *mammoth* and other extinct animals, predetermined to share their fate because they had seemingly failed to "improve" the North American wilderness. This rhetoric only promoted the notion of *Manifest Destiny* and provided a convenient excuse to further the agenda of settler colonialism. The vision of the Native American as a "Noble Savage" selected for extinction thusly almost became a self-fulfilling prophecy.[630] On the contribution scientific thinking had on this process Semonin writes:

> [t]he idea of extinction was adapted to doctrines of white supremacy long before social Darwinism made these ideas fashionable in the late nineteenth century. In Joseph Richardson's *The American Reader*, published in 1813, children were taught that the extinction of the American Indian was ordained by nature's god: 'The religion of nature, the light of revelation, and the pages of history, are combined in the proof, that God has ordered that nations shall become extinct, and that others shall take their place.' Such beliefs were not yet anchored in geological science or the fossil record, but they lend themselves easily to similar interpretations when theories of scientific racialism were later used to substantiate beliefs in racial superiority and the doctrine of Manifest Destiny.[631]

Walter Hixson further elaborates on the "passing of a noble race". He links the emergence of the "Noble Savage" stereotype in early nineteenth century literature with the nation's endeavor to fulfill its *Manifest Destiny*, the war with Mexico, the Civil War, and all the genocidal violence these conflicts brought.[632]

This conception changed somewhat after the Civil War, when the Great Plains were settled (or invaded) by US Americans in ever greater numbers. Now the Native Americans of the West (especially the Lakota) became enemies and obstacles to

630 Kevin Hutchings: Transatlantic Extinctions and the "Vanishing American", in: Kevin Hutchings; John Miller (eds.): Transatlantic Literary Ecologies. Nature and Culture in the Nineteenth-Century Anglophone Atlantic World, London 2017, pp. 58–72. For a comprehensive but short overview of US-American settler colonialism and the "winning of the West" by Euro-American settlers, backed by the US Army, see: Hixson: American Settler Colonialism, pp. 113–144.

631 Semonin: American Monster, p. 365.

632 Hixson: American Settler Colonialism, pp. 85–111.

the progress of civilization and the expansion of the United States. This perspective spread quickly, propagated in countless, very cheap and therefore widely read, dime novels. This form of pulp literature boomed due to lower paper and printing costs, thanks to technical innovations.[633]

While the difference between the Native and the Euro-Americans were justified at first through religious theories, the theory of evolution and the growing scientific enthusiasm in the second half of the nineteenth century contributed an apparently scientific race theory, and scientific racism.[634] The fact that epidemics had decimated various Native American populations throughout the continent was understood as yet more proof that the Euro-Americans were superior to the long-term inhabitants of North America. The quest to prove this racial superiority was the reason that the fate of Native Americans was often linked to the fate of the African Americans, especially after the Civil War.[635]

Slotkin further elaborates on the link between science, evolution, and Social-Darwinist rhetoric, and on how said themes were exploited to justify a "savage war" against the Native Americans. The theory of evolution had revealed the underlying principles of the struggle for survival in nature. It seemed that the same inevitable struggle was fought out in the West. Many contemporaries suggested that this struggle would regenerate the primeval and pure characteristics of the respective "race," and because the "Celtic/Teutonic race" displayed the noblest and strongest features, the "frontier" experience would advance the US-American nation (further implying that said nation's "true" citizens were of Celtic/Teutonic ancestry).[636] Consequently, the closing of the "frontier" presented the nation with a dilemma: "The problem of a post-frontier American is how to preserve and develop those leadership virtues that were fostered by hunting and Indian-fighting in a world without wilderness or savages."[637] The self-image of a nation that so strongly identified with the "frontier"-struggle was to be deeply confused by the disappearance of the "frontier".

Slotkin describes the goals of the Grant administration's "Peace Policy" as follows:

> The reform of Indian policy was intended to raise a dark and victimized race of primitives from barbarism [...] to the light of Christianity and economic

633 Berkhofer: The White Man's Indian, pp. 86–102.

634 Berkhofer: The White Man's Indian, pp. 49–61. For an elaboration on the link between scientific racism, the boom of natural history in the US, and Social Darwinism in the latter half of the nineteenth century, see: Thomas: Skull Wars, pp. 38–70.

635 For further elaboration on the scientific racism concerning African and Native Americans, see: Dippie: The Vanishing American, pp. 82–106.

636 Slotkin: Gunfighter Nation, pp. 38–51.

637 Slotkin: Gunfighter Nation, p. 56.

> self-reliance. Economic progress and the civilization process would go hand in hand: as the Indian acquired the skills and economic philosophy of the white yeoman, mechanic, or small businessman, his manners would alter and the religious philosophy of Christianity would come home to him.[638]

5.5.1 Naturalizing Genocide: Extinction Narratives

In light of a series of very costly "Indian Wars," the government in Washington D.C. (the Grant administration, to be precise) devised a peace policy to deal with the Native Americans of the Great Plains. A series of negations and treaties established new reservations in which the Native Americans were to be resettled and fed by the US government. In the end this was a much cheaper alternative to a continuation of the wars.[639] Walter Hixson judges that

> [...] the 'peace policy' demanded unconditional surrender of Indian homelands and hunting grounds and relocation onto reservations [...]. The 'peace policy' and the subsequent movement for Indian assimilation signaled an effort to institute internal colonial rule once Indians had been militarily subdued.[640]

Klaus Frantz provides a broad study of Native American life during the twentieth century. He also provides a short overview of US–Native American relations from the eighteenth to the twentieth century. He attests that after 1871 Native Americans were no longer regarded as independent peoples with whom treaties would be signed and other diplomatic relations would be conducted. Instead, most Native Americans were now restricted to their corresponding reservations, which were managed by the Bureau of Indian Affairs (BIA). As a part of Grant's "Peace Policy," the BIA sought the "cultural assimilation" (other – more poignant – terms might be cultural genocide or ethnocide) and reeducation of the Native Americans under their management. In addition, the bigger part of the land Native Americans possessed was redistributed to Euro-Americans under the supervision of the BIA.[641] Assimilation seldom proved ben-

---

638 Slotkin: The Fatal Environment, p. 315.

639 Berkhofer: The White Man's Indian, pp. 166–175. Slotkin also describes the "Peace Policy" of the Grant administration in some detail. He puts an emphasis on General Custer, the various Native American nations, and the conflict concerning the Black Hills, see: Slotkin: The Fatal Environment, pp. 316–476.

640 Hixson: American Settler Colonialism, p. 119.

641 Klaus Frantz: Die Indianerreservationen in den USA. Aspekte der Territorialen Entwicklung und des Sozio-Ökonomischen Wandels, Stuttgart 1993, pp. 27–39, 53. For more reflection on the mechanisms by which the Native Americans were dispossessed and the establishment of the reservation

eficial for Native Americans, even tribes that had adopted most aspects of US-American agriculture were forced from their land in the interest of Euro-American colonization.

The BIA is one of the oldest federal agencies. It was created as a unit of the Department of War in 1824 and tasked with implementing federal policy concerning Native Americans. In 1824 it was placed under the jurisdiction of the newly founded Department of the Interior.[642] After the Civil War the reservation system was first established. Improvements in the infrastructure (especially the completion of the transcontinental railroad in 1869) allowed a more thorough exploitation and colonization of the western territories. First Native Americans were restricted to reservation land to avoid clashes with settlers. In the early 1870s provisions of rations were delivered to the reservations to ensure that the Native Americans would stay on their allotted land. This was understood to be a provisional solution until the reservations' economies were developed enough to stand on their own. In many cases this never happened. Congress never provided adequately for the rations and so they were seldomly delivered as promised. Provisioning was to be run by agents of the BIA, but at first they were highly autonomous and in full control of the implementation of provisioning. Many agents were very susceptible to corruption, lining their own pockets with funds that should have gone to the inhabitants of the reservations. Later the reservation system was abandoned in favor of Native-American reeducation and their assimilation into mainstream US-society.[643]

Only after 1874, when Marsh led his very public campaign against the corrupt agents of the Red Cloud Agency (see below), serious anti-corruptions efforts were made. In the late 1870s the institutional structure of the BIA was revised, permanent inspectors for internal control established.[644]

Even celebrities like Henry David Thoreau (1817–1862) argued for the establishment of "preservation zones," in which the true beauty of the natural American landscape, its fauna, and even the Native American cultures could be preserved: "Natural preserves [...] in which the bear and panther, and some even of the hunter race, may still exist [...] not for idle sport or food, but for inspiration and our own true re-creation."[645]

Like George Catlin, Marsh traveled to Washington D.C. to plead on behalf of Red Cloud. Catlin was a romantic artist who painted western landscapes and above all the

---

system, see Dippie: The Vanishing American, pp. 141–182. On Native American reeducation also see: Hixson: American Settler Colonialism, pp. 140–142.

642 Paul Stuart: The Indian Office. Growth and Development of an American Institution, 1865–1900, Ann Arbor, MI 1979, p. 5.

643 Stuart: The Indian Office, pp. 15–26

644 Stuart: The Indian Office, pp. 63–64, 77–85.

645 Quoted after Pomeroy: In Search of the Golden West, pp. 90–91.

Native Americans who still lived in the West. He became acquainted with the cultures of some tribes, and was an advocate for the vanishing cultures.[646] Since at least the first half of the nineteenth century, the Native Americans were understood as a people doomed to extinction, inevitably falling in the wake of Euro-American "progress and civilization". This way of thinking was interwoven with the nostalgia and melancholia which quickly became a part of the mythical West. Thus, the exploration of extinct lifeforms and prehistoric worlds long gone could be thematically linked to the impending doom that was, as many US-Americans presumed, in store for the Native Americans in the near future: "The races of the mammoths and mastodons, and the great sloths, came and passed away: the red man of America is passing away!"[647]

In 1819 statesman Henry Clay (1777–1852) rhetorically linked the state and foreseeable demise of Native American nations to the fate of the *mammoth*:

> We are powerful and they are weak [...] to use a figure drawn from their own sublime eloquence, the poor children of the forest have been driven by the great wave which has flowed in from the Atlantic ocean to almost the base of the Rocky Mountains, and, overwhelming them in its terrible progress, has left no other remains of hundreds of tribes, now extinct, than those which indicate the remote existence of their former companion, the mammoth of the New World.[648]

In an 1889 article Otto Meyer, Marsh's former assistant, also likened the "melancholic" destiny of many Native Americans to that of the now-extinct species. He goes one step further and writes that a Native American in 1889 would look onto the past of his people in the same way the descendants of the now-extinct reptiles of prehistory would:

> INDIANS of to-day, who are well acquainted with the history of their race, may often think with melancholy of the olden times, when their forefathers were the only masters of the country. Numerous and powerful tribes occupied the vast territory between two oceans, some hunting the deer in the forests of the East, others ruling supreme in the plains and mountains of the West. [...] The majority of the tribes, and among them the most powerful ones, have been extinguished entirely; while others, sadly diminished in numbers, linger here and there, and the pale-face is met everywhere. The same feelings of melan-

646 Goetzmann: Exploration and Empire, pp. 184–191. For more on Catlin and his fascination with the "vanishing" tribes, see: Dippie: The Vanishing American, pp. 25–31. For a comprehensive biography, see: Brian W. Dippie: Catlin and his Contemporaries. The Politics of Patronage, Lincoln, NE 1990.

647 James H. Carleton on July 25, 1865, quoted after: Dippie: The Vanishing American, pp. 130–131.

648 Quoted after: Dippie: The Vanishing American, p. 8.

> choly must enter the mind of an alligator of geological education, when, during a *siesta* in the sun, he thinks of the good old Mesozoic times and compares them with the pitiable present.[649]

Joseph Henry, first secretary of the Smithsonian Institute and president of the NAS, wrote an open letter perfectly demonstrating the linkage between the perceived destiny of the Native American people and natural history. He conjures up an image in which Native Americans, whom he calls a "disappearing race," are seen as living fossils, ready to be studied to learn more about the pre-historic ancestors of Europeans and Euro-Americans:

> I would respectfully urge the importance of purchasing these valuable records of the previous inhabitants of North America, which, if not secured at this time, will be dissipated and lost to the world. They will grow in importance with advancing years, and when the race of which they are representations shall have entirely disappeared their value will be inestimable. [...] No scientific subject of the present day is exciting more interest than that of the past history of the world [...]. It is proved by cumulative arguments the most irresistible that the ancestors of the most civilized races of the present day were at one time savages, of whom the manners and customs can only be understood by a comparative study of the lives of savages now existing in different parts of the world. Comparative ethnology forms the basis of pre-historic science.[650]

In 1911 a Native American man emerged from the "wilderness" of California. The man was named Ishi and he was the last survivor of his tribe, the Yahi. Before 1911 Ishi had had no or only limited contact with US-American society and civilization, and was therefore seen as a relic, some kind of vestige, almost a living fossil, a sensation to be studied by anthropologists. Dippie describes the whole situation as follows: "[...] Ishi received the attention that would be lavished on a dinosaur that happened to stumble into a paleontologists' convention"[651]

---

649 Otto Meyer: Giant Reptiles of a Past Age, in: Popular Science Monthly, vol. 34, no. 4 (Feb. 1889), pp. 466–473. Quote on page 466.

650 Joseph Henry: Letter of Prof. Henry, of the Smithsonian Institute, Washington, DC, 13 December 1873, MS 343, Series I. Correspondence, Box 16, Folder 646.

651 Dippie: The Vanishing American, p. 208.

5.5.2 Native American Legends and Guidance

In the US, Native Americans were always thematically linked to extinct life forms: Thomas Jefferson displayed Native-American artifacts alongside fossilized bones in his Monticello estate. On his quest to disprove Buffon's theory on American inferiority (see chapter 2.6.) he also had several Native burial mounds unearthed to prove the worthiness of the ancient American people.

Thomas even writes that, at the closing of the century, Native Americans (and Inuit) were seen as living fossils, and were thus studied vigorously at anthropological museums and exhibitions.[652]

Native Americans played an important role in the development of US-American paleontology, and even its very beginning. Native-American legends partly inspired Thomas Jefferson's instructions to Lewis and Clark. Furthermore, the guidance of Abenaki scouts let Baron Longueuil to the discovery of the Big Bone Lick in 1739 (see chapter 2. 6.). These interactions are instances of knowledge transfer, and the importance of local knowledge, which led to the discovery of fossils, cannot be overstated.

Both Cope and Marsh employed the help of local guides, who often were Native Americans, in their fossils-hunting endeavors. Cope employed such help in his expeditions to New Mexico in 1874 and 1883, though it is not known whether his guides were Native or Euro-Americans.[653] On his first major expedition in 1870 Marsh also employed Native American scouts, but he did not record any of their myths or legends which might have been inspired by the gigantic fossils bones.[654] Cope and Marsh recognized the Native Americans and their stories as interesting curiosities, but seldom employed said stories in their searches, or in any other scientific way.[655] One notable exception is Marsh's cooperation with Chief Red Cloud, whom the scientist met in 1874. He listened intently to the Oglala accounts of the bones of "thunder beasts," and was inspired to name *brontotherium* and *brontosaurus* after the legendary thunderous monsters.[656]

---

652 Thomas: Skull Wars, pp. 77–90.

653 Mayor: Fossil Legends of the First Americans, p. 135.

654 According to Mayor, the Native Americans of that region knew about the bones since at least the eight century, for they crafted trinkets from them. Mayor: Fossil Legends of the First Americans, pp. 165–167.

655 Mayor: Fossil Legends of the First Americans, pp. 154, 182–184.

656 Mayor: Fossil Legends of the First Americans, pp. 240–241. Since then some other Native American myths and words have entered the paleontological vocabulary via naming conventions, see: Mayor: Fossil Legends of the First Americans, p. 242. Furthermore, Marsh acquired an ammonite that had been worn by a Native American as jewelry and brought it to New Haven. Mayor: Fossil Legends of the First Americans, pp. 269–270.

Native Americans also appear as antagonists in stories remembering expeditions, either told by the members of the expeditions themselves or by eager journalists, both looking to bolster their stories with excitement and adventure. In these tales Native Americans, in their role as "primitive" peoples who are "in touch with nature," become a part of the adverse lands, similar to hostile weather conditions and wild animals; an obstacle to be overcome by the heroic explorer and "civilization."

Cope seldom asked for or received the help of army escorts,[657] and while Osborn writes that the pacifistic Quaker went on his expeditions unarmed, Davidson argues against that.[658]

Sometimes the differing cultural approaches concerning the fossils led to minor dispute, as some Native Americans found the unearthing and transportation of the bones precarious, or even insulting. In addition, Cope and Marsh excavated countless human skulls to be studied for science, also a practice that offended the spiritual sentiments of some.[659]

5.5.3 Red Cloud and Marsh

Though Sitting Bull was the most famous and infamous Native American at the closing of the nineteenth century, Red Cloud was the most notorious leader of the Lakota prior to Sitting Bull: "[...] he [Sitting Bull] was not the most famous or feared chief of the early 1870s – that distinction probably goes to Red Cloud."[660]

Red Cloud was known to newspaper-reading audiences for his victorious war[661] against the United States, and his subsequent journey to Washington, D.C. to renegotiate the details of the peace treaty. His meeting with President Grant was widely publicized in various newspapers.[662]

In 1874, after his successful fossil-hunting expedition into the Black Hills had returned to the Red Cloud Agency, Marsh met up with Red Cloud and the chief discussed with him the issue of the shoddy supplies they had received, handing the professor some samples. When Marsh attended a convention of the NAS in Washington

---

657 Mayor: Fossil Legends of the First Americans, pp. 261–264

658 Davidson: The Bone Sharp, pp. 10–11.

659 Mayor: Fossil Legends of the First Americans, pp. 298–302.

660 Coward: The Newspaper Indian, p. 162. For a comprehensive biography of Red Cloud see: James C. Olson: Red Cloud and the Sioux Problem, repr. ed. Lincoln, NE 1975. Also see: John D. McDermott: Red Cloud. Oglala Legend, Pierre, SD 2015.

661 For a very detailed description of the war see: John D. McDermott: Red Cloud's War. The Bozeman Trail 1866 – 1868, vol. I+II, Norman, OK 2010.

662 McDermott: Red Cloud, pp. 50–57.

the following April, he showed the samples to the Commissioner of Indian Affairs, E. P. Smith. The next day the professor met with President Grant and told the soldier of the mistreatment of the Oglala. Later, while still in the capital, Marsh met with the Board of Indian Commissioners, and the whole affair was printed in the newspapers, where it loudly resonated and was reprinted countless times.[663] Joseph Ferdinand Keppler (1838–1894) depicted the "Indian-ring affair" in a cartoon in Frank Leslie's Illustrated Newspaper on September 18, 1875. The cartoon is titled "Our Indian Policy" and depicts the inside of a military fort in which the officers of the BIA offer substandard blankets, rifles, and obviously rotting beef to Red Cloud's Lakota. O. C. Marsh can be seen in the background peering over the stockade, planning to report the fraudulent procedure to the president (see figure 5).[664]

Figure 5: Joseph Ferdinand Keppler: Our Indian Policy, in: Frank Leslie's Illustrated Newspaper, vol. 41, no. 1042 (18. Sep. 1875), p. 1.

663 The story featured in more than 111 newspaper articles, see: Wendler: Die "Knochenkriege", pp. 42–43.

664 Joseph Ferdinand Keppler: Our Indian Policy, in: Frank Leslie's Illustrated Newspaper, vol. 41, no. 1042 (18. Sep. 1875), p. 1.

In May Red Cloud and other chiefs arrived at Washington to plead their case and negotiate the selling of territory, keeping the story alive in the newspapers. In July Marsh had the charges printed as a short pamphlet and sent some copies to the president.[665] The following investigation of the affair would result in the sacking of agent Saville (1830–1910) and the resignation of Secretary of the Interior Delano (1809–1896).

In May 1875 Red Cloud traveled to Washington to discuss with President Grant the lacking quality of the supplies the US-government sent to the Oglala, but went home disappointed.[666] Olson suggests that the samples Red Cloud presented to Marsh might have been "fraudulent" and not representative of the overall quality of the supplies. As for the professor's motivation to get involved: on the one hand Marsh needed Red Cloud's approval for his fossil-hunting expeditions to the Black Hills in South Dakota, on the other hand he was no friend of Agent Saville, who was charged with providing the substandard provisions: Saville had tried to deny Marsh access to a military escort for the expedition.[667] A third reason might have been that it was very fashionable to publicly complain about the Grant administration because it constituted a good way to generate publicity for oneself. It worked for Marsh, for his accusations and the whole affair were widely published in various newspapers. But Marsh also went to Washington, spoke to Grant in person, and published a 38-page pamphlet with the accusations.[668] The president had received the pamphlet via mail (Marsh had sent the pamphlet to Secretary Delano, who forwarded it to Grant)[669], as evidenced by his written reply:

> I have rec'd and read your letter and the accompanying pamphlet, dated July 10th, and am taking steps to verify or refute the statement you made in regard to the bad management of the Red Cloud Agency. The charges and statements you made are sufficiently explicit either to be substantiated or to be refuted – to prove the fraud and bad management, or incompetent observation. Assuring you of my earnest desire for an honest administration in every department of the government and willingness to find out and punish fraud wherever found.[670]

---

665 Schuchert; LeVene: O. C. Marsh, pp. 145–155.

666 Olson: Red Cloud and the Sioux Problem, pp. 175–198. Also see: McDermott: Red Cloud, pp. 73–74.

667 Olson: Red Cloud and the Sioux Problem, pp. 179, 183.

668 Othniel Charles Marsh: A Statement of Affairs at Red Cloud Agency. Made to the President of the United States, 1875.

669 Library of Congress, Digital Collections, Ulysses S. Grant Papers: Series 2, Letterbooks, 1869–1877; Vols. 2 (cont.)-5, p. 494.

670 Ulysses S. Grant, Long Branch, NJ to Othniel Charles Marsh, New Haven, CT 16 July 1875, MS 343, Series I. Correspondence, Box 13, Folder 538.

Only after Marsh had complained to the president, an investigation of the agents of the Red Cloud Agency was initiated. As a result, the cattle that was delivered to the Oglala was weighed from then on. Before, the agents had just guessed the weight of the cattle delivered as part of the promised supplies, which usually meant that much less meat than promised was delivered to the Oglala. As further consequence of the investigation, Agent Saville was replaced.[671] "In this instance, Red Cloud's technique of persuading men from various professions and parts of the country to carry his message to government officials and politicians had proven beneficial."[672]

Red Cloud and Marsh remained friends. The chief wrote some letters to the professor in the subsequent year. The O. C. Marsh correspondence provides three letters written by Red Cloud rather by his interpreters, or some other person, for Red Cloud never really learned to speak or write English). In all his letters to Marsh Red Cloud refers to his friendship with the professor and references his aid in the "Red Cloud affair." Furthermore, Red Cloud tried to employ Marsh's help in his political ambitions and negotiations with the US government, which was his typical approach to politics (see below).

In 1877 Red Cloud had his peace-pipe sent to Marsh. In the accompanying letter Lieutenant William Lewis Carpenter (1844–1898), then stationed at Ford Robinson, Nebraska, wrote that Red Cloud told him that he remembered Marsh, whom he (or, to be more precise, his interpreter) fondly called the "wise chief."[673] Because "He told the Great Father [President Grant] everything just as he promised he would, and I think he is the best white man I ever saw. I like him." LeVene and Schuchert quote this part of the letter, but omit the rest,[674] in which Red Cloud again asks Marsh for help against the mistreatments in the name of the US government: "I want you to tell him this, and also that Genl. Mc Kenzie [Col. Randal Ranald Slidell Mackenzie, of Fort Robinson ?] came and took 100 horses which belonged to me, and the soldiers burned some of my things. I am now very poor, and without friends. This is all." It is not known if Marsh could help Red Cloud in this matter as well. But the two were still friends in 1883, when Red Cloud came to New Haven.

In 1883 Red Cloud sent a letter to Marsh in which he calls him a "[d]ear friend."[675] He wanted to update the professor on his and his people's situation. He told him that

---

671 McDermott: Red Cloud, pp. 80–81.

672 McDermott: Red Cloud, p. 81.

673 William Lewis Carpenter, Fort Robinson, NE to Othniel Charles Marsh, New Haven, CT 31 January 1877, MS 343, Series I. Correspondence, Box 06, Folder 216.

674 See: Schuchert; LeVene, p. 167.

675 Red Cloud, Pine Ridge Agency, Dakota Territory to Othniel Charles Marsh, New Haven, CT 2 May 1883, MS 343, Series I. Correspondence, Box 27, Folder 1121.

he had made "many warm friends to Stand by the Poor Indian and healp [sic!] them." He further stated his new friends were "good white m[e]n."

The next surviving letter was later dated by pencil to 1890. Red Cloud again invokes his amiable relation to Marsh by calling him his "kind friend."[676] He told Marsh that he would like to visit the east again, but lacked sufficient funds for such a journey, then asked the professor for money: "I would very much like to go, but have no money, If [sic!] you can assist me I would like it very much." It seems Marsh never lent his friend the money.

The last letter Red Cloud wrote to Marsh was sent in 1897 from Washington, D.C., on Red Cloud's last visit to that city (he resided in the National Hotel, as stated in the handwritten letterhead). In this last letter he makes reference to Marsh's assistance during the "Red Cloud Affair:"

> I shall never forget my troubles during the earlier days of reservation life. Your kind assistance rendered at that time I shall never forget. The [illegible] association of those days come back to me when I arrive here and wish to write you a few lines. I have not taken much part in the affairs of the reservation [in] recent years, but I desire to make this my last appeal to the government to fulfill all the provisions of the treatis [sic!] that were made with us. I can not say how long I shall be here, but probably two weeks or so. I trust that you may be able to render me some assistance again, That [sic!] is, I would like to have you write to your senators [?] here to help me all they can.[677]

5.5.4 Preliminary Conclusion

Many Native American ethno-cultural groups were associated with the "frontier myth." On the one hand, they were perceived as obstacles to the *Manifest Destiny* of a nation bound to rule over and bring "civilization" to a whole continent. In short: savage enemies of civilization. On the other hand, their various lifestyles were romanticized as more "natural" than that of Euro-Americans, more "in tune with nature" (a bias still prevalent to this day). They were perceived as a part of nature and sometimes even the natural land. As the nineteenth century came to a close, the significance of the "frontier" for US- American culture and identity became apparent, or rather its significance was propagated by historians, politicians, and other influential mem-

676 Red Cloud, Pine Ridge Agency, SD to Othniel Charles Marsh, New Haven, CT 25 October 1890, MS 343, Series I. Correspondence, Box 27, Folder 1121.

677 Red Cloud, Washington, DC to Othniel Charles Marsh, New Haven, CT 1 May 1897, MS 343, Series I. Correspondence, Box 27, Folder 1121.

bers of US society. When US paleontology became part of this great narrative, it was, almost inevitably, linked to Native Americans, especially the inhabitants of the Great Plains. Here they are almost exclusively constructed to be enemies, an ever-looming threat to the heroic scientists and bone-hunters. In some cases, Native Americans, whose ancestral legends told of thunderbirds and giants, were helpful guides to the expeditions. But Native Americans and paleontology were interconnected in the minds of nineteenth-century contemporaries in another way: they were perceived to be a group of people seemingly selected for extinction; they were perceived as another example for the survival of the fittest. Genocidal acts would therefore often be excused as an unavoidable fact of nature. On a more individual level, actors with individual agendas were part of this greater narrative, as exemplified by the interactions of Red Cloud and Marsh. The professor needed the permission of Red Cloud (and other chiefs) for his expeditions into the Black Hills. He was therefore very willing to lobby in Washington for a more just and humane treatment of the Oglala. He did so on Red Cloud's behalf to repay a favor to him, maybe because he believed this course of action to be the right thing to do, maybe because it was very much in style to publicly criticize the corrupt Grant administration, and it greatly contributed to the reputation of the professor. Red Cloud knew how to champion his agenda with the BIA and the national government in Washington, often imploring the help of prominent allies within the high society such as Marsh.

5.6 Chapter Conclusion

While highly entertaining, the bone-hunting expeditions and the concomitant "discovery" of the West are but a fraction of the history of the genesis of US-American paleontology:

> Published histories of vertebrate paleontology concentrate too often upon the adventures of the bone hunters. The rivalry of Marsh and Cope and the fanciful battles between their collecting parties; expeditions of students from eastern colleges, armed to the teeth [...] – all these make interesting reading but give a woefully incomplete picture of the development of this subject and its contributions to related branches of science.[678]

Still, the connection between the "Bone Wars" and the "frontier myth" is the main reason why the story of Marsh and Cope is remembered and retold to this day. The "fron-

678 Gregory: North American Vertebrate Paleontology, p. 307.

tier" made paleontology a (US-) American venture, it proved to be a treasure-trove of the most spectacular fossils. The scientists who described these most fascinating life-forms had a distinct edge over their colleagues. Moreover, the "Bone Wars" were part of the "winning of the West," to this day *the* US-American epic. Therefore, it can be argued that paleontology was "Americanized" on the "frontier."

Paleontology owes a great deal of its mainstream success to the depictions of extinct lifeforms in popular culture. When cheap magazines and dime-novels began to constitute a popular culture in the United States, a heroic and often fictionalized narrative of the "conquest" of the American West was born. Roughly at the same time dinosaurs and other extinct lifeforms began to captivate the imagination of a general audience in the US. It is no wonder that dinosaurs were popularized at the same time the "frontier myth" caught on. Like no others before, the dinosaurs of the West, like *triceratops, brontosaurus,* and *stegosaurus,* came to inspire public imagination, and became the inspiration for many works of (paleo) art and fictional stories.

> The relationship between science and society in mid-century America involved more than parallels. Developments in American life during that period directly affected American science. The national pride that supported wars to win or to keep territory also spurred scientists and gave them a claim to public support. Their ambitions for American science had overtones of Manifest Destiny and the War for the Union. Territorial expansion drew them back for a time toward descriptive natural history and gave them government employment. The strengthening of democratic ideals and representative government forced American scientists to cultivate the general public.[679]

As seen in chapter 2. 6., in the minds of US patriots the imagined prehistoric past of the continent was, beginning in the early days of the republic, linked to savagery. Domination of nature and its inhabitants, animals as well as Native Americans, had been linked to a savage past that could now be symbolically dominated through scientific de-mystification. But because the imagined prehistoric creature had to be as fierce as possible, some herbivores were imagined as carnivores and the bloodthirsty martial nature of prehistoric creatures was underlined in scientific literature as well as in popular culture.

Westward expansion and the associated "frontier" life became *the* defining American feature, and the most spectacular dinosaurs were found in the western territories. Dinosaurs became American, and paleontologists and other fossils hunters became imbued with the same whiff of adventure that characterized all western exploration and "frontier"-life. Mitchell also comes to the conclusion that

679 Robert V. Bruce: The Launching of Modern American Science 1846–1876, New York 1987, pp. 5–6.

[D]inosaurs rightly belong in the picture with cowboys – and Indians and buffalo and outlaws and railroads and cavalry – in short, in the world of the American frontier, understood as a blend of fact and fantasy, a real place and a Hollywood invention.[680]

680 Mitchell: The Last Dinosaur Book, pp. 30–31.

6

“M.’s activity is simply low” – Marsh and his German Assistants

The transatlantic scientific network of the nineteenth-century paleontologists is a continuation of the Republic of Letters, dating back to the seventeenth century. After his study visit to Europe (see chapter 2), O. C. Marsh returned to the United States where he became a professor for paleontology at Yale and made a name for himself via the paleontological exploration of the Western territories (see chapter 5. 4.). His excellent fieldwork and scientific descriptions (especially his contributions to the theory of evolution, see chapter 7. 2.) earned Marsh international renown. While Marsh and many other aspiring US-American scientists had to visit Europe to further or complete their education during the last quarter of the nineteenth century, European paleontologists often visited US-American centers of knowledge and education, such as Philadelphia and New Haven. Within the scientific correspondence network new discoveries and theories were exchanged. In the case of paleontology, fossils are valuable scientific data, and therefore fossils were part of this international exchange. When visiting another country, many paleontologists tried to acquire new specimens and take them back home. In other cases, professional fossil-vendors were instructed to hunt down certain skeletons and send them across the Atlantic. One of the leading German paleontologists, Karl Alfred von Zittel of Munich, visited the United States in 1883.[681] He was invited by Marsh to stay at his home in New Haven. After his return to Germany, Zittel remained in contact with the US-American professor, helped him to buy Bavarian fossils, and introduced him to two young German paleontologists. These two young paleontologists were Max Schlosser and Georg Baur; they traveled to New Haven in 1884 and became Marsh's assistants. At Yale they met Otto Meyer, another German migrant working for Marsh. While Schlosser soon returned to Munich and Meyer quit his job, Baur stayed with Marsh in New Haven for several years; later he became a professor at Clark University (1890–1892) and at the University of Chicago (1892–1898). Marsh's transatlantic paleontological network after 1865 and his contacts to the aforementioned German scientists are analyzed and explained in this chapter.

Schuchert and LeVene write that some of Marsh's assistants became scientists of renown after they had parted ways with their tutor and that their help had therefore been a major influence on Marsh's own work.[682] They devote a whole chapter to the relationship between Marsh and his laboratory assistants.[683] Schuchert and LeVene propose that Marsh had some fifty assistants working with him between 1873 and

681 For a comprehensive analysis of Zittel's position in the German paleontological community and his legacy see: Marco Tamborini: "If the Americans Can Do It, So Can We". How Dinosaur Bones Shaped German Paleontology, in: History of Science, vol. 54, no. 3 (Sep. 2016), pp. 225–256, https://doi.org/10.1177/0073275316671526.

682 "The Group of assistants who worked with Marsh in his laboratories at various times included men who were later to become leaders in their own branches of science, and for this reason they must be considered in any discussion of his work". Schuchert; LeVene: O. C. Marsh, p. 290.

683 Schuchert; LeVene: O. C. Marsh, pp. 290–312.

1898, twenty-three of whom could be traced and written about. The longest employed, and presumably most loyal assistant was Thomas Attwater Bostwick (1857–1923), who was hired by Marsh in 1873 and stayed until Marsh's death. Bostwick was charged with managing the vast collections of Marsh and is said to have kept minute and detailed accounts of the specimens. Others were hired to do secretarial work, illustrations, engravings, restorations, etc. The first German employed by Marsh was Adam Hermann (before 1878 – after 1909), who prepared fossils for study. Hermann had worked in a brass manufactory and as a taxidermist. After Marsh had noticed Hermann's specimens displayed in a store window and the German had received a recommendation from his employer Henry B. Sargent (1851–1927, owner of the brass manufactory and a member of the first Yale expedition), he hired him in 1876 to prepare specimens. Hermann stayed with Marsh for ten years and then moved on to the American Museum of Natural History in New York (AMNH).[684] Additionally, there was a group of assistants who were employed to help Marsh with his laboratory work; they probably had the most influence on his scientific work. Among these were well-known paleontologists like George Brid Grinnell (1849–1938), Oscar Harger (1843–1887),[685] Samuel Wendell Williston (1851–1918) and Erwin Hinckley Barbour (1856–1947). The last two would become very outspoken enemies of Marsh (see below).[686]

Schuchert and LeVene draw a connection between Marsh's ties to the United States Geological Survey (USGS) and federal funds, his experiences as a student in Germany, and the employment of the three German assistants who were most important to this thesis:

> In 1882–83, when Federal money was becoming available, Marsh began to look around for still other scientific assistants, and recalling the helpful Privatdozenten whom he had seen at work in the German laboratories, he appealed to Professor Karl A. von Zittel of Munich, a guest at his home in the fall of 1883, to help him find such. Zittel, enthusiastic over Marsh's hospitality and his collections, felt sure that he knew of several promising young men. Not long after his return to Munich, he wrote Marsh that one of his own students, Max Schlosser, was 'pleased to go to America,' and that another budding paleontologist, George Baur, would probably join him. When these two young Germans

684 Schuchert; LeVene: O. C. Marsh, p. 293.

685 Despite his German name, Harger was born in the US and had no special ties to the German scientific community. Besides that, and though he was very well read and an excellent student, Marsh did not allow him to publish about the specimens of the Museum. Schuchert; LeVene: O. C. Marsh, pp. 298–299.

686 Schuchert; LeVene: O. C. Marsh, pp. 297–302.

> reached the Museum in the spring of 1884, they found a third, Otto Meyer, already on the ground.[687]

The institutional influence of German scientific practices on US-American higher education, including the adoption of "Privatdozenten," is the subject of chapter 8. Lanham also writes that Marsh's experiences in Germany directly led to the employment of his German assistants:

> Remembering the Prussian laboratories from his student days in Europe, where a single professor ruled as undisputed autocrat over assistants who loved to be subservient to capricious and arbitrary authority, Marsh imported three laboratory assistants from Germany – Max Schlosser, Otto Meyer, and Georg Baur. Marsh bungled the job. Two of the three left him soon after their arrival, and the other, Georg Baur, was a monumentally bad choice for an assistant, since he came from a long line of aristocratic German professors. Baur negotiated an agreement with Marsh to be allowed to publish under his own name, and during the six years he was at Yale published seventy-five papers. The only hold that Marsh had over him was to keep him in a state of carefully modulated poverty. Like a coal miner in a company town, Baur was continually in debt to his employer, and it took a great deal of maneuvering for him to get clear of Marsh and find another job. One of his final publications based on his experiences in the Yale laboratory was an exposé of Marsh published in the *American Naturalist*. Convinced of his own superiority over Marsh, he tore the reputation of his one-time employer to shreds in a paper that is perhaps unique in American scientific literature.[688]

This chapter details the careers of Baur, Schlosser, Meyer, and their changing relations with Marsh. Since Baur and Schlosser were introduced to Marsh by Zittel, who continued to be a broker in the paleontological scientific network, the first part of this chapter is dedicated to Zittel.

---

687 Schuchert; LeVene: O. C. Marsh, p. 302.

688 Lanham: The Bone Hunters, p. 247.

6.1 Karl Alfred von Zittel

There is no extensive biography of Zittel. Records of his life are incomplete because most of Zittel's correspondence fell victim to the bombs of the Second World War. The only comprehensive account of his life was published in Munich in 1989 by paleontologist Helmut Mayr, commemorating Zittel's 150th birthday.[689] Beginning in 1856 Zittel studied science with an emphasis on geology and paleontology at Heidelberg. He was educated by Blum and Bunsen, similarly to Marsh, who visited Heidelberg seven years later (chapter 3. 2.). Zittel was granted a doctorate in 1860 and sought to complete his geological education in Vienna afterwards. Before he could move there, Bronn introduced Zittel to a US American named Dale, who lived in Paris and planned to employ a travel companion for his fifteen-year old son, Thomas Nelson Dale (1845–1937). Zittel accompanied Thomas Nelson Dale and another American named Norman Sprang on a journey through Scandinavia, where Zittel collected various fossils and minerals in addition to his social duties. Thomas Nelson Dale seemed to be impressed by Zittel's geological knowledge and went on to become a geologist himself.[690] After he studied the paleontological collection of the Sorbonne in Paris in 1861, Zittel resumed his studies in Vienna in 1862, where he was employed at the Imperial Mineralogical Cabinet ("Mineralogisches Hofkabinett."). One year later he was habilitated (qualified as a professor). After serving two years as a professor for mineralogy and geology at the technical school of Karlsruhe ("Polytechnikum zu Karlsruhe"), Zittel moved to Munich and became a professor for paleontology and the director of the paleontological collection.[691] In 1869 Zittel became a member of the Bavarian Academy of Sciences and Humanities ("Bayerische Akademie der Wissenschaften") and in 1870 he joined the Geological Society of Germany ("Deutschen Geologischen Gesellschaft"). Roemer and Beyrich, both acquaintances of Marsh, had advocated Zittel's membership in the Geological Society, making the nineteenth-century scientific world seem rather small.[692] In 1880 Max Schlosser, a student of Zittel's, was granted a doctorate in paleontology. In the same year Zittel started to correspond with O. C. Marsh, thanks to Thomas Nelson Dale, who had established contact with the American professor on behalf of Zittel. Marsh was interested in trading the casts of the dinosaurs of the American

689 Helmut Mayr: Karl Alfred von Zittel zum 150jährigen Geburtstag (25.9.1839–5.1.1904), in: Mitteilungen der Bayerischen Staatssammlung für Paläontologie und histor. Geologie, vol. 29 (Dec. 1989), pp. 7–51.

690 For the later scientific achievements and career of Dale, see: Peggy Champlin: Raphael Pumpelly. Gentleman Geologist of the Gilded Age, Tuscaloosa, AL 1994, pp. 150–151.

691 Mayr: Karl Alfred von Zittel, pp. 10–19.

692 Mayr: Karl Alfred von Zittel, p. 22.

West for casts of European paleontological specimens found in Pikermi, Greece.[693] In 1883, thanks to the intercession of Zittel, Marsh was elected a member of the Bavarian Academy of Sciences. Later that year Zittel embarked on a journey to the United States, where he met Agassiz in Boston and Marsh in New Haven. He was then invited by Henry Villard (1835–1900), president of the Northern Pacific Railway, to visit the Yellowstone National Park, who would later also be named as a reference in the job application Meyer sent to Marsh.[694] The natural landscape of Yellowstone left a lasting impression on the Bavarian paleontologist: in 1885 he published a short pamphlet, detailing his experiences at the “Yellowstone-Wonderland” (“Das Wunderland am Yellowstone”). There he praised the grandiose nature of the American landscape, where the history of nature could be studied in “broader strokes” than in Europe. Zittel states that nothing in Europe could compare to the “immeasurable prairies,” the “terrific beauty of the Columbia River,” the “adventurous Badlands,” the “grove of the gods” that was Colorado, the “fantastic terraces of the rocky deserts of Arizona,” or the Grand Canyon.[695] After his return to Munich, Zittel arranged for Max Schlosser and Georg Baur to become assistants to Marsh in New Haven. After he returned to Germany, Schlosser became Zittel’s assistant in 1886. Dana, Agassiz, and Cope were elected corresponding members of the Bavarian Scientific Academy during the same year.[696] Zittel attended the International Geological Congress in Washington D.C. in 1891. He was elected vice-president of the congress alongside Carl Hermann Credner (1841–1913),[697] and visited Yellowstone again.[698] Between 1876 and 1893 Zittel’s textbook on paleontology (“Handbuch der Palaeontologie”) was published in five volumes. Under Zittel’s guidance Munich became one of the most important centers of paleontology.[699]

On over 4000 pages the textbooks detailed all then known aspects of paleontology and proved to be an international success. He expanded on the textbooks with a book

693 Mayr: Karl Alfred von Zittel, p. 28.

694 On Villard and his career see: Thomas Childs Cochran: Railroad Leaders, 1845–1890. The Business Mind in Action, Cambridge, MA 1953, pp. 49–51.

695 “In der neuen Welt ist die Geschichte der Urzeit in grober Fractur, in Europa in zierlicher mit Schnörkeln und Arabesken überladener Miniaturschrift aufgezeichnet […] Wo finden wir in Europa eine Ebene, die sich den unermeßlichen Prärien des amerikanischen Westens vergleichen ließe; wo einen Strom der an grandioser Schönheit den Columbiafluß überträfe? Auch den abenteuerlichen Landschaften der Bad-lands von Montana und Wyoming, dem Götterhain von Colorado, den phantastisch gegliederten Terrassen der Felswüsten von Arizona und gar den in die Ebene eingeschnittenen Riesenschluchten des Gran Cañon [sic!] in Colorado haben wir in Europa nichts Ebenbürtiges zur Seite zu stellen.” Karl Alfred von Zittel: Das Wunderland am Yellowstone, Berlin 1885, p. 1.

696 Mayr: Karl Alfred von Zittel, p. 29.

697 Credner was professor for geology in Leipzig, corresponded with Marsh, and taught Meyer.

698 Mayr: Karl Alfred von Zittel, p. 30.

699 Tamborini: “If the Americans Can Do It, So Can We”, pp. 230–233.

on the main features of paleontology ("Grundzüge der Palaeontologie") in 1895. Zittel had written a comprehensive abstract on more than 100 years of geological and paleontological discoveries, encompassing the endeavors of countless scientists, working on all continents. The English translation of Zittel's textbook (published in three volumes between 1899 and 1925) became the standard reference textbook for paleontological education in the United States. When Zittel's book on the history of paleontology ("Geschichte der Paläontologie")[700] was published in 1899, it was the first comprehensive history of this scientific discipline. Zittel was a member of the Boston Society of Natural History, the National Academy of Sciences of the United States (NAS), and the New York and Philadelphia Academies of Science.[701]

A very flattering obituary called attention to the fact that in Zittel's mind science was furthered by international cooperation, in which he himself readily participated. He is quoted to have said that the twentieth century would be the century of international scientific cooperation and that the nineteenth century unfortunately had seen the rise of "nationalistic" and "confessional" science;[702] a curious verdict given he had already formed extensive international connections within the scientific world during the nineteenth century. Henry Fairfield Osborn, who had been taught by Zittel during his stay in Munich in 1886, wrote in another obituary that Zittel was a truly international scientist and that he may have done more for paleontology than any other scientist of the nineteenth century, first and foremost through his coherent and comprehensive textbooks:

> Although a German by birth, Professor von Zittel belonged to every country, and through his remarkable work 'Handbuch der Palaeontologie' his influence extended everywhere. It is probably not an exaggeration to say that he did more for the promotion and diffusion of paleontology than any other single man who lived during the nineteenth century [...] Immediately after the completion of this work the author began the preparation of a condensed treatise upon the whole subject, entitled 'Grundzüge der Palaeontologie,' [...] We mention this monumental work first, because it was chiefly through this that the

700 Karl Alfred von Zittel: Geschichte der Geologie und Paläontologie bis Ende des 19. Jahrhunderts, Munich 1899.

701 For a full bibliography of Zittel, see: Mayr: Karl Alfred von Zittel, pp. 40–45, 49. For a more detailed and contemporary review of the quality and scale of Zittel's books, see: August Rothpletz: Gedächtnisrede auf Karl Alfred von Zittel, gehalten in der öffentlichen Sitzung der K. B. Akademie der Wissenschaften zu München zur Feier ihres 146. Stiftungstages am 15. März 1905, Munich 1905.

702 "Das 19. Jahrhundert hat die unglücklichen Begriffe von nationaler und konfessioneller Wissenschaft hervorgebracht; die Tätigkeit der Akademien im 20. Jahrhundert steht unzweifelhaft unter dem Zeichen der Internationalität". Karl Theodor von Heigel: Zum Andenken an Karl von Zittel. Rede in der öffentlichen Festsitzung der K.B. Akademie der Wissenschaften am 14. März 1904, Munich 1904.

> influence of von Zittel was exerted. The prodigious progress of paleontology in the nineteenth century was scattered through thousands of monographs and special papers, a hopeless labyrinth to the student, and an extremely difficult field even to the expert investigator; it had ceased to be possible to gain a perspective view of the whole subject, not to speak of the difficulty of mastering the details. With remarkable clearness and fullness, with imperial justice to workers in every country, with especially warm appreciation of the work done in America, von Zittel devoted himself for twenty years to this great task.[703]

6.1.1 Zittel and Marsh

On February 20, 1880, Zittel wrote his first letter to Marsh in which he stated that Thomas Nelson Dale had informed him that Marsh was looking for casts of European vertebrates, especially for the cast of a *compsognathus*[704] specimen, which had been discovered in the lithographic limestone of Bavaria. Zittel stated he would commission a cast of *compsognathus* and would furthermore include a list of other specimens that could be of interest to Marsh in his letter. Zittel then asked if Marsh could send him some American specimens in return. He praised the "marvelous paleontological treasures of the American West" which Marsh had discovered. Said treasures were accessible to the Europeans exclusively through Marsh's and Cope's descriptions and depictions. Zittel then elaborated that he would love to remedy the situation by acquiring American specimens, or at least their casts.[705] In the next letter, written on May 8 of that same year, Zittel thanked Marsh for sending him a catalogue of the specimens the Peabody Museum would like to trade, and ordered some of them for his work on the handbook of paleontology. He furthermore wrote that the *compsognathus* cast was in the care of a Mr. Trübner of London and on its way to New Haven.[706]

---

703 Henry Fairfield Osborn: Karl Alfred von Zittel, in: Science, new ser., vol. 19, no. 474 (Jan. 29, 1904), pp. 186–188. Quote on pages 186–187

704 Compsognathus is a small bipedal carnivorous dinosaur, probably a close relative of archaeopteryx, and a minor cast member of the "Jurassic Park" movies.

705 "Die wunderbaren, von Ihnen entdeckten palaeontologischen Schaetze des Amerikanischen Westen[s] sind uns Europaeern [sic!] lediglich nur durch Ihre und Prof Cope's Beschreibungen und Abbildungen bekannt und diese erwecken den dringendsten Wunsch, etwas von diesen wunderbaren Dingen auch in Natura oder doch wenigstens im Abguss zu besitzen." Karl von Zittel, Munich to Othniel Charles Marsh, New Haven, CT, 20 February 1880, MS 343, Series I. Correspondence, Box 36, Folder 1570.

706 Karl von Zittel, Munich to Othniel Charles Marsh, New Haven, CT, 8 May 1880, MS 343, Series I. Correspondence, Box 36, Folder 1570.

Zittel wrote the next letter to Marsh in Newport, Connecticut on his 1883 tour through the United States. Dale told Zittel that Marsh had returned to New Haven and would be delighted to take him in as a guest. He would give him the exact information of his arrival after his trip to Boston, where he would meet Agassiz. Note that Zittel wrote this letter, and this letter alone, in English:

> I just hear by my friend M. Nelson Dale, that you have arrived in Newhaven [sic!] and that you will have the kindness to offer to myself your hospitality while my staying in Newhaven [sic!]. I intend to go to Boston this afternoon and to stay with Prof. Agassiz until Friday evening. If this is no inconvenience for yourself I wish to come to Newhaven [sic!] Saturday morning or Friday evening. I would give you by telegraph the exact time of my arrival.[707]

Zittel had a comfortable stay in New Haven. During this time Marsh must have informed him that he would like to employ some German assistants in New Haven and that he would like to purchase a *pterodactyl* skeleton for sale in Bavaria. A few weeks later, and back in Munich, Zittel wrote to Marsh, thanked him enthusiastically for the "extraordinary hospitality" and assured Marsh that the visit to New Haven ranked amongst "the most pleasant memories" of his journey. He would be hard pressed to name another place of higher "stimulation" and "instruction". He furthermore had talked to Max Schlosser, who would be willing to move to the United States and was anticipating a letter from Marsh with further details about his employment. Schlosser had published various scientific papers that he would gladly send to Marsh. At this moment he had had no opportunity to talk to Georg Baur, the assistant of Professor Kupffer (1829–1902),[708] but Kupffer stated that Baur would most likely also be happy to become Marsh's assistant at New Haven for the next two or three years in order to enhance his education, and that Baur had published an essay on bird limbs that he would send to Marsh as soon as possible. Zittel promised furthermore to ask around for a young German paleontologist specializing in invertebrates and to send him to Marsh as well. To the author's knowledge no third student of Zittel's was sent to Marsh, maybe this position was filled by Otto Meyer (see below). Zittel then promised Marsh to look into the acquisition of the desired *pterodactyl*.[709] Later it turned out that the *pterodactyl* was to be sold for an outra-

707 Karl von Zittel, Newport, CT to Othniel Charles Marsh, New Haven, CT, 10 October 1883, MS 343, Series I. Correspondence, Box 36, Folder 1570.

708 Karl Wilhelm von Kupffer was professor of anatomy at the University of Munich from 1880 to 1901.

709 "Nach meiner glücklichen Rückkehr draengt es mich Ihnen meinen innigsten Dank auszusprechen für die ausserordentliche Gastfreundschaft, welche ich bei Ihnen gefunden habe. Die Tage in New Haven gehören zu den schoensten Erinnerungen meiner amerikanischen Reise und kaum wüsste ich einen anderen Ort zu nennen wo ich eine reichere Fülle von Anregung und Belehrung gefunden hätte [...] mit Herrn Dr Schlosser habe ich gesprochen. Es ist gerne bereit, nach Amerika überzusiedeln und wartet

geous sum of money (4,000 marks). Zittel blamed a man named Ernst Otto Häberlein (1819–1896),[710] who functioned as a middleman between the owner of the fossil (a quarry owner named Fritz Ehrenberger) and the potential buyers, for the racketeering. Zittel included letters detailing the business offer made to Marsh through Schlosser and Zittel in his letter.[711] The matter was further complicated by the fact that Alexander Agassiz had made up his mind to purchase the specimen for his museum of comparative zoology at Harvard. Häberlein had offered the *pterodactyl* to Agassiz for £ 170 in September, before Zittel made an offer on Marsh’s behalf. Zittel stated that he would have to comply with Agassiz’ request, but he would try to find another *pterodactyl* for Marsh.[712] On a more personal note Zittel sent a photo of himself to Marsh as a token of his “grateful adoration.” Zittel was returning a favor, for it can be deduced from his letter that Marsh had sent him a portrait photograph of himself sometime prior. Zittel furthermore communicated that Schlosser and Baur were planning their immediate professional futures and would be very thankful to hear from Marsh about his employment offer.[713] Details

---

nur auf Ihren Brief, worin Sie ihm die Bedingungen mittheilen, unter denen er engagiert werden soll […] Herrn Dr Bauer habe ich noch nicht gesehen, wohl aber Professor Kupffer, bei dem er eine Aßisteneten Stelle bekleidet. Letzterer glaubt, daß Bauer gleichfalls gerne für 2–3 Jahre nach Amerika gehen würde, um seine Kenntnisse daselbst zu erweitern […] Für einen jungen Palaeontologen, welcher mi den Invertebraten vertrat ist, glaube ich Ihnen im Verlaufe des Winters sicher sorgen zu können, so daß Sie bis 1ten April auf 3 meiner Schüler rechnen können. In nächster Woche werde ich wegen Ankauf des Pterodactylus Schritte thun und Ihnen das Resulthat meiner Handlungen mittheilen.” Karl von Zittel, Munich to Othniel Charles Marsh, New Haven, CT, 2 November 1883, MS 343, Series I. Correspondence, Box 36, Folder 1570.

710 Ernst Häberlein was a pharmacist from Pfaffenhofen, Bavaria. He continued the business of his father Carl Friedrich Häberlein (1787–1871), who had bought fossils directly from the quarry owner of Solnhofen and sold them subsequently to the highest bidder. Among these fossils was the first archaeopteryx skeleton ever found (BMNH 37001), which Carl Häberlein sold to the Natural History Museum in London.

711 Fritz Ehrenberger, Pappenheim, to Max Schlosser, Munich, 12 November 1883; Ernst Häberlein, Pappenheim, to Max Schlosser, Munich, 13 November 1883, attached to: Karl von Zittel, Munich to Othniel Charles Marsh, New Haven, CT, 17 November 1883, MS 343, Series I. Correspondence, Box 36, Folder 1570.

712 “Herr Ehrenberger, der Besitzer des Pterodactylus (Haeberelein ist nur der Zwischenhaendler) beansprucht laut beiliegendem Brief an Dr Schlosser den enormen Preis von 4000 M und ist noch zweifelhaft, ob der das Stück isoliert verkaufen wolle. Seinen in Aussicht gestellten mündlichen Verhandlungen wird heute durch einen Brief von Prof. Alexander Agassiz ein Ende gemacht. Haeberlein hatte dem Museum of comparative Zooglogy den Pterodactylus am 29ten September für £ 170 angeboten und da das Museum keinen Repräsentanten der Flugsaurier besitzt, so schreibt mir Prof. Agassiz soeben, dass er das Anerbieten Haeberlein’s anzunehmen geneigt sei und bittet mich ein in diesem Sinne abgefasstes und meinem Briefe beigefügtes Schreiben an Haeberlein gelangen zu lassen. Ich kann loyaler Weise nichts Anderes thun, als dem Wunsche Prof. Agassiz‘ zu entsprechen, werde mich aber bemühen Ihnen sobald als möglich einen Pterodactylus zu verschaffen, dessen Preis nicht durch Haeberleins Vermittlung zu so exorbitanter Höhe hinauf geschraubt wurde.” Zittel to Marsh, 17 November 1883.

713 “Erst heute bin ich in der Lage Ihnen meine Photographie in Erwiederung [sic!] der Ihrigen zu übersenden. Mögen Sie dieselbe als Beweis meiner dankbaren Verehrung annehmen. Meinen Brief bezüglich der Herrn Dr Schlosser und Baur haben Sie wohl erhalten. Da beide Herrn in der Lage sind, sich in den nächsten Monaten über die Gestaltung ihrer Zukunft zu entscheiden, so wäre ich Ihnen für

on the job offer arrived attached to a letter to Zittel in February 1884 (Marsh had written the letter on January 30). Zittel relayed the information to Schlosser and Baur and replied to Marsh that the two young scientists would accept the offer and had agreed to become Marsh's assistants for the next three years. He then asked Marsh whether the date of the commencement of employment could be postponed to May 1, because Schlosser and Baur were in the process of publishing scientific essays. Zittel proposed that the assistants would be of great service to Marsh's "grand examinations." He praised their qualifications, stating that Baur had a "sound scientific mind" and Schlosser was a "scrupulous, dependable, and diligent worker". Finally, Zittel stressed his involvement in the procedure, maintaining that this deal would further his professional relationship with Marsh.[714] It seems that the relationship between Marsh and Zittel indeed flourished during the following decade, for in his next letter, preserved at Yale (dated September 8 1895), Zittel no longer called Marsh "Dear Sir" ("Hochgeehrter Herr") or "Dear professor Marsh" ("Lieber Professor Marsh"), as he did in all the letters referenced above, but called him more endearingly his "Most Revered Friend" ("Hochverehrter Freund"). Marsh must have sent a letter to Zittel, informing his friend that he was to embark on a journey to Europe and would like to meet Zittel in Leyden or Stuttgart. Zittel replied that he would very much like to meet Marsh, but was accompanying his sick wife to a health resort in Baden-Baden; this would greatly complicate the meeting.[715] Indeed they did not meet on Marsh's journey, which Zittel greatly regretted in a letter written in October of the same year. In this letter he informed Marsh that he was working on translating his "Grundzüge der Palaeontologie" into English and would have loved to discuss a few details of the translation with Marsh. In the meantime he had received Marsh's "wonderful" essay on the classification of dinosaurs, which he hoped would be of use to his handbook. Zittel delighted in Marsh's efforts to make the treasures of the Peabody Museum publicly accessible through his publications. He then called attention to the

---

eine baldige Nachricht in dieser Sache zu Danke verpflichtet." Karl von Zittel, Munich to Othniel Charles Marsh, New Haven, CT, 7 January 1884, MS 343, Series I. Correspondence, Box 36, Folder 1570.

714 "Ihr Schreiben vom 30ten Jan. kam vor einigen Tagen hier an. Ich habe die beiden eingeschlossenen Briefe an Dr Schlosser und Dr Baur übergeben und mit denselben gesprochen. Beide Herrn sind bereit Ihr Anbieten anzunehmen und für 3 Jahre als Assistenten bei Ihnen einzutreten. Sie bitten jedoch den Termin ihres Eintrittes auf den 1ten Mai zu verschieben, da sowohl Dr Schlosser als auch Dr Baur Abhandlungen vollendet haben, deren Druck erst bis Mitte April abgeschlossen sein wird [...] Ich halte ihn [Baur] für einen gut veranlagten wissenschaftlichen Kopf, Dr Schlosser für einen sehr gewissenhaften zuverlässigen und sehr fleissigen Arbeiter. Es würde mir eine besondere Freude sein, wenn die beiden neuen Assistenten Ihren Wünschen und Anforderungen genügten und wenn sich dadurch unsere in so erfreulicher Weise angeknüpften Beziehungen noch lebhafter entwickelten." Karl von Zittel, Munich to Othniel Charles Marsh, New Haven, CT, 23 February 1884, MS 343, Series I. Correspondence, Box 36, Folder 1570.

715 Karl von Zittel, Munich to Othniel Charles Marsh, New Haven, CT, 8 September 1895, MS 343, Series I. Correspondence, Box 36, Folder 1570.

"wonderful discoveries that were made in America" and called the European discoveries "meager," likening European scientists to "beggars" in this regard.[716]

In 1901 Zittel learned from Edward S. Dana (1849–1935), the son of James Dana, that he was to receive an honorary doctorate from the "venerable" and "distinguished" Yale University. Zittel replied that his failing health and other commitments would prohibit him from accepting the doctorate in person. He nonetheless assured Dana that he remembered New Haven and his stay at Marsh's home (and a meeting with James Dana, whom he called America's greatest geologist) very fondly.[717]

6.1.2 Zittel and Osborn

In 1883, while Zittel was traveling through America, he was invited by Henry Fairfield Osborn to come to Princeton. Osborn had written him a letter on October 8;[718] Zittel responded on October 14, thanking the former for "amiably encouraging"[719] his visit, but told him that he would not be able to come to Princeton. He had spent more time in Cambridge and New Haven than he had initially planned and had no time to come

---

716 "Ich hätte gerne Verschiedenes bezüglich der Uebersetzung meiner Grundzüge der Palaeontologie mit Ihnen besprochen [...] Inzwischen habe ich Ihre wundervolle Abhandlung über die Classification der Dinosauria erhalten und vor wenigen Tagen gelangte ich auch in Besitz Ihrer schoenen Restauration ausgestorbener Wirbelthiere. Von beiden Publicationen hoffe ich für mein Lehrbuch Nutzen ziehen zu können. Ich bewundere Ihre unermüdliche Thätigkeit und den unerschöpflichen Reichthum Ihres Museums, dessen Schätze Sie nun allmählich zum Gesammtgut der wissenschaftlichen Welt machen. Wenn ich mit den wunderbaren Entdeckungen in Amerika die dürftigen neuen funde in Europa vergleiche, so sehe ich recht deutlich, welche Bettler wir Ihnen gegenüber sind." Karl von Zittel, Munich to Othniel Charles Marsh, New Haven, CT, 22 December 1895, MS 343, Series I. Correspondence, Box 36, Folder 1570.

717 "Mit freudiger Ueberraschung erfahre ich aus Ihrem Schreiben vom 23ten Februar, dass mir die alte und berühmte Universität in New Haven die hohe Ehre zugedacht hat, mich bei Gelegenheit ihren 200 jährigen Jubilaeums zum Doctor der Rechte zu ernennen [...] die Tage, welche ich in Ihrer schoenen Stadt bei meinem verstorbenen Freunde Professor O. Marsh zugebracht habe, gehören zu meinen schönsten Erinnerungen und mit freude denke ich daran zurück, dass es mir damals vergönnt war in Ihrem Vater den grössten Geologen America's kennen zu lernen. Leider muss ich aber, trotz Ihrer lockenden Einladung darauf verzichten, meine Beziehungen zum Yale College in diesem October persönlich zu erneuern, da mich um diese Zeit amtliche Pflichten an München fesseln und überdies meine Gesundheit in diesem Jahr besondere Schonung erheischt." Karl von Zittel, Munich to Edward S. Dana, New Haven, CT, 28 March 1901, Yale University, Sterling Memorial Library, Dana Family papers (MS 164), Series II. Edward Salisbury Dana, Box 21, Folder 222.

718 Osborn's letter was not preserved, but Zittel expresses his gratitude for the letter in his response, see below.

719 "Zu meinem lebhaften Bedauern muss ich meinen beabsichtigten Besuch in Princeton zu dem Sie mich durch Ihren freundlichen Brief vom 8ten in so liebenswürdiger Weise ermuntert hatten, wegen mangelnder Zeit aufgeben." Karl von Zittel, New York to Henry Fairfield Osborn, Princeton, NJ, 14 October 1883, American Museum of Natural History, VPA 1/108, General Correspondence Yo–Z, von ZITTEL, Prof. KARL A. Munich 1883–1904, Folder 47.

and visit Osborn. But he had had such a good time in America that he hoped he would one day return to the country where science was "ascending so rapidly" ("in wissenschaftler Hinsicht so rasch aufsteigendem Lande"), and then would find the time to spend a few days in Princeton.

Osborn was born to a wealthy and prominent New York family, his father William Osborn (1820–1894) being a well-known and wealthy figure in the shipping and railroad business.[720] Within this framework of New York high society, Henry was taught to cultivate the social connections necessary for his business ventures and publicity. Osborn entered the college of New Jersey (later Princeton) in 1873, a socially and intellectually conservative environment. Ancient extinct life was brought to Princeton in the form of a cast of Leidy's and Hawkins's reconstruction of *hadrosaurus* and seventeen watercolor paintings, depicting lost fauna and flora. Osborn and his fellow student – and later colleague – William Berryman Scott (1858–1947) started their scientific careers in this environment, partaking in Princeton's scientific expedition of 1877, and entering Princeton's first postgraduate class. Scott and Osborn remained at Princeton and joined the faculty in 1881, Scott teaching geology, Osborn comparative anatomy, often collaborating in their teachings. In 1885 Osborn turned his attention to paleontology, establishing a close working relationship with Cope. In that same year he ventured to Munich where he studied under Zittel. During his stay in Munich, he also befriended Max Schlosser, Zittel's assistant. He invited the taxidermist Rudolph Weber to Princeton, establishing another German-American network. Beginning in 1885, Osborn intensified his contact with Cope, becoming his disciple, and sided with him during the "Bone Wars." Cope employed Osborn and Scott in his efforts to ally with Marsh's frustrated assistants.[721] While Scott traveled to New Haven to interview the assistants, Osborn learned from Schlosser in Munich what he could about Marsh's supposed failings and skullduggeries. From Zittel and Baur's parents Osborn learned about Baur's misery, his financial situation and dissatisfaction with his situation in New Haven.[722] It seems that Cope willfully instigated the discontent of Marsh's assistants, visited the laboratory, met with them conspicuously, and tried to fathom what Marsh was working on. As a reaction Marsh forbade visitor access to the facility. Cope had convinced four of Marsh's associates to testify against their former employer in 1885, when he involved Osborn in his plot. Some of the material

---

720 On William Osborn's railroad career see: Cochran: Railroad Leaders, pp. 44–46.

721 Note that Rainger states that Zittel sent Meyer along with Baur and Schlosser to New Haven, conveying that Zittel had send Meyer directly to Marsh. While Zittel is an all-important lynchpin in the Meyer-Marsh relationship, Meyer joined Marsh on his own accord and not directly via Zittel, see below.

722 Ronald Rainger: Vertebrate Paleontology as Biology. Henry Fairfield Osborn and the American Museum of Natural History, in: Ronald Rainger et al. (eds.): The American Development of Biology, pbk. ed., New Brunswick, NJ 1991 (orig. publ. 1988), pp. 219–256.

gathered in this manner found its way to the pages of the “New York Herald” in 1890. Statements of Marsh’s assistants, mainly Williston and Baur, were published by the “New York Herald.”[723]

Osborn returned to New York in 1890. Through his family he had ties to the wealthy New York elite. These contacts helped Osborn to lobby successfully for the establishment of a department of biology at Columbia College, which was just now growing into a real university. Besides various other responsibilities, Osborn taught vertebrate morphology and paleontology at Columbia. A position at the American Museum of Natural History (which was also very much a product of the philanthropic elite of the city) promised great opportunities. To further the willingness of the museum to acquire and display paleontological specimens, Osborn invested heavily – and out of his own pocket – in its paleontological department and the expeditions that were to provide the museum with new fossils. Osborn became a trustee and the vice president of the museum in 1901; in 1908 he was appointed president of the museum.[724]

> Osborn was a wealthy man who was fully convinced of his own self-importance and often treated others in the museum, including scientists, in a condescending manner. Osborn held his own scientific work in high regard. As indicated, he took virtually no part in the mundane aspects of vertebrate paleontology; others did the collecting, cleaning, and preparing of fossil specimen.[725]

To make a name for himself in the field of paleontology, Osborn became a competitor of Marsh’s, who dominated the field in the 1880s. This inspired his efforts to incite Marsh’s assistants against the professor.

> Nevertheless Osborn, unlike Marsh, was able to maintain a large and flourishing program in vertebrate paleontology. At Yale, Marsh’s program had broken down as a result of dissatisfaction among his assistants, continued problems from Cope and Osborn, and political and economic troubles associated with the Geological Survey. Osborn’s program never experienced such problems. In part that was true because of the vast financial and political resources at Osborn’s disposal. Through his connections and ability to gain support from a network of influential and wealthy patrons, Osborn was able to almost continuously expand the size and scope of his program.[726]

---

723 Lanham: The Bone Hunters, pp. 247–256.

724 Rainger: An Agenda for Antiquity, pp. 44–66.

725 Rainger: An Agenda for Antiquity, p. 74.

726 Rainger: An Agenda for Antiquity, p. 77.

As had been the case with the Cope-Marsh dispute, professional disagreements mixed with personal ones as the Osborn-Marsh feud grew. Osborn soon tried to hire some of Marsh's fossil hunters and started to undermine his position at the USGS in 1892, which finally led to Marsh's resignation from the USGS. Marsh's downfall in the 1890s proved to be a boon for Osborn, who hired some of Marsh's collectors and gained much public attention for his program. Osborn employed Barnum Brown, who revisited the fossil hunting grounds of Como Bluff, exhuming many more specimens which had been overlooked by Marsh's teams and claiming the fossil beds of Como Bluff for the AMNH in the process. In 1902 Brown discovered the first fragments of *tyrannosaurus rex*, which was to become the uncontested super star amongst all extinct life (after Brown had found an almost complete *tyrannosaurus* skull in 1906), making the AMNH the leading institution in the collecting of dinosaur bones.[727] Osborn might have realized what the previous generation of paleontologists, Marsh, Cope, and Leidy, had not: the enormous popularity of dinosaur exhibitions promised profit and public prestige, meaning the golden future of vertebrate paleontology (and its funding) lay with public education, not academic discourse.

Unfortunately, no letters detailing Osborn's journey to Munich in 1886 have been preserved, neither of Osborn to Zittel, nor vice versa. However, it can be deduced that they met and became friends for Zittel addressed Osborn in his letter of October 1883 as "Dear Sir" ("Geehrter Herr"), but in his subsequent letters Zittel called him "Revered Friend" ("Verehrter Freund"), indicating they had definitely met in person and for a longer period of time. In a letter written on July 28, 1890, Zittel thanked his friend for his article on "The Mammalia of the Uinta Formation,"[728] which shed light on some of the mammalian fauna of the American West. Zittel praised Osborn and William B. Scott, the co-author of the article, for bringing some orderliness to a field which had been in "sorrowful confusion" due to the "discrepancies" between Cope and Marsh ("die jammervolle Confusion [...] welche durch den Zwiespalt zwischen Marsh u. Cope entstanden ist") and the resulting overhasty naming of the species. Zittel furthermore stated that only Osborn and Scott could have done this, because they were not only experts on the mammals of America but also Europe and could thusly provide the required context. He would promptly incorporate the discoveries of the article in his handbook. Being able to think and write internationally and to acquire location-independent knowledge of the species were great assets, and sometimes the prerequisite for paleontology.[729] In a letter of March 27, 1891, Zittel again thanked Osborn

727 Rainger: An Agenda for Antiquity, pp. 80–84, 94–95.

728 William B. Scott; Henry Fairfield Osborn: The Mammalia of the Uinta Formation, in: Transactions of the American Philosophical Society, new ser., vol. 16, no. 3 (1890), pp. 461–572.

729 "Für Ihre Mamalia of the Uinta Formation meinen allerbesten Dank. Es ist eine wahre Erlösung, dass Sie sich im Verein mit Prof. Scott entschlossen haben, die jammervolle Confusion bei den fossi-

for the clarification on Marsh's "Cretaceous Mammalia,"[730] praised Osborn's professional authority on the matter and promised to update his handbook in accordance with the new information ("Nun kann ich mich auf eine Autorität stützen und diesen Theil meines Handbuchs gründlich reformieren."). He stated that he had become a "thankful student" of Osborn's because the latter had brought light to some aspects even the efforts of Cope and Marsh had left in the dark ("bin [...] Ihr dankbarer Schüler geworden. Wie vieles haben Sie jetzt trefflich aufgeklärt, was trotz Cope und Marsh für den ferner Stehenden dunkel geblieben war."). He then told Osborn that he was in the process of finishing the corresponding chapters in his textbook and would then prepare for his journey to Washington to attend the international geological congress later that year; that he would afterwards very much like to study the various American collections of extinct mammals, and was especially looking forward to meeting Scott and Osborn (and was hoping that the two could enlighten him on some open details on the American mammals); this time he would make time for a visit to Princeton.[731] Zittel did visit Osborn in Princeton, as can be surmised from a letter he sent to Osborn on May 4, 1892, in which he refers to his "unforgettable" visit. Likely Osborn had promised Zittel his guidance on the American mammals for the handbook when the two met in Princeton, for Zittel now enlisted Osborn to review his drafts on the American mammals he had included in the letter. He had furthermore employed the help of Marsh, who sent him an illustration depicting *protoceras* and clarifying the evolutionary position of *meniscotherium*.[732] In the same letter he wrote that he felt elat-

len Säugethieren N. America's, welche durch den Zwiespalt zwischen Marsh u. Cope entstanden ist, aufzuklaeren. Niemand hätte diese Aufgabe besser lösen können, als Sie und Prof Scott, da ihnen nicht nur die amerikanischen sondern auch alle Europäischen Gattungen und Arten auf das genaueste bekannt sind. Ich habe in den letzten Wochen mit der Bearbeitung der Säugethiere für mein Handbuch begonnen und wie willkommen mir dabei Ihre neueste Abhandlung ist, brauche ich Ihnen wohl nicht erst zu versichern. Mit der Bitte mich Ihrer Frau Gemahlin und Prof Scott empfehlen zu wollen, verbleibt" Karl von Zittel, Munich to Henry Fairfield Osborn, Princeton, NJ, 28 July 1890, VPA 1/108, General Correspondence Yo–Z, von ZITTEL, Prof. KARL A. Munich 1883–1904, Folder 47.

730 Most likely Zittel was referring to: Henry Fairfield Osborn: A Review of the "Discovery of the Cretaceous Mammalia", in: American Naturalist, vol. 25, no. 295 (Jul. 1891), pp. 595–611.

731 "so dass ich mit leidlich guter Vorbereitung nach Washington reisen und nach dem Congress die verschiedenen herrlichen Sammlungen Amerikanischer Mammalia studieren kann. Ich freue mich dabei ganz besonders auf eine Zusammenkunft und gemütliche Aussprache mit Ihnen und mit Prof. Scott und hoffe, daß ich dabei über eine Menge von zweifeln, welche ich auf dem Herzen habe, Aufklaerung und Belehrung erhalte. Hoffentlich sehen wir uns in Washington, damit ich dort von Ihnen erfahren kann, ob und zu welcher Zeit ich Sie in Princeton aufsuchen darf." Karl von Zittel, Munich to Henry Fairfield Osborn, Princeton, NJ, 27 March 1891, VPA 1/108, General Correspondence Yo–Z, von ZITTEL, Prof. KARL A. Munich 1883–1904, Folder 47.

732 "Sie hatten während meines Besuches in America und während der mir unvergesslichen Tage in Garrison und Princeton die Güte mir Ihren Beistand für die Klarstellung der Amerikanischen Formen zu versprechen. Sie werden es darum auch nicht als Unbescheidenheit auslegen, wenn ich jetzt von jener Zusage Gebrauch mache und Sie erwirke, etwaige Irrthümer in den Ihnen vorliegenden Fahnen zu cor-

ed that Baur had secured the tenure as professor for paleontology at the University of Chicago, which would allow him to devote himself completely to his scientific work.[733] Undoubtedly, Zittel knew about Baur's discontent with New Haven and Marsh. Apparently Baur had sent some letters describing his misery to Zittel in Munich, asking his old tutor for help (see below). Osborn had complied with Zittel's request, for Zittel conveyed his thanks in a series of letters written between June 8 and November 2, 1892, in which he also discussed further details on the classification of the American mammals.[734] Zittel's completed handbook was greatly appreciated by Osborn, who employed it in his classes at the museum. He furthermore promised to write a review of the handbook for the magazine "Science" (he would not get around to writing the review in 1894, see the letter of February the 12, 1896, below):

> I am taking great pleasure in writing a long review of your Palaeontology for 'Science', at the suggestion of Mr. Scudder of Cambridge.[735] We are using your volume daily in the museum, and find it simply invaluable. It is so richly stored with references. I am also using it with my classes of students. I want to extend my heartiest gratulations to you and upon the splendid manner in which you have completed this work.[736]

In the next preserved letter dated February 12, 1896, Osborn invited Zittel to visit a new exhibition at the museum, apologized since he had still not written the promised review, and stated his intention to exchange duplicates of fossils with Zittel in Munich:

> I shall then not be happy until we have had a visit from you and I can have an opportunity of showing you the arrangement of our collection for study, research, and exhibition. I write especially however, at present to ask whether

---

rigieren […] Von Prof. Marsh habe ich vor einigen Tagen eine gute Abbildung von Protoceras erhalten, sowie die Extremitäten einer neuen Condylarthren Gattung / Hyracops / erhalten, welche die Stellung von Meniscotherium klärt." Karl von Zittel, Munich to Henry Fairfield Osborn, New York, 4 May 1892, VPA 1/108, General Correspondence Yo–Z, von ZITTEL, Prof. KARL A. Munich 1883–1904, Folder 47.

733 "Daß G. Baur an der neuen Universität in Chicago die Professur für Paläontologie erhalten hat, werden Sie ohne Zweifel wißen. Ich freue mich ausserordentlich darüber. Er befindet sich jetzt in einer gesicherten Stellung, welche ihm gestattet ruhig und ohne Ueberstürzung zu Arbeiten und seine bedeutenden Fähigkeiten im Dienste der Wißenschaft zu bethätigen." Zittel to Osborn 4 May 1892.

734 Karl von Zittel, Munich to Henry Fairfield Osborn, New York, 8 June 1892; Karl von Zittel, Munich to Henry Fairfield Osborn, New York, 11 October 1892; Karl von Zittel, Munich to Henry Fairfield Osborn, New York, 2 November 1892, all in: VPA 1/108, General Correspondence Yo–Z, von ZITTEL, Prof. KARL A. Munich 1883–1904, Folder 47.

735 Samuel Hubbard Scudder (1837–1911), paleontologist and pupil of Agassiz.

736 Copy of a letter from Henry Fairfield Osborn, New York to Karl von Zittel, Munich, 13 April 1894, VPA 1/108, General Correspondence Yo–Z, von ZITTEL, Prof. KARL A. Munich 1883–1904, Folder 47.

> we can effect some exchange with you. We have a large number of duplicates which are very well preserved and we should be very glad to exchange them for similar duplicates in your collection. I shall be perfectly willing to do what we did with Princeton; namely, to make yo u [sic!] a shipment and ask you to send what you considered an equivalent exchange [...] We are constantly using your Palaeontology in our classwork so that I feel I have been very ungrateful in not reviewing your Hand-book in ‘Science’ as I promised to do. I shall take an early opportunity of writing this review. Please give my best greetings to Dr. Schlosser and my thanks for his admirable Bibliography.[737]

In his letter of February 28, Zittel congratulates Osborn on the completion of the exhibition that was going to be the most complete collection of Northern American Mammals in existence and should prove to be of great value to paleontologists from all around the world.[738] He furthermore informed Osborn that Schlosser would compile a list of specimens from Munich suitable for the proposed exchange, which Zittel then sent to Osborn in his next letter. In this letter he praised the abundance of fossilized mammals in America while denouncing the collection of the museum in Munich. Zittel might have done this out of modesty, or to arrange a more favorable exchange.[739] Zittel used almost the exact same words to compare the abundance of American paleontological discoveries to their European counterparts in a letter to Marsh in 1895, likening the European scientists to “beggars”.[740]

The American duplicates arrived in September 1896, to the great delight of Zittel and Schlosser.[741] On December 5 a list cataloging the plaster casts of the fossils of Munich crossed the Atlantic, in hopes of initializing a new exchange.[742]

---

737 Copy of a letter from Henry Fairfield Osborn, New York to Karl von Zittel, Munich, 12 February 1896, VPA 1/108, General Correspondence Yo–Z, von ZITTEL, Prof. KARL A. Munich 1883–1904, Folder 47.

738 “Nach dem, was Sie mir früher mitgetheilt haben, dürfte New York nunmehr die vollstaendigste Sammlung Amerikanischer fossiler Mammalia besitzen und wird für alle Palaeontologen der Welt einen grossen Attracktinospunkt bilden. ” Karl von Zittel, Munich to Henry Fairfield Osborn, New York, 28 February 1896 in: VPA 1/108, General Correspondence Yo–Z, von ZITTEL, Prof. KARL A. Munich 1883–1904, Folder 47.

739 “Anbei sende ich Ihnen das Verzeichnis der von Herrn Dr Schlosser ausgesuchten Dubletten von Säugethieren und Vögeln aus tertiären und diluvialen Ablagerungen Europas [. . .] Im Vergleich mit dem wunderbaren Reichtum an fossilen Vertebraten in N. America, sind wir in Europa ja wahre Bettler.” Karl von Zittel, Munich to Henry Fairfield Osborn, New York, 13 May 1896 in: VPA 1/108, General Correspondence Yo–Z, von ZITTEL, Prof. KARL A. Munich 1883–1904, Folder 47.

740 Zittel to Marsh, 22 December 1895.

741 Karl von Zittel, Munich to Henry Fairfield Osborn, New York, 18 September 1896 in: VPA 1/108, General Correspondence Yo–Z, von ZITTEL, Prof. KARL A. Munich 1883–1904, Folder 47.

742 Karl von Zittel, Munich to Henry Fairfield Osborn, New York, 5 December 1896 in: VPA 1/108, General Correspondence Yo–Z, von ZITTEL, Prof. KARL A. Munich 1883–1904, Folder 47.

Between 1896 and 1899 Osborn must have visited Zittel and his family in Munich. On this occasion he must have shown a series of photos depicting the fossil collection of the American museum to his German colleague. In a letter dated January 3, 1899, Zittel thanked Osborn for his kind visit and asked him where he could buy copies of the photographs (taken by Charles Knight) for the museum in Munich. On a more personal note, he conveyed the appreciation of his wife and daughter for Osborn's visit and asked him whether his son Ernst had visited Osborn in New York. Ernst Zittel had lost his wife in the late fall of 1898 and had now moved in with a friend of his in Brooklyn.[743] After Karl Zittel had died in 1904, Osborn sent a letter to Zittel's newly widowed wife, further underlining the private level of this transatlantic relationship.[744]

Alongside scientific news and professional requests, private and sometimes quite personal information was exchanged via correspondence. While traveling abroad the scientists were usually invited to stay at the homes of colleagues and depended on the help of local acquaintances. In this manner not only professional but also personal bonds were forged, which would commonly last many years. An unusual and heartwarming example centers on the widow of Karl von Zittel and is conveyed in a series of letters from 1920.[745] It demonstrates how the bonds established through scientific exchange could also serve the scientists and their families in personal matters and extended beyond the professional community.

Ida Schirmer had married Zittel in 1865; they had three children and various grandchildren. After Zittel's death in 1904 she left Munich, only to return a few years later. A grandson of hers and her son in law had been killed in the Great War, after which Ida Zittel must have had a hard time making a living in war-torn Germany. In 1920 Richard von Hertwig (1850–1937), professor of zoology in Munich and friend to Ida Zittel, wrote to a letter to John Y. Graham, professor of biology at the University of Alabama, informing his colleague of the plight of Ida Zittel and requesting some aid on her behalf. Graham had studied in Germany for three years and must have met Hertwig in Munich, where he received a Ph.D., magna cum laude.[746] He must have

743 "Mein Sohn Ernst hat Sie wohl noch nicht aufgesucht. Er verlor im Spätherbst ganz plötzlich seine reizende Frau, mit welcher er nur ein kurzes Jahr in glücklichster Ehe verbunden war und hatte sich diesen entsetzlichen Verlust so zu Herzen genommen, dass er Niemanden sehen wollte. Jetzt ist er wieder etwas getröstet und hat sich mit einem Freund in Brooklyn eine stille Häuslichkeit gegründet." Karl von Zittel, Munich to Henry Fairfield Osborn, New York, 3 January 1899 in: VPA 1/108, General Correspondence Yo–Z, von ZITTEL, Prof. KARL A. Munich 1883–1904, Folder 47.

744 Henry Fairfield Osborn, New York to Ida von Zittel, Munich, 13 February 1904 in: VPA 1/108, General Correspondence Yo–Z, von ZITTEL, Prof. KARL A. Munich 1883–1904, Folder 47.

745 Note that all letters sent by Osborn are (unsigned) carbon copies of the originals. Graham's letters, written by typewriter, are signed by hand. Hertwig's letter of June 6 is also a copy, most likely typed by Graham; Hertwig's letter of August 25 is in the original handwriting.

746 James B. Sellers: History of the University of Alabama, vol. 1: 1818–1902, Tuscaloosa, AL 1953, p. 377.

kept amiable ties to Hertwig, for he had sent him a food parcel and a friendly letter in 1920. In his letter to Hertwig Graham thanked him for the parcel and explained that he and his family were coping relatively well with the food shortage, which was due to rapidly rising inflation. He wrote that he would share some of the food (which he calls “Kostbarkeiten,” meaning treasures) with friends like Ida Zittel; she had returned to Munich but had a hard time making ends meet because her widow’s pension had not been adjusted to the massive inflation plaguing post-war Germany.[747] Hertwig also gave a brief insight to his political (and emotional) state of mind, stating that he despised the men who sought to “destroy the industrious and peace-loving [German] people” (“die Maenner, die es versucht haben [...] ein arbeitsfrohes, friedfertiges Volk zu vernichten”) and those “traitors” who used the distress of their country to further their own political agendas (“die Vaterlandsverraeter, die in der Stunde der Noth die Revolution inscenierten”). Graham then sent a copy of the letter to Henry Fairfield Osborn in New York. After reintroducing himself as a former student of Osborn, he correctly supposed that the misery of Ida Zittel might be of interest to Osborn. He describes Hertwig as “one of the most lovable men it has ever been my good fortune to know”. He then explains Hertwig’s rather harsh political tone and implores Osborn to help the colleague and Ida Zittel:

> Of course, they are all royalists and will doubtless remain so to the end of the chapter, just as the old timers down here still think they were right at the time of the civil war. But that people of such fine personal characteristics and whose interests outside of politics are identical with ours should be in want of a little flour and bacon stirs my sympathies and is my excuse for sending you this letter.[748]

Osborn replied promptly and confesses his great empathy for Hertwig and Ida Zittel, to both of whom he had become personally acquainted during his own stay in Munich:

> I am indebted to you for your letter of July first, calling my attention to the sad condition of my former friend and colleague Professor von Hertwig and

747 “Ich werde von Ihrer Erlaubnis Gebrauch machen und auch anderen von den von Ihnen gesandte Kostbarkeiten abgeben, so der Wittwe [sic!] Zittels, dem Sie wohl auch ein freundliches andenken bewahrt haben. Sie lebt nach laengerer Abwesenheit wieder in Muenchen und ist bei ihren knappen Einkuenften schlecht dran.” Copy of a letter from Richard von Hertwig, Munich to John Y. Graham, Tuscaloosa, AL, 6 June1920, attached to: John Y. Graham, Tuscaloosa, AL to Henry Fairfield Osborn, New York, 1 July 1920, American Museum of Natural History, Henry Fairfield Osborn Papers, Correspondence: Individuals, Box 23, Welling to Zittel, Folder 35.

748 John Y. Graham, Tuscaloosa, AL to Henry Fairfield Osborn, New York, 1 July 1920, Henry Fairfield Osborn Papers, Correspondence: Individuals, Box 23, Welling to Zittel, Folder 35.

> of Frau von Zittel. It is indeed terrible that such people should be suffering, because both were extremely prosperous and well-off when I was studying in Munich in 1886.[749]

On the same day he wrote a letter to Vernon Kellogg (1867–1935) in Washington, who had been the director of the Commission for Relief in Belgium in 1915/16, and had previously corresponded with Osborn about a food parcel to Vienna:

> Referring to your recent letter to me regarding relief in Vienna, I write to ask how I may send relief to certain of my former friends in Munich? I learn through an old student of mine that Professor von Hertwig, but more especially the widow of the very distinguished Professor Karl A. von Zittel, are in very sad circumstances. I would like particularly to do something for Frau von Zittel, who was extremely kind to me while I was in Munich as a student in 1886.[750]

Another ten days later Graham wrote to Osborn and informed him how and where to purchase a food-parcel and what it would most likely contain. He then described the logistic details in Germany and how the parcel would travel from Hamburg, where the parcels were sent, to Munich. Apparently, the delivery of the parcels within Germany still relied on the help of local contacts, and would not reach their final destination when left to the postal service:

> You asked me to suggest how best to make a contribution. My wife and I have been doing that sort of thing partly by means of Hoover food drafts and partly by parcel post. The food drafts are very simple. We buy the draft at the bank [...]. On the draft is written the name and address of the one to whom you wish the food to be sent. The draft must be presented by the recipient or his legal representative to the store house in Hamburg and he then receives a package containing flour, bacon, beans, cotton-seed oil and condensed milk, - not exactly what we would think of as Kostbarkeiten, (that Professor Hertwig should apply that term is to me another of the pathetic touches), but certainly the most necessary things and doubtless the greatest number of calories that the sum named can furnish them with. [...] Since we do not know Frau von Zittel's address it will probably be necessary to send to her through Professor v. Hertwig, Schackstr. 2/iii, Munich. Prof. Hertwig you will note in his letter secured the

749 Henry Fairfield Osborn, New York to John Y. Graham, Tuscaloosa, AL, 16 July 1920, Henry Fairfield Osborn Papers, Correspondence: Individuals, Box 23, Welling to Zittel, Folder 35.

750 Henry Fairfield Osborn, New York to Vernon Kellogg, Washington, DC, 16 July 1920, Henry Fairfield Osborn Papers, Correspondence: Individuals, Box 23, Welling to Zittel, Folder 35.

> help of a friend of his in Hamburg. I am sure that Dr. Malcolm C. Burke, care of U. S. Consular Officer, Hamburg, Germany, would be glad to act in this capacity. He took his degree in Munich (Latin) and was a member of our faculty until the war broke out when he went into the service. --- I do not suppose that either of us will live long enough to see the end of this work of clearing up the wreckage of war.[751]

On August 4 Osborn replied that he would write to Burke in Hamburg; he once again emphasized that Ida Zittel had been very kind to him.[752] On the same day he did indeed send a letter to Burke, informing him of his intentions to send a fifty-dollar food parcel to Ida Zittel.[753] Graham then thanked Osborn for his letter and the promised help for Ida Zittel: “Your letter of August 4th is a source of great satisfaction to me and I wishto [sic!] thank you sincerely for what you are doing to relieve the need of Frau von Zittel.”[754] Osborn also sent a letter to Hertwig directly:

> My dear Professor von Hertwig: I recently learned through our mutual friend, Professor John Y. Graham, of the University of Alabama, of the straightened conditions in Munich. This is the first news I have had of the kind, with the exception of a letter from Professor Broili.[755] I would like to inquire especially in regard to the widow of Professor Karl von Zittel, whether she is still living and in need of friendly assistance. When I last heard from the family her son was living and was quite prosperous. In case I can be of any assistance to an old friend I hope you will let me know.[756]

Hertwig received the letter and replied to Osborn promptly on August 25. He informed Osborn of the death of Ida Zittel’s grandson and son in law, that her widowed daughter had remarried and just a few days ago had had a child of her own. Still, Ida Zittel suffered from the food shortage and Osborn’s proposed food parcel would be

751 John Y. Graham, Tuscaloosa, AL to Henry Fairfield Osborn, New York, 26 July 1920, Henry Fairfield Osborn Papers, Correspondence: Individuals, Box 23, Welling to Zittel, Folder 35.

752 Henry Fairfield Osborn, New York to John Y. Graham, Tuscaloosa, AL, 4 August 1920, Henry Fairfield Osborn Papers, Correspondence: Individuals, Box 23, Welling to Zittel, Folder 35.

753 Henry Fairfield Osborn, New York to Malcom C. Burke, Hamburg, 4 August 1920, Henry Fairfield Osborn Papers, Correspondence: Individuals, Box 23, Welling to Zittel, Folder 35.

754 John Y. Graham, Tuscaloosa, AL to Henry Fairfield Osborn, New York, 8 August 1920, Henry Fairfield Osborn Papers, Correspondence: Individuals, Box 23, Welling to Zittel, Folder 35.

755 Ferdinand Broili (1874–1946), German paleontologist, studied under Zittel and worked on Zittel’s handbook. Professor at the University of Munich.

756 Henry Fairfield Osborn, New York to Richard von Hertwig, Munich, 9 August 1920, Henry Fairfield Osborn Papers, Correspondence: Individuals, Box 23, Welling to Zittel, Folder 35.

greatly appreciated.[757] There are no more letters concerning the matter to be found within the Osborn Papers. While there is no proof that Osborn had followed through and sent the food parcel, it is likely, given the attention and effort he invested. While this episode is not part of the scientific knowledge exchange, it is an example of how private matters and interpersonal relationships were also a substantial part of these correspondences and demonstrates how networks established through scientific exchange where then utilized in more personal ways by the scientists involved and their family members and acquaintances.

6.1.3 Preliminary Conclusion

In 1883 Zittel, a German paleontologist of great diligence and renown, had come to the United States to study the extensive fossil collections of Northeastern scientists, most of whom had traveled through Europe in a similar endeavor. He was fascinated by the geological features and sheer magnificence of the countryside, especially Yellowstone National Park. Zittel had corresponded with some of his US-American colleagues before his journey to America, but in 1883 he had met many of the men in person, enhancing their relationships. Marsh and Osborn were among these newfound friends. Zittel helped Marsh with his efforts to hire German assistants. Encouraged by Zittel, Baur and Schlosser went to New Haven to work for the great O. C. Marsh. This mirrors Marsh's own scholarly journey to Europe roughly twenty years before. Then again, Marsh was never employed by his European teachers, remained independent, and was bankrolled by his uncle Peabody. Osborn on the other hand went to Munich in 1886 and worked with Zittel in a purely scholarly way; he was never employed by the Bavarian paleontologist. Zittel and Osborn kept in touch and arranged a prolific scientific exchange between New York and Munich (mostly through Schlosser, who continued his work with Zittel after his return to Germany). Osborn's efforts to help Ida Zittel in 1920, more than sixteen years after her husband Karl had died, show that the paleontological bonds sometimes exceeded professional relationships and became deeper, amiable ties.

757 Richard von Hertwig, Munich to Henry Fairfield Osborn, New York, 15 August 1920, Henry Fairfield Osborn Papers, Correspondence: Individuals, Box 23, Welling to Zittel, Folder 35.

6.2 Georg Baur

Out of all the assistants to Marsh, this study focuses on Georg Baur the most. He came to New Haven alongside Schlosser and remained employed there until 1890, when Cope initiated the most public phase of the "Bone Wars." Baur, who was long since disappointed by Marsh, his work ethics, and terms of employment, finally managed to secure a position at Clark University and later at the University of Chicago, which enabled him to leave Yale and Marsh, whom he despised very much at this point, behind. In contrast to Meyer and Schlosser, Baur had stayed with Marsh for years and continued his scientific career within the United States.

The only comprehensive biography of Baur is an obituary written by William Morton Wheeler (1865–1937), which was published in the "American Naturalist" in January 1899.[758] The Wisconsinite Wheeler had deep ties with Germany since childhood; he had attended the German Academy of Milwaukee and he could read French, German, Greek, Italian, Latin, and Spanish fluently. He was an entomologist, myrmecologist (the study of ants), had received a Ph.D. at Clark University in 1892, and was employed by the University of Chicago from 1892 until 1899.[759] These parameters explain his personal friendship to Baur.

According to Wheeler, Baur was born in the small town of Weißwasser in Bohemia (nowadays Bělá pod Bezdězem, Czech Republic). His father and three of his uncles were university professors, constituting what Wheeler calls "a family noted for its learning."[760] After having studied at the "Realgymnasium" of Stuttgart from 1873 to 1877, he began studying geology and paleontology at Munich in 1878. Baur then spent a year in Leipzig (1880–1881), studying under Credner and Carus.[761] After his return to Munich he continued his paleontological studies under Zittel and received a doctorate for his dissertation titled "Der Tarsus der Vögel und Dinosaurier. Eine Morphologische Studie" (The Tarsus of Birds and Dinosaurs, a morphological study) in 1882. He was called to New Haven in March 1884 (in 1890 he lived in 31 York Square) and left in 1890 due to what Wheeler calls "certain difficulties" with his employer Marsh. In summer of the same year, he collected fossilized reptiles and fish in Western Kansas

---

758 William Morton Wheeler: George Baur's Life and Writings, in: The American Naturalist, vol. 33, no. 385 (Jan. 1899), pp. 15–30.

759 At Clark University Baur and Wheeler worked for Charles Otis Whitman (1842–1910). When Whitman accepted the call to Chicago, where he became the head of the biology and zoology departments, he was accompanied by Baur and Wheeler, amongst others. Jane Maienschein: Whitman at Chicago. Establishing a Chicago Style of Biology?, in: Ronald Rainger et al. (eds.): The American Development of Biology, pbk. ed., New Brunswick, NJ 1991 (orig. publ. 1988), pp. 151–182.

760 Wheeler: George Baur's Life and Writings, p. 15.

761 Julius Victor Carus (1823–1903), professor of comparative anatomy and director of the Zoological Museum in Leipzig.

and sent them to Zittel. Baur then taught osteology and paleontology at Clark University. It seems his scientific undertakings now really got into gear after being freed from the oppressive Marsh regime. In 1891 Baur partook in a well-publicized expedition to the Galapagos Islands, made possible by the generous funding and support of Osborn, amongst others. In 1892 he became assistant professor of comparative osteology and paleontology at the University of Chicago. Though having spent more than a decade in the US Baur never really developed a full grasp of the English language, for Wheeler judges (quite harshly for an obituary): "His classes were never large, owing partly to the advanced and highly specialized nature of the subjects presented and partly to his inability to express himself in a clear and attractive manner in the English language."[762]

He later doubles down on this observation: "This unpleasantness of expression was unintentional, however; being due to a certain abruptness in the use of the English language."[763] Contrarily, Oliver Perry Hay (1846–1930), who wrote another obituary to his colleague Baur for the magazine "Science," stated: "His [Baur's] ideas were expressed in clear and simple language, quite in contrast with the usual German style, and one reads after him with pleasure and profit, even if one does not agree with him."[764]

In late 1897 Baur's health quickly deteriorated when he developed a nervous condition Wheeler calls "general paresis." He took leave for a year and visited his relatives in Munich. Instead of improving, Baur's condition took a turn for the worse; he was institutionalized in an asylum and died there in 1898. Wheeler describes Baur as being enthusiastic and investigative.

Notably, most of Baur's papers were published in the "American Naturalist" and the "Zoologischer Anzeiger." Some of his papers seem to be mere translations, published almost simultaneously in the American and the German journal.[765] Albeit being an expert on the comparative anatomy of (non-avian) dinosaurs and birds, Baur seems to have been a Neo-Lamarckist, much like Cope, whom he greatly admired. Wheeler seemed to resent his colleague's convictions:

762 Wheeler: George Baur's Life and Writings, p. 18.

763 Wheeler: George Baur's Life and Writings, p. 19.

764 Oliver Parry Hay: George Baur, in: Science, new ser., vol. 8, no. 185 (Jul. 15, 1898), pp. 68–71. Quote on page 69.

765 For example, an article called "Der älteste Tarsus (Archegosaurus)" was released in the "Zoologischer Anzeiger," while "The Oldest Tarsus (Archegosaurus)" was published in the same year in the "American Naturalist," see Georg Baur: Der älteste Tarsus (Archegosaurus), in: Zoologischer Anzeiger, vol. 9, no. 216 (Feb. 22, 1886), pp. 104–106.; George Baur: The Oldest Tarsus (Archegosaurus), in: The American Naturalist, vol. 20, no. 2 (Feb. 1886), pp. 173–174. For further examples and a comprehensive bibliography of Baur's see Wheeler: George Baur's Life and Writings, p. 23–30.

> Dr. Baur would probably have dissented from this crude view in after years, but he never altogether abandoned the assumption that variations in living organisms are traceable to the inherited effects of the environment.[766]

Another, shorter obituary published in the “Naturalist” a few issues before gives the highest praise to Baur and his paleontological achievements:

> Science in America has met with a severe loss. In fact, since the death of Professor Cope there has been no one in our country who had a more extensive and a more accurate knowledge of vertebrates, living and fossil, than he. [...] The amount of work which he accomplished in his early years in this country is known only to few, but these few are fully aware that his contributions, especially to the study of fossil reptiles, were both numerous and of the highest importance. He was in reality the victim of that system against which this journal has always protested – he was not allowed to publish his discoveries over his own name. When the release came, Baur at once stepped into prominence, and had time spared him, he would soon have stood, in popular esteem, among the world’s first paleontologists.[767]

Schuchert and LeVene call Baur the “most gifted” of Marsh’s three German assistants, elaborating that “in the six years that he was with Marsh, he published no fewer than 64 short papers and 11 long ones, totaling about 430 pages.” These publications would “show the brilliance of Baur’s mentality, his marked productivity, and his great store of information.”[768]

Note that it was Baur who first told Schuchert about Marsh when Schuchert visited New Haven in 1889. He writes: “I went to New Haven for the winter holidays that year [1889], and we spent Christmas Eve at the home of another of Marsh’s assistants, George Baur. During this visit Professor Marsh was absent, but I learned much of him from both Beecher and Baur.”[769]

Wheeler does not delve too deeply into Baur’s family life, he merely describes that he married Auguste Wachter in 1884, never even mentioning any children. The first short obituary, published in the “Naturalist,” speaks of four children who survived him.

---

766 Wheeler: George Baur’s Life and Writings, p. 19.

767 N.N.: Georg Baur, in: The American Naturalist, vol. 32, no. 381 (Sep. 1898), pp. 717–718. Quote on page 717.

768 Schuchert; LeVene: O. C. Marsh, p. 303.

769 Schuchert; LeVene: O. C. Marsh, p. xi.

6.2.1 Baur and Marsh

After he had grown disenfranchised with his employer, Baur lamented his financial indebtment to Marsh. Most documents written by Baur and archived at the Sterling Library in New Haven consist of receipts detailing loans from Marsh. Baur had to frequently take advances on his salaries to finance himself and his family in New Haven. Between April and June of 1884, he accumulated $ 250 in debt, in July of the same year Baur wrote two receipts for $ 100 each. In September he took out an advance of $ 100, on October 8 $ 100, and another $ 250 at the end of the month; in December he borrowed another $ 50.[770]

In a letter dated December 15 Baur argues for the establishment of a department at Yale that would join paleontology with embryology, the letter is typewritten and not signed, so it might be a copy:

> Palaeontology as assisted by Embryology.
>
> Palaeontology of Vertebrates is becoming more and more a morphological science, forming a branch of comparative Osteology, and the best support for the phylogeny of vertebrates. There is another morphological science, Embryology, which is also an eminent support for the phylogeny.
>
> Everybody knows to-day, that Ontogeny of an Organisme [sic!] is a short recapitulation of its phylogeny. This related to the skeleton of Vertebrates: Morphogeny of the skeleton is a short recapitulation of the skeleton of its ancestors. Conf. my paper: On the tarsus of Birds and Dinosaurs.
>
> It would be of the highest importance if Palaeontology and its sister-science, Embryology, would go hand in hand. I will give an example: Dinosaurs are now generally considered as the ancestors of Birds; if this view is correct, we must find in embryos of Birds, especially the older birds, the Ratitae, characters of the Dinosauria. That this is really the case, I have proved in my above cited paper.

---

770 Georg Baur, New Haven, CT to Othniel Charles Marsh, New Haven, CT, 3 June 1884; Georg Baur, New Haven, CT to Othniel Charles Marsh, New Haven, CT, 7 July 1884; Georg Baur, New Haven, CT to Othniel Charles Marsh, New Haven, CT, 22 July 1884; Georg Baur, New Haven, CT to Othniel Charles Marsh, New Haven, CT, 5 September 1884; Georg Baur, New Haven, CT to Othniel Charles Marsh, New Haven, CT, 8 October 1884; Georg Baur, New Haven, CT to Othniel Charles Marsh, New Haven, CT, 24 October 1884; Georg Baur, New Haven, CT to Othniel Charles Marsh, New Haven, CT, 15 December 1884, MS 343, Series I. Correspondence, Box 2, Folder 73.

> Such a manner of working would fulfil all the claims of a true scientific palaeontology, and could be possible with relatively few expenses.
>
> 2000# per annum would be sufficient for the construction of such a department of Palaeontology.
>
> 1500# must be paid to a man quite familiar with all embryological practicime [sic!], and 500# for the laboratory and embryological material.
>
> Such an institution would be the first in this kind and will certainly find imitation.
>
> 15 Dec., ’84.
> Dr. G. Baur.[771]

Obviously Baur had himself in mind when he recommended “a man quite familiar with all embryological practicime” should be the head of the proposed department. This would have meant that his salary would have increased dramatically, to $ 1500 a year (in 1884 he received a salary of $ 650 a year). A list, dated January 1, 1885, details Baur’s loans and salary; on January 2 he took another $ 200 in advance. Marsh had paid him $ 1150 in total, his agreed upon salary being $ 650 meant that he was now $ 500 indebted.[772] Baur’s debt to his employer grew even further in the following months and years. On another receipt of October 23, Baur stated his total debt as being $ 700 and pledged to repay Marsh “at the rate of no less than $ 50 per month.”[773] Despite this, Baur still owed Marsh $ 390 at the beginning of 1887.[774] The receipts were written by Marsh and cosigned by Baur.

Nonetheless, Baur renewed his contract with Marsh on July 1, 1885. The document states that he would remain in Marsh’s employ for three more years (until July 1, 1888) for the annual payment of $ 1600. According to this contract, Baur was obliged to work for seven hours a day, or forty hours per week with one month “vacation.” The contract would be void if the USGS did not pay the “division of Paleontology” the aforemen-

771 Georg Baur, 15 December 1884, MS 343, Series I. Correspondence, Box 2, Folder 73.

772 Georg Baur, New Haven, CT to Othniel Charles Marsh, New Haven, CT, 1 January 1885; Georg Baur, New Haven, CT to Othniel Charles Marsh, New Haven, CT, 2 January 1885, MS 343, Series I. Correspondence, Box 2, Folder 73.

773 Georg Baur, New Haven, CT to Othniel Charles Marsh, New Haven, CT, 23 October 1885, MS 343, Series I. Correspondence, Box 2, Folder 73.

774 Handwritten bill titled “Dr. Baur, 1886”, MS 343, Series I. Correspondence, Box 2, Folder 73.

tioned sum.[775] The contract was drafted by Marsh. The next contract between Marsh and Baur is dated April 27, 1888. In it Baur obligated himself to work for Marsh until their co-authored book on the osteology of reptiles had been published by Gustav Fischer of Jena (see below). The new contract sets Baur's yearly salary at $ 1800, granted the USGS was willing to pay for his service.[776]

In 1887 Baur travelled through Europe, visited colleagues, studied fossil collections, and acquired fossils for Marsh. According to Schuchert and LeVene he was accompanied by Charles Emerson Beecher (1856–1904), another one of Marsh's assistants. They assert that "Baur's letters to Marsh during this period have a very friendly attitude."[777]

He arrived in Hamburg on February 23 and travelled to Munich, where he met Zittel and Kupffer. He wrote Marsh that he would visit Italy next and France after that. He had procured a letter of recommendation from the German consulate. Baur then told Marsh that "there seems to be no danger with France at the moment and I will stay in Paris and certainly [illegible] any trouble." It seems the relationship between France and Germany was still strained almost sixteen years after the conclusion of the Franco-Prussian War. As a postscript Baur asked for more money: "P. S. Please send with the next check one hundred dollars for drawings and purchase of fossils that I do not get in trouble."[778] Marsh did send the money (Baur noted it to be 783 Marks), it arrived in Munich while Baur was in Italy. He considered his journey "a very great success," and elaborated on the details of the journey in his letter to Marsh, written after his return to Munich on March 25.[779] The highlight of the excursion seems to be Milan and the collection of Francesco Bassani (1853–1916):

> But the most important pieces of the Mailand [Milanese] collection are not less than five well preserved specimens of a small Triassic Ichthyosaur. [...] the limbs are entirely different from those of the Liassic Ichthyosauria. They resembled very much the limbs of the Mosasauria. Radius and Ulna are separated in the middle and are entirely different from the distal phalangeal bones. Therefore, it is a fact that the Liassic Ichthyosauria went from forms with extremities of a land living ancestor. The vertebrate are much less in number and

775 Georg Baur, New Haven, CT to Othniel Charles Marsh, New Haven, CT, 1 July 1885, MS 343, Series I. Correspondence, Box 2, Folder 73.

776 Georg Baur, New Haven, CT to Othniel Charles Marsh, New Haven, CT, 27 April 1888, MS 343, Series I. Correspondence, Box 2, Folder 74.

777 Schuchert; LeVene: O. C. Marsh, p. 304.

778 Georg Baur, Munich to Othniel Charles Marsh, New Haven, CT, 5 March 1887, MS 343, Series I. Correspondence, Box 2, Folder 74.

779 Georg Baur, Munich to Othniel Charles Marsh, New Haven, CT, 25 March 1887, MS 343, Series I. Correspondence, Box 2, Folder 74.

> different from those of the Liassic forms. Prof. F. Bassani is just now occupied with a monograph on those forms; but having received a call to Naples, he may not be able to continue his researches and so he offered in that case to send the whole collection to New Haven for a morphological description. I tried to receive a specimen for money or exchange but it was not possible. [...] Nobody even not Prof. Zittel has an idea of the splendid material in Mailand.

Besides his endeavors to obtain fossil specimens, Baur was on a mission to propagate the latest US-American paleontological discoveries (made by Marsh and his assistants) in Europe:

> I think it would be very good, if you could send another box with your monographs and the plates of restorations which were forgotten, to my address in Munich. Prof. Bassani of course must have the books.

He ended the letter with the usual postscript asking for more money to obtain drawings and fossils, but this time added a more personal note: “Mrs. Baur has improved very much and is all right now. I hope that there will be no relapse.”

Baur sent the next letter to Marsh from Stuttgart on April 10; he described the specimens of the Museum for Natural History (nowadays known as “Staatliches Museum für Naturkunde Stuttgart”) and proposed that Marsh might want to hire Eberhard Fraas (1862–1915), the son of “the old Prof. Fraas,” Oscar Fraas (1824–1897); both were paleontologists working for the museum:

> The people in the Museum are of the greatest kindness, especially the old Prof. Fraas and the young Dr. He would like very much to be an assistant of yours for a year or two, and would be very glad to heare [sic!] your conditions. He is the best man I have found so far on my trip and understands the Museum work first rate.[780]

Fraas was never hired by Marsh, it is unclear if Marsh had no interest in his help, or Fraas did not agree with the conditions, or whether he declined employment for another reason altogether.

On April 21, while staying in Rheims, he wrote back to New Haven (in a very clean and neat handwriting):

> Rheims, April 21. 1887.

780 Georg Baur, Stuttgart to Othniel Charles Marsh, New Haven, CT, 10 April 1887, MS 343, Series I. Correspondence, Box 2, Folder 74.

My dear Professor

I just return from Prof. Lemoine.[781] He was very kind and showed me all his specimens. He sends you his best regards and thanks you for your volumes. From Stuttgart I started to Metzingen for the 'Geologenversammlung'

I purchased some very splendid specimens of Ichthysaurus, Teleosaurus, [illegible], etc. …, which are on the way to New Haven.

Please send the following sums to the named dealers:
1. Mark 185, 50 to Bernhard Hauff,[782] Holzmaden, Kirchheim, u.T. Württem berg
2. Mark 180, – to Friedrich Oberdoerfer, Praeparator in Plieningen bei Stuttgart.

There are two large Ichtyosauria, which are worked out with the greatest care. If we had at New Haven one of these two men, it could be done a good deal in fine work. Mr. Hauff would perhaps accept a position at the Museum. I could only recommend him to you.[783]

Despite Baur's recommendations, Hauff was not hired by Marsh. The next letter Baur sent to Marsh from London on May 20. He described his travels through France and England. As always, money troubles were not far off and he wrote to Marsh that he had received a check but employed Marsh to send him another $ 200 to Munich: "I had different expenses for drawings and so on, that I am pretty short of money, until I am back in Munich."[784]

There is a handwritten résumé of Baur's journey, dated December 31, 1887. The report details the locations and peculiarities of many European fossils specimens but

781 Victor Lemoine (1837–1897), physician and paleontologist, not to be confused with the florist and flower breeder of the same name. For more details and Lemoine's function in the transatlantic network of paleontologists see: Eric Buffetaut: From Giant Birds to X-Rays. Victor Lemoine (1837–1897), Physician and Palaeontologist, in: Christopher J. Duffin et al. (eds.): Geology and Medicine. Historical Connections, London 2017, pp. 115–131 https://doi.org/10.1144/SP452.

782 Bernhard Hauff (1866–1950), who collected and prepared fossils found in his father's quarry, the most important one being the outline of an ichthyosaurus, complete with visible fins, which in most other finds are not preserved. See: Paul A. Selden; John R. Nudds: Evolution of Fossil Ecosystems, 2nd ed., London 2012, pp. 135–138.

783 Georg Baur, Reims to Othniel Charles Marsh, New Haven, CT, 21 April 1887, MS 343, Series I. Correspondence, Box 2, Folder 73.

784 Georg Baur, London to Othniel Charles Marsh, New Haven, CT, 20 May 1887, MS 343, Series I. Correspondence, Box 2, Folder 73.

leaves out Baur’s employment efforts on Marsh’s behalf. The report further describes that Baur journeyed from London to the Netherlands on May 24, visited Leiden and Harlem, and continued to Munich on May 28. On June 18 he went to Prague, and then went on to Dresden on June 19 to study the collection of Geinitz; afterwards he travelled to Leipzig and Berlin. On June 26 he boarded the steamer “Bohemia” in Hamburg and arrived in New York on July 12 (no letters sent to Marsh during this period are archived).[785]

In November 1887 Marsh sought to publish a book about the comparative bone structures of reptiles in Jena, Germany. He employed Baur to handle this transaction, as can be deduced from a contract signed by Gustav Fischer (1845–1910), the proprietor of the “Gustav Fischer Verlag,” a publishing house specialized in publishing scientific, medical, and juridical texts. The contract is arranged between “Dr. O. C. Marsh und Dr. G. Baur in New Haven, Conn. und der Verlagsbuchhandlung [publishing house] von Gustav Fischer in Jena.” In the first section of the contract the proposed book is called “Grundzüge der vergleichenden Osteologie der Reptilien” [outlines of comparative reptile osteology]. In further sections the printing costs and the copies of the first edition (there were to be 1,000 copies) are specified. Section seven states that the book would be published in English by the same publisher one year after the publication of the German edition. Marsh and Baur signed the contract on December 12, 1887.[786] Though the book never got published, there is a receipt of $ 500 for the preparation of the publication, dated July 20, 1887.[787]

By January 1891 Marsh considered Baur an enemy, as evidenced in a letter of his to Leidy. In the same letter he describes how Osborn purchased fossils from Marsh’s contractors and asks Leidy whether Osborn had to be considered an enemy to Marsh. The copy of the letter is typewritten but seems to be signed by Marsh. The top of the page is marked as confidential:

> I learn that Osborn has been imitating Cope in trying to secure specimens from one of my men. If the facts are, as stated to me, he has acted most dishonorably. Is it true that he read a paper before the Philadelphia Academy this week, and showed specimens of Cretaceous mammals recently obtained from the West? You will do me a great favor, if you will let me know all you can about this. I will regard anything you say as strictly confidential. I do not want to believe that Osborn would do anything wrong intentionally, but he is so com-

---

785 Baur, Georg: Dr. Baur Report 1887, MS 343, Series I. Correspondence, Box 2, Folder 73.

786 Gustav Fischer, Jena to Othniel Charles Marsh, New Haven, CT, n.d., MS 343, Series I. Correspondence, Box 11, Folder 469.

787 Georg Baur, New Haven, CT to Othniel Charles Marsh, New Haven, CT, 20 July 1887, MS 343, Series I. Correspondence, Box 2, Folder 74.

pletely under the influence of Cope and Baur, my enemies, that they may have led him into doing what he would not otherwise have done. Did he give the name of the man from whom he obtained the fossils?[788]

Schuchert and LeVene assert the relationship of Baur and Marsh as follows:

It would seem, from this evidence, that Marsh recognized Baur's great ability, and that he went much further to keep him satisfied than he had ever done with any of his other assistants. In spite of this, Baur's feelings toward him grew increasingly bitter. That Baur was brilliant there is no denying, but hand in hand with this brilliance went the lack of balance that so frequently accompanies it. Baur had, moreover, the sense of superiority that belongs to his class in Germany (although he came from South Germany, his looks suggest the Prussian), and it was not long before he began to feel that Marsh's mental equipment and knowledge were much inferior to his own, particularly in Embryology, and his talk of these and other shortcomings helped to position the minds of an already restive staff. This was the belief of Gibb and Westbrook.[789] This feeling of Baur's was further aggravated by his desire to become the holder of a professorship at Yale, and he tried to have Marsh take steps to have a chair established for Anatomy and Embryology of the Vertebrates.[790]

On the next page they assert Baur's character in contrast to that of Marsh:

Baur and Marsh – one young, brilliant, dogmatic; the other past middle age, slower to reach conclusions, equally dogmatic, hard to convince. Probably harmony between such divisive personalities was too much to expect. That they could not agree was really a tragedy, for had they been able to work harmoniously together, they might have attained greater results than were achieved by either separately, and science would have been the gainer.[791]

788 Othniel Charles Marsh, New Haven, CT to Joseph Leidy, Philadelphia, 24 January 1891, The Academy of Natural Sciences of Drexel University, Coll#1, Box#7, Folder#157.

789 Richard Ward Westbrook (1865–1934) was hired by Marsh in 1879 to do secretarial work. He also studied medicine and graduated in 1891. Schuchert (or LeVene) interviewed him about Marsh and his work in 1931; he is one of the most important sources on Marsh. He stated himself that he was very close to the paleontologist. Schuchert; LeVene: O. C. Marsh, p. 291. Hugh Gibb (1860–1932) joined the museum staff in 1882, worked as a very talented fossil preparator. He remained very loyal to Marsh and stayed with him to the end. Schuchert; LeVene: O. C. Marsh, p. 293.

790 Schuchert; LeVene: O. C. Marsh, p. 305.

791 Schuchert; LeVene: O. C. Marsh, p. 306.

6.2.2 Baur and Cope

While no correspondence between Baur and Cope survived, Cope met Baur in person on several occasions. It is safe to assume that he employed the disillusioned assistant in his crusade against Marsh and fanned the flames of discontent. Cope met Baur in the summer of 1886,[792] and again in November 1888 when he attended the meeting of the NAS in New Haven; on this occasion he boarded with Baur.[793]

One of the obituaries mentions that Baur "corresponded probably with every man of note in his particular branch of science, and many of them were his personal friends." One of these friends and correspondents was Henry Fairfield Osborn.

6.2.3 Baur, Osborn, and Hatcher

The first letter Baur ever wrote to Osborn is dated October 1, 1884. It is written in a quite formal and polite manner, and it is in German. Baur addressed Osborn as "Dear Sir" (Sehr geehrter Herr) and sent with the letter a copy of his dissertation on the tarsus of birds and dinosaurs. Baur told Osborn he was working on the morphology of reptile and amphibian limbs, then asked him to send embryological specimens for his studies, crocodilian embryos to be specific, for the collection of the Yale Museum was short on any embryological material.[794] In the next letter of October 18 Baur thanked Osborn for his response (which unfortunately is not preserved) and mentioned that he would be honored to meet him in person.[795] In 1885 they met, presumably in Princeton, for Baur thanked Osborn for his kind reception after his return to New Haven in a letter dated August 9.[796] He wrote that he was now reunited with his wife and child and that his wife reckons that ready-furnished apartments could be rented in Munich. Osborn must have discussed his intended journey to Munich with Baur, who had lived there for some time. His wife recommended one such apartment called "Pension Washeim," located at "Carls. Str. 10," close to the "Pinakotheken" (picture galleries) called the "Maximilians Platz," which she reckoned the most beautiful public place in Munich. They could discuss the matter further when Osborn would visit Baur and his

792 It is uncertain if they had met before the summer of 1886 but Baur had started to complain about his position at Yale before that, for example in letter he wrote to Osborn in August 1885, see below.

793 Osborn: Cope, pp. 374, 384.

794 Georg Baur, New Haven, CT to Henry Fairfield Osborn, Princeton, NJ, 1 October 1884, American Museum of Natural History, VPA 1/8, General Correspondence Bas–Bel, BAUR, GEORGE (Prof.) Univ. of Chicago 1884–1897, Folder 36.

795 Georg Baur, New Haven, CT to Henry Fairfield Osborn, Princeton, NJ, 18 October 1884, VPA 1/8, General Correspondence Bas–Bel, BAUR, GEORGE (Prof.) Univ. of Chicago 1884–1897, Folder 36.

796 The date on the letter is blurry and could either be "Aug." or "Apr."

wife in New Haven in the near future. Baur further promised to write to a friend of his, Dr. Gerster, to instruct him to look for similar lodgings as well and recommended that Osborn should contact Gerster as soon as he had arrived in Munich. He told Osborn the addresses of the family physician Dr. Moddes and Prof. Kupffer (both located at Hess Straße), and mentioned that his parents (who also lived at Hess Straße) would be of little help, for they were on vacation in the mountains and would not return to Munich before the middle or the end of September.[797]

On September 18 Baur wrote to Osborn, who at that time would have moved to Munich together with his wife, that he would like to leave New Haven and seek employment at another institution, the National Museum for example; he would be very grateful if Osborn could help him with this endeavor since he would be aware of Baur's resentments towards his current employment.[798]

Indeed, Osborn had arrived in Munich and had already sent a letter to Baur, which arrived on September 21, three days after Baur had written the aforementioned letter. Baur mentioned this letter in his reply, in which he stated that he was glad that Osborn had safely arrived in Munich and already accommodated himself to the Bavarian capital. He then wrote that he would get homesick thinking about having a cool drink with friends in a Bavarian inn but would have to bear what could not be altered ("doch man etraegt was man nicht aendern kann"). This might be another nod to his dissatisfaction with Marsh and his employment at Yale. Marsh was still in Europe and would not return before October 20. Before he left Marsh had spread the news that Osborn would become his assistant, which were indeed bewildering news for Baur (and probably Osborn as well). In addition, Baur mentioned that he had received a letter from

797 "Ich kam gestern Abend hier an und traf Frau und Kind ganz wohl. Für Ihre freundliche Aufnahme sage ich Ihnen nochmals meinen besten Dank. Meine Frau sagt mir, dass es verschiedene Wohnungen in München giebt [sic!], die moeblirt [sic!] sind und wo Sie zur gleichen Zeit essen können. Eine Derselben ist [an der] Ecke von Barer Str. und Carls-Str., (Pension Washeim Carls. Str. 10.) eine sehr gesunde Gegend. In der Nähe der Pinakotheken und zunächst dem Maximilians-Platz, dem schoenstem Platz in München, diese Wohne waere sehr zu empfehlen. Wenn Sie hierher kommen, koennen wir mehr darüber sprechen. Ich schreibe außerdem an meinen Freund Dr. Gerster Promenadeplatz 14 damit er sich nach allen derartigen Wohnungen in gesunder Lage erkundige, das beste waere dann, wenn Sie ihn sofort in München aufsuchen würden. Unser Arzt in München heisst Dr. Moddes, wohnt Hess-Str. 1a, neben Prof. Kupffer. Meine Eltern wohnen Hess-Str. 32 2ter Stock. Dieselben sind, wie mir ein heute angekommenen Brief meldet im Gebirge und werden nicht vor Mitte oder Ende September zurückkommen." Georg Baur, New Haven, CT to Henry Fairfield Osborn, Princeton, NJ, 9 August [?] 1885, VPA 1/8, General Correspondence Bas–Bel, BAUR, GEORGE (Prof.) Univ. of Chicago 1884–1897, Folder 36.

798 "Offen gestanden, waere es, mir auch das liebste, wenn ich von hier weg, an einen anderen Ort gelangen koennte, z.B. National Museum. Vielleicht koennten Sie etwas in dieser Hinsicht thun, ich waere ihnen sehr dankbar dafür. Sie wissen ja selber, dass hier vieles nicht so ist wie es sein sollte. Hoffentlich sind Sie und Familie glücklich und gesund in München angekommen, und verleben gute Zeit." Georg Baur, New Haven, CT to Henry Fairfield Osborn, Munich, 18 September 1885, VPA 1/8, General Correspondence Bas–Bel, BAUR, GEORGE (Prof.) Univ. of Chicago 1884–1897, Folder 36.

Cope in which the latter had told him that he intended to leave Philadelphia. Baur was corresponding with the enemy.[799]

The next letter to Osborn is dated October 30. Baur wrote that Scott had arrived in New Haven and would write to Osborn directly and tell him what he had accomplished there. It is unclear whether Scott's visit was of a purely scientific nature, or whether he agitated Marsh's disenfranchised assistants and plotted against their employer. Be that as it may, Baur told Osborn that he himself, Williston, and Meyer had coauthored an article about Marsh's misconduct which would be published as soon as Baur had left Marsh's employment. He calls Marsh's conduct a "base fraud" ("dieser niedertraechtige Schwindel"); alas he was bound to Marsh by his $ 700 debt. He explained that he was paid merely $ 600 in his first year of employment, that he was paid quarterly, and that he had to borrow the money for his livelihood in the first quarter of his employment. Marsh had lent him the money but did so only to "ensnare" Baur. Scott had informed Cope about the assistant's grievances and Cope was very interested in the publication of the abovementioned article. He then asked whether Osborn could loan him $ 600 and emphasized that he would do him a very great favor in doing so. He would then come to Germany as soon as he was released by Marsh, and his wife would maybe travel to Germany before that. He instructed Osborn not to tell his father about the financial troubles, that he would have to see the situation through on his own (or rather with Osborn's help). In a postscript he told Osborn that the accusatory article would be published in the "Naturalist" and the "Zoologischer Anzeiger" at the same time.[800] The article was never published in that form, but may have contrib-

799 "Heute erhielt ich Ihren Brief vom 21. Sept. Ich freue mich, dass Sie sich in München rasch eingewohnt haben und dass Sie und die Ihren wol [sic!] sind. Der Gedanke, mit guten Freunden auf dem Spatenbräukeller bei einem kühlen Trunke zu sitzen; ist allerdings Heimweherregend doch man ertraegt, was man nicht aendern kann. [...] Hier gibt es nicht viel Neues. Prof. Marsh ist immer noch in Europa und wird nicht vor dem 30. Oct. Zurückkommen. [...] Prof. Cope schrieb mir heute, dass er die Absicht habe Philadelphia zu verlassen. [...] Es wird Sie vielleicht interessieren zu hoeren, dass sich hier im Museum die Nachricht verbreitet hat Sie würden Assistent bei Marsh werden!! Marsh hat es verbreitet ehe er wegging. Was sagen Sie dazu?" Georg Baur, New Haven, CT to Henry Fairfield Osborn, Munich, 7 October 1885, VPA 1/8, General Correspondence Bas–Bel, BAUR, GEORGE (Prof.) Univ. of Chicago 1884–1897, Folder 36.

800 "Gestern war Herr Scott hier; er wird Ihnen selber schreiben, wie er Alles hier vorgefunden hat. Dr. Williston, Dr. Meyer und ich haben einen Articel verfasst gegen Marsh, welcher zum Druck kommt, sobald ich von hier abgehe. Mein Weggehen nun ist durch einen Umstand erschwert; durch den Umstand, dass ich Marsh 700 $ schulde. Wie diese Summe moeglich wurde, werden Sie begreifen wenn ich Ihnen mitteile, dass mein Gehalt als ich hierher kam nur 600 $ jährlich betrug und dass ich vierteljährlich und zwar [illegible] bezahlt wurde. Das erste Vierteljahr schon musste ich mir das ganze Geld für meinen Lebensunterhalt borgen. Prof. Marsh that dies auch mit grossem Vergnügen, natürlich aber nur deshalb, um mich ihm gegenüber zu binden. Prof. Cope u Scott sind vollkommen mit den ganzen Verhaeltnissen durch mich vertraut gemacht worden; und es waere enorm viel daran gelegen, wenn unser Articel noch dieses Jahr zum Druck kaeme denn dieser niedertraechtige schwindel muss an den Tag kommen. Wenn es Ihnen also moeglich waere mir auf einige Zeit 600 $ zu borgen, so würden Sie mir

uted to the articles published in the "New York Herald" and the "Naturalist" in 1890 (see below).

In his next letter to Osborn, dated February 19, 1886, Baur lamented that he was still stuck with Marsh, while Meyer had been employed by Aldrich in Cincinnati, and Williston was now the assistant editor of "Science."[801] He had written to Whitman and Ryder to seek employment with them at Cambridge or Washington, a fruitless effort, but Whitman seemed to be dissatisfied with Cambridge and planned to leave.[802] Fortunately for Baur, Whitman would employ him at Clark University once he finally managed to leave New Haven. Meanwhile, Scott had visited again but Marsh had prevented him from meeting Baur. He further wrote that he would very much like to leave New Haven but that there was simply no opportunity for that yet. He even thought about returning to Germany but without an imminent job opportunity there his return would be impossible. Baur told Osborn that he was working on a book about the skeletons of reptiles and that he would love to leave Yale for Princeton and finish the book there. He ends his letter imploring Osborn to get in touch, further illustrating his desperation.[803]

---

und uns Allen en einen sehr grossen Dienst leisten. [...] ich komme mit meiner Familie nach Deutschland, sowie ich von Marsh frei bin. Meine Frau kommt vielleicht schon vorher. Bitte sagen Sie meinem Vater nichts von der Geldangelegenheit, ich moechte dieses allein abmachen. [...] P. S. Unser Articel soll zur gleichen Zeit im 'Naturalist und Zoolog. Anzieger' publicirt werden." Georg Baur, New Haven, CT to Henry Fairfield Osborn, Munich, 30 October 1885, VPA 1/8, General Correspondence Bas–Bel, BAUR, GEORGE (Prof.) Univ. of Chicago 1884–1897, Folder 36.

801 Truman Heminway Aldrich (1848–1932) was a businessman, paleontologist, and politician. He represented Alabama's 9th district in the US House of Representatives in 1896 and 1897 and said he was the first paleontologist to serve in Congress since Thomas Jefferson. Together with Meyer he published on the subject of geology and fossils of Alabama. Charles Willison Johnson: Truman Hemin[g]way Aldrich, in: The Nautilus, vol. 46, no. 1 (Jul. 1932), pp. 34–35.

802 John Adam Ryder (1852–1895) was a zoologist and in 1886 worked in for the United States Fish Commission. Edward Drinker Cope et al. (eds.): In Memoriam John Adam Ryder, Philadelphia 1895.

803 "Hier giebt [sic!] es nichts Neues. Mayer [Meyer] ist von Aldrich in Cincinnati engagiert. Williston ist Hilfs-Herausgeber der Science. Ich bin immer noch hier. In den letzten Tagen habe ich an Whitman und Ryder, wegen einer eventuellen Stellung in Cambridge oder Washington geschrieben. Aber ohne Erfolg. Whitman scheint mit den Cambridger Verhältnisse[n] auch nicht sehr zufrieden zu sein und gedenkt den Ort zu verlassen. Prof. Scott war vorgestern hier, Marsh verstand es aber so zu wenden, dass wir uns nicht sprechen konnten. Ich waere sehr froh, New Haven verlassen zu koennen, bis jetzt aber sind noch gar keine Aussichten hierzu vorhanden. Ich arbeite an der Bibliographie für eine [illegible] Osteologie der Reptilien. [...] Ich hatte schon die Absicht nun dieses Opus zu vollenden, nach Deutschland zurückzukehren, aber ohne eine Stellung dort, ist es auch eine miesliche Sache. Ich moechte Ihnen nun einen unbescheidenen Vorschlag machen. Waere es nicht moeglich in Princeton eine Stelle zu bekommen[?] [...] Auch wenn ich lieber ein Lebenszeichen von ihnen zu erhalten hoffe bin ich mit den besten Grüssen Ihr ganz ergebener G. Baur." Georg Baur, New Haven, CT to Henry Fairfield Osborn, Princeton, NJ 19 February 1886, VPA 1/8, General Correspondence Bas–Bel, BAUR, GEORGE (Prof.) Univ. of Chicago 1884–1897, Folder 36.

The next surviving and rather lengthy letter Baur wrote to Osborn is dated October 26. Within the first paragraph Baur apologized that he could just now reply to Osborn’s “kind” (“liebenswürdig”) letter of October 23. Baur’s youngest daughter had died that week. This news set the tone for the rest of the letter. He described emotionally and embittered his grievances with Marsh. Baur recounts that he was very happy when working as an assistant to Kupffer in Germany (Osborn had met Kupffer himself). Then in the autumn of 1883 Zittel had told him of Marsh’s desire to hire assistants and of his marvelous paleontological collection. Enthusiastic about this once in a lifetime opportunity and encouraged by Zittel he agreed to get in touch with Marsh. The professor’s reply arrived in Munich in February 1884. Baur reproduced Marsh’s reply (the original is not recorded) as follows:

> Prof. Zittel has recommended you to me as an assistant and also explained to you fully, what services I require in my Museum. I therefore offer you a situation here for three years from April 1st 1884. The salary will be fifty Dollars ($ 50) per month, or six hundred Dollars ($ 600) per annum with one month vacation on each year, without any loss of pay. I expect (8) eight hours service per day, but you will have Saturday afternoon to yourself.

The style of this excerpt matches the conditions of Marsh’s contract with Baur, mentioned above, while some stylistic and word choices imply that Baur was paraphrasing. Zittel had told Baur that this salary would be more than sufficient, that $ 600 would amount to 2500 Marks, which was “quite a lot of money” (“eine ganze Menge Geld”), and that his salary in Munich had been a meager 1000 Marks. His prime motivation being scientific rather than monetary opportunity anyway, he took the job and moved to New Haven, where he arrived on April 1, 1884. Here he (and presumably Schlosser as well) were surprised to hear that he would receive his first payment only three months after he had started his work. This was quite surprising to him because in Germany one was paid in advance. Thus, from his first day in New Haven he had no choice but to borrow money from Marsh. He told Osborn that he was sure that Marsh had plotted his assistants’ indebtedness from the beginning to “indenture” them (“Aber das war sicher gerade das, was er wollte, er wollte, dass wir von ihm abhaengig würden”). Marsh had not even reimbursed them for their journey to America. Schlosser soon had enough of this, had someone send him 600 Marks, and returned to Germany. He had endured Marsh’s working conditions for just one year. Baur remarks that Schlosser had had little opportunity to learn anything of scientific value in New Haven and that Marsh had forced him to promise not to talk about any scientific work done in New Haven. Baur then sarcastically remarks that Schlosser had paid c. 2000 Marks to learn about Marsh’s true character. But he himself had been forced to remain at Yale. On July 1, 1885, Baur’s salary had been raised to $ 1600 (this corresponds with the new contract, see above),[804] but he stilled owed $ 750 to his employer. He had managed to save and repay $ 450 through "enormous personal sacrifice" ("Durch die allergrößte Einschränkung"), but Marsh now demanded that he pay the remaining $ 300 as soon as possible and constantly urged him to pay up. To make matters worse, Baur's (and anyone else's) salary was not paid on time but often one or two weeks late. He wrote that his salary for August had been paid on September 20. Williston had filed a complaint with Powell, but he of course was "in cahoots" with Marsh ("die Herren stecken ja unter einer Decke"), and Powell had given the letter of complaint to Marsh. Due to these circumstances Baur would have loved to leave New Haven sooner rather than later. At least he could focus on his scientific work and was left to his own devices in this regard. The only things he had done for Marsh were compiling a bibliography about dinosaurs and writing the last chapter on the "Dinocerata" for the professor.[805] Left to his own devices Baur worked at the embryological laboratory and helped Marsh as little as possible. He wrote that he felt he was treated as a "milk cow" by his employer ("ich bin so zusagen die Kuh, welche gemelkt werden soll"), who sought to exploit his work for his own gain. Therefore he kept Marsh in the dark ("aber wie unwissend dieser Marsh ist wissen Sie gar nicht!") and thought it "scandalous that such an ignorant man would be at the top of American science" ("Es ist geradezu ein scandal, dass ein solcher Ignorant an der Spitze der amerikanischen Wissenschaft steht"), but Baur predicted that "his time would come and his reign would end in terror" („Doch seine Zeit wird auch noch kommen und er wird ein Ende nehmen mit Schrecken!"). Paleontological specimens were not shown to Marsh's assistants and kept under lock and key. Baur had not learned anything of paleontological value thus far. What little he had learned he was planning to publish with Whitman's help and without telling Marsh.[806]

---

804 Baur to Marsh, 1 July 1885.

805 See: Othniel Charles Marsh: Dinocerata. A Monograph of an Extinct Order of Gigantic Mammals, Washington, DC 1886.

806 "Erst heute bin ich im Stande auf Ihren liebenswürdigen Brief vom 23. zu antworten. Meine Jüngste starb donnerstag Abend und wurde gestern begraben; daher die Verzögerung. Nun will ich Ihnen Alles schreiben, wie es hier aussieht und warum ich mich so sehr bemühe von hier weg zukommen. Mein lieber Freund A. Böhm und ich waren Assistenten bei Kupffer, dass diese Zeit eine äusserst angenehme für mich war, koennen Sie sich vorstellen nachdem Sie Böhm und Kupffer persönlich kennen gelernt haben. Es war Herbst 1883, als Zittel mir mitteilte, dass Marsh deutsche Assistenten suche und dass er mir rate mich als solchen zu melden. […] Mitte Februar 1884 kam dann auch Marsh's Antwort. […] 'Prof. Zittel das recommended you to me as an assistant and also explained to you fully, what services I require in my Museum. I therefore offer you a situation here for three years from April 1st 1884. The salary will be fifty Dollars ($ 50) per month, or six hundred Dollars ($ 600) per annum with one month vacation on each year, without any loss of pay. I expect (8) eight hours service per day, but you will have Saturday afternoon to yourself.' Zittel meinte man koenne mit diesem Gehalt ganz gut leben; Schlosser und ich natürlich kannten die Verhältnisse nicht und übersetzten die 600 Dollar in 2500 Mark, was uns eine ganze Menge Geld schien. (Mein Gehalt in München war 1000 M.) Das Geld war aber Nebensache, wir dachten nur an die Wissenschaft und

Baur’s next letter to Osborn is dated October 23. He thanks Osborn for his letter, which had arrived one day prior and had contained a note written by Ryder. He informed Osborn that he had written a letter to Professor Baird,[807] asking him for a job. Even though the aforementioned job (which unfortunately is never specified in the letter) would most likely not be very fulfilling to him, news of his job-hunting might prompt Marsh to further raise Baur’s pay. He now planned to publish a book on the skeletons of reptiles. Baur’s wife and daughter (accompanied by his sister) would go to Germany in March of next year. Baur himself would stay in the US until July to visit the collections of Cambridge and the Smithsonian, Cope’s collection, and Princeton. Then he would embark for London and his European tour, as described above. He would be looking for a new job in Europe, and only if he could not find one would he return to America. In the meantime, and in Europe, he would continue his work on his book about reptile osteology.[808]

---

die Erfolge, die wir durch einen Aufenthalt beim grossen Marsh erringen würden. Am 1. April kamen wir hier an, und waren sehr überrascht zu hören, dass wir erst nach 3 Monaten den ersten Gehalt bekommen sollten, das waere hier so Usus. […] Also vom ersten Tag an waren wir genöthigt von Herrn Marsh uns Geld zu borgen. Aber das war sicher gerade das, was er wollte, er wollte, dass wir von ihm abhaengig würden. Nachtraeglich muss ich noch bemerken, dass wir für die Reise keinen Cent Entschädigung bekommen haben; […] Schlosser wurde diese Sache bald zu bunt, er liess sich 600 Mark schicken und fuhr ab, nachdem er es gerade ein Jahr hier ausgehalten. […] Gesehen hat er nichts hier und das Wenige was er gesehen, kann er nicht gebrauchen, da ihm Herr Marsh vor seinem Weggange das Ehrenwort abnahm von nichts, was er hier gesehen etwas zu sprechen oder zu publiciren. Nun Schlosser ist wenigstens die Genugthung durch eine Ausgabe von circa 2000 M. kennengelernt zu haben, was Marsh für ein Herr ist. Ich war gezwungen zu bleiben. Am 1. Juli 1885 erhöhte sich mein Gehalt bis auf 1600 Dollar, zeitgleich war aber meine Schuld an Marsh auf 750 Dollar angeschwollen. Durch die allergrösste Einschränkung nur habe ich diese Summe um 300 Dollar vermindert, mit dem Rest draengt Marsh bestaendig, und will so schnell wie moeglich Alles zurückbezahlt haben. Dies ginge alles noch, nun aber kommt die Unpünktlichkeit der Bezahlung. Ich kenne auch nicht einen einzigen Fall, dass irgend einer im Museum seinen Gehalt zur richtigen Zeit bekommen hätte, immer vergeht eine Woche, oft 14 Tage, bis man endlich sein Geld in die Hände bekommt. Williston beklagte sich letztes Jahr deshalb bei Powell, doch die Herren stecken ja unter einer Decke, Powell übergab Williston’s Brief an Herrn Marsh. Meinen Gehalt für August erhielt ich am 20. Sept. […] Was nun meine wissenschaftliche Thätigkeit betrifft, so kann ich mich hierüber nicht beklagen, ich arbeite eben für mich osteologisch-embryologisch. Das Einzige, was ich für Marsh gethan habe, ist eine Bibliographie der Arbeiten über Dinosaurier und den Schlusstheil der Dinoceraten den ich ihm so ziemlich dictirt habe. […] Wenn es sich um einen wichtigen Punkt handelt werde ich ausgefragt, ich bin so zusagen die Kuh, welche gemelkt werden soll. Aber wie unwissend dieser Marsh ist, wissen Sie gar nicht! Es ist geradezu ein scandal, dass ein solcher Ignorant an der Spitze der amerikanischen Wissenschaft steht. Doch seine Zeit wird auch noch kommen und er wird ein Ende nehmen mit Schrecken! […] Das [paläontologische Material] ist Alles wolweisslich abgesperrt und verstellt.” Georg Baur, New Haven, CT to Henry Fairfield Osborn, Princeton, NJ 26 September 1886, VPA 1/8, General Correspondence Bas–Bel, BAUR, GEORGE (Prof.) Univ. of Chicago 1884–1897, Folder 36.

807 Spencer Fullerton Baird (1823–1887), a very prolific naturalist and the first curator of the Smithsonian Institution.

808 “Ihren Brief mit der Einlage von Ryder habe ich gestern erhalten. Ich sage Ihnen meinen besten Dank für Ihre Bemühungen. Ich habe an Professor Baird geschrieben. Man kann es ja einmal versuchen;

While he was in London Baur wrote a letter to Osborn, which is dated March 8, 1887. This has to be the first letter Baur wrote to Osborn from Europe, for he recapped his journey thus far. He included some drawings of paleontological specimens. Marsh or any other troubles remained unmentioned and Baur promised further cooperation with Osborn.[809] His next letter (as well as most of the subsequent letters) to Osborn is written in English. The subject matter is purely professional.[810] Schlosser was part of the professional exchange, this is further documented in a letter dated January 15, 1888.[811]

In a letter dated January 13 the next chapter in Baur's quest for independence unfolds. He reported to Osborn that:

> Hatcher is back from the West,[812] again with an enormous collection. He intends to give up collecting for the Survey, especially for Marsh. He is going to collect independently, and I intend to go with him, and to leave my position here. We shall collect successively in all the formations of the West, beginning with the Withe River. We hope to sell collections and single specimens to the different Museums in this Country and in Europe. You have seen, what Hatcher brought together last year. Of course, collecting in these regions is very expensive, as you know; we are looking now for somebody, or some Institution, who would be willing to advance a sum, high enough to permit a start. Hatcher informs me that $ 2000 would be sufficient. If the Princeton Museum could advance this amount; we would collect for it a fine set of fossils, worth this sum; and would give to it for future times the first chance to purchase specimens. I

doch glaube ich nicht, dass diese Stellung sehr befriedigend für mich sein wird. Vielleicht aber wird Marsh dadurch veranlasst meinen Gehalt etwas zu erhöhen. Ich habe nemlich [sic!] nun andere Plaene. Meine Absicht, eine grosse vergleichende Osteologie der Reptilien zu schreiben soll verwirklicht werden. Meine Frau und die Kleine gehen im Maerz mit meiner Schwester nach Deutschland, ich bleibe hier bis etwa Juli, besuche zuerst die Museen in Cambridge, und Smithson. Inst; sodann Cope's Sammlung und wenn Sie es zulassen Princeton; namentlich um osteologische Studien zu Machen. Dann gehe ich nach London, Leyden, Harlem, Bruxelles, Paris, Stuttgart, Tübingen, München. Dies sind die Hauptplätze für fossile Reptilien. Mitte September hoffe ich in München zu sein. Wenn sich unterdessen keine Stellung in Europa gefunden hat kehre ich nach Amerika zurück." Georg Baur, New Haven, CT to Henry Fairfield Osborn, Princeton, NJ 23 October 1886, VPA 1/8, General Correspondence Bas–Bel, BAUR, GEORGE (Prof.) Univ. of Chicago 1884–1897, Folder 36.

809 Georg Baur, London to Henry Fairfield Osborn, Princeton, NJ 8 March 1887, VPA 1/8, General Correspondence Bas–Bel, BAUR, GEORGE (Prof.) Univ. of Chicago 1884–1897, Folder 36.

810 Georg Baur, New Haven, CT to Henry Fairfield Osborn, Princeton, NJ 13 December 1887, VPA 1/8, General Correspondence Bas–Bel, BAUR, GEORGE (Prof.) Univ. of Chicago 1884–1897, Folder 36.

811 Georg Baur, New Haven, CT to Henry Fairfield Osborn, Princeton, NJ 15 January 1888, VPA 1/8, General Correspondence Bas–Bel, BAUR, GEORGE (Prof.) Univ. of Chicago 1884–1897, Folder 36.

812 John Bell Hatcher (1861–1902), a paleontologist and fossil hunter, who worked for Marsh until his contact with Osborn provided him a job at the AMNH and later Princeton.

> thought it a good idea to talk the whole matter over [?] personally with you, and if you can make some arrangement, I should be greatly obliged to you. I shall be here 5 years the 1st of April, and I think this is enough. There are no chances for me with Marsh, and the 'Reptilian-Work' would never be finished in 'Company' with him. I never liked his ways nor him, and I think it better to stop with him. Please consider this letter as private, I do not want to let Marsh know anything before the whole is fixed. Of course you can tell Prof. Scott about it.[813]

Baur was secretly plotting against his employer, and, maybe even more importantly, he was planning to enter the fossil hunting business and not collect for Marsh, or Cope, but independently, promising the first choice to Osborn and Princeton. Baur's anticipation for a reply from Osborn must have been palpable, for the next letter Baur sent to him is dated January 22, and he opens by writing: "I am anxious to hear from you, being obliged to decide about my future arrangements in a short time." Understandably Baur wanted his situation sorted out as soon as possible, but Osborn did not intend to make a rash decision. Maybe did he not trust Baur and Hatcher with that much money, or maybe could not raise the money that easily? Still, business went on as usual for in the same letter Baur functions as a broker between Osborn and Schlosser, the latter wanting to ask Osborn for some sketches. Baur still worked with Marsh, helping with paleontological reconstructions: "You will find in the next number of the Am. Journ. a restoration of 'Brontops robustus' Marsh.[814] When his 'monograph' will be published nobody knows." Note the emphasis on the word "monograph," implying that it really was another work of collaboration Marsh claimed for himself without giving any credit to his coauthors.[815] On January 27 the proposed venture with Hatcher was once again the main gist of the Baur-Osborn correspondence. There must have been additional correspondence with Osborn in in the meantime via telegram, proposing a meeting in New York, for Baur wrote:

> I am not able to be in New York tomorrow, as I telegraphed you yesterday; the boy and Mrs. Baur being not quite well. I am very sorry that I can not meet you; thinking it very much better to talk the whole matter over with you.

He still intended to follow through with his plan, trying to secure Osborn's backing:

---

813 Georg Baur, New Haven, CT to Henry Fairfield Osborn, Princeton, NJ, 13 January 1889, VPA 1/8, General Correspondence Bas–Bel, BAUR, GEORGE (Prof.) Univ. of Chicago 1884–1897, Folder 36.

814 Othniel Charles Marsh: Restoration of Brontops Robustus, from the Miocene of America, in: The American Journal of Science, ser. 3, vol. 37, no. 218 (Feb. 1889), pp. 163–165.

815 Georg Baur, New Haven, CT to Henry Fairfield Osborn, Princeton, NJ, 22 January 1889, VPA 1/8, General Correspondence Bas–Bel, BAUR, GEORGE (Prof.) Univ. of Chicago 1884–1897, Folder 36.

> I do not go in this business, without having a guarantee of its success; and I am waiting therefore for news from Europe [?]. That we can collect material enough to supply many museums I do not doubt whatever. Hatcher is an absolute competent and reliable man. He has a place at Long Pine, Nebr., a very central point for the expeditions. He has built this summer a large barn, which could be used, he says, as a very good place to work up the fossils and to store them. The expenses out there would be very small, and I do not doubt that we could make it a success. A systematic scientific collecting in the different localities would be of the greatest importance, and our principal notion is to collect series as complete as possible. A complete series of the Phenacodus – Equus Line for example, could be of the greatest value. That I could not find a better companion as [sic!] Hatcher, there is no doubt. Hatcher of course will undertake nothing as long as he is connected with the Survey. He only wants to have a certain guarantee before he gives up his position.

Baur ended his letter urging Osborn to a conspiratorial face to face meeting: "I should be very glad if we could arrange a rendezvous, if possible here in New Haven."[816]

The next letter, dated February 14, shows that Baur had written concerning his defection idea to a couple of other colleagues, most likely asking them to support his endeavor. His elation on the prospect of success and cutting loose his ties to Marsh is evident:

> I have answers from Woodward, Gaudry, Fraas.[817] Gaudry is very pleased with the idea, but has not offered a definite sum. Woodward, Fraas and Zittel however will buy for $ 2220, allthogether [sic!], i.e. at present. I have not heard yet from Winkler of the Teyler Museum.[818] So far things look all right. Have you heard or done anything more in this matter? If we beginn [sic!] in August, we may perhaps arrange to get along without the advance of $ 2000. I need about

816 Georg Baur, New Haven, CT to Henry Fairfield Osborn, Princeton, NJ, 27 January 1889, VPA 1/8, General Correspondence Bas-Bel, BAUR, GEORGE (Prof.) Univ. of Chicago 1884–1897, Folder 36.

817 Most likely Henry Woodward (1832–1921), curator of the Geology Department of the Natural History Museum of London and friend of Marsh (see chapter 3. 3.). Arthur Smith Woodward (1864–1944), who was not related to Henry Woodward, replaced him as curator in 1901 but had worked at the museum since 1882, so it is hard to tell which Woodward Baur corresponded with. Jean Albert Gaudry (1827–1908), a French paleontologist.

818 Tiberius Cornelis Winkler (1822–1897), curator of paleontology and geology at the Teylers Museum in Haarlem, Netherlands.

> $ 500, to arrange everything before I leave. Do you know of any way how I could get this sum? I should pay it back either in money or in fossils.[819]

In April Marsh had returned to New Haven and Baur had tried to fathom his career-opportunities at New Haven, again without a result. Desperate about his indebtedness, he asked Osborn for the money to purchase his "freedom":

> I did not have any talk with Marsh until now; he has been away. I just had some conversation with him but without result. He said he could not say anything at present about my future chances here. The whole thing is the result of my dependence from being in money-matters; if this would not be, the whole thing would be different. If I only were free! Could you make it possible to let me have $ 300 for a few months? I would pay him off and would return the money to you in monthly rates. I could ask my father for the money but I do not want to do this; he has trouble enough. I feel very badly, being much worried by this and other matters. You would do me a very great favor and relieve me very much if you could do something for me.[820]

Baur's situation still worsened, as evidenced in his next letter to Osborn (he apparently could not wait for a reply):

> I am anxiously waiting for an answer from you. I am in a bad condition at present and really do not know what to do. M. has given me only $ 60 the first of April, and when I asked him for more, he said, he had none. This he has never done before. I have sold one part of my library and I have to let go another part. It is very disagreeable to me to write to you about these things; but you are really the only one here, to whom I dare to entrust my present situation. I hope you may not take amiss this letter, which is dictated by the circumstances.[821]

In his very short subsequent letter, dated April 18, he informed Osborn, that the situation had not changed at all, and that he was still quite unhappy: "M. did go to New York Saturday afternoon; and will be back probably Monday. Nothing has changed in my conditions yet." Only in a postscript he informs Osborn that he had managed to get

---

819 Georg Baur, New Haven, CT to Henry Fairfield Osborn, Princeton, NJ, 14 February 1889, VPA 1/8, General Correspondence Bas–Bel, BAUR, GEORGE (Prof.) Univ. of Chicago 1884–1897, Folder 36.

820 Georg Baur, New Haven, CT to Henry Fairfield Osborn, Princeton, NJ, 2 April 1889, VPA 1/8, General Correspondence Bas–Bel, BAUR, GEORGE (Prof.) Univ. of Chicago 1884–1897, Folder 36.

821 Georg Baur, New Haven, CT to Henry Fairfield Osborn, Princeton, NJ, 8 April 1889, VPA 1/8, General Correspondence Bas–Bel, BAUR, GEORGE (Prof.) Univ. of Chicago 1884–1897, Folder 36.

some scientific work done, continuing the trend of mixing personal and professional correspondence: "P.S. I have dissected a Megalobatrachus (Cryptobr.) from Japan, it is a different genus from Menopoma; Menopoma is the older stock."[822]

If anything, Baur's financial situation had worsened during April since his next letter to Osborn, dated May 7, reads even more drastically (obviously Osborn had not offered money):

> My condition is becoming more serious every day. The only way out was to write home, what I have done today. But until I hear from home I do not know what to do. Marsh does not give me enough money to live one week after I have paid my necessary monthly expenses. I still owe him $ 190. What can I do? My cousin in New York has gone to Europe there is nobody here, whom I know, who could give me some money to live on this month. M.'s activity is simply low. I just now had a talk with him; I wanted him to advance me 100 $ until I hear from home. He asked me, whether 25 would not be enough, than [sic!] he raised to 30, than [sic!] to 35, and his last offer was 40, when I left the room, without saying any thing [sic!]. I prefer to pay 20% to a Polish Jew; than to have one other cent advance from him. I am ashamed to write to you, but perhaps you know of any way to get out of this; I do not.[823]

By June 12 Baur still had to beg Osborn for help. This time he asked for his help in securing a position at the American Museum for Natural History. In addition to his grievances with Marsh, it seems that Baur had fallen very ill:

> Would it not be possible, that I could receive the position at the New York Museum? Perhaps it would be a good idea, to see the people there in person. I do not know exactly what a man they want. I have to leave New Haven, i.e. M. it does [not] do me any good, to stay longer with him. Besides that the climate here is very bad, and I suffer again from malaria since some weeks, in such degree [?] that I am nearly unable to do any scientific work, it is even difficult for me to write this letter. I should like to have control of a department in a Museum, and to make this special department as instructive as possible.[824]

---

822 Georg Baur, New Haven, CT to Henry Fairfield Osborn, Princeton, NJ, 18 April 1889, VPA 1/8, General Correspondence Bas–Bel, BAUR, GEORGE (Prof.) Univ. of Chicago 1884–1897, Folder 36.

823 Georg Baur, New Haven, CT to Henry Fairfield Osborn, Princeton, NJ, 7 May 1889, VPA 1/8, General Correspondence Bas–Bel, BAUR, GEORGE (Prof.) Univ. of Chicago 1884–1897, Folder 36.

824 Georg Baur, New Haven, CT to Henry Fairfield Osborn, Princeton, NJ, 12 June 1889, VPA 1/8, General Correspondence Bas–Bel, BAUR, GEORGE (Prof.) Univ. of Chicago 1884–1897, Folder 36.

The next day he sent a letter to Osborn, asking for a meeting in New York and mentioned that he would like to be introduced to some of the decision-makers at the museum.[825] Even though Baur's financial and occupational problems dominate his correspondence with Osborn in 1889, some professional, scientific information was also included. On June 18, for example, he wrote that an issue of the Quarterly Journal of the Geological Society had just been published and gives a very compact review of Seeley's article in said issue.[826] But in his very next letter Baur again argued for his employment by the New York Museum, again reaffirming his willingness to take any position as long as it would spare him from renewing his contract with Marsh:

> July 1st is near, on which day I probably have to sign a new contract with M. I should be very glad therefore, if you could give me some news in regard to New York. As I understood you, they want a man there to work out and bring to exhibition the fossil vertebrata; and who would be able to collect new material during a part of the year. Could it not be arranged that the money to be spent for this purpose could be divided in the forthcoming way. One part for a fixed Salary, and the other for the purchase of fossils from a collector. I could bring the collection in order and would also be glad to receive charge of the living Reptilia. Hatcher, who is going to collect for different Museums, would give to the N. Y. Mus. the first chance to buy. I should like to see you and Prof. Bickmore to talk things over fully,[827] if there is any possibility for me to get a place.[828]

Baur was still in contact with Hatcher, who was still willing to help the depressed German by giving the first choice of fossils to whatever institution Osborn was attached to. Osborn had answered in a letter written on June 26, as can be deduced from a letter written by Baur on June 28:

---

825 "Can I meet you in New York some day this or the coming week? I should like to have a talk with you about the New York Museum, and to see some of the people." Georg Baur, New Haven, CT to Henry Fairfield Osborn, Princeton, NJ, 13 June 1889, VPA 1/8, General Correspondence Bas–Bel, BAUR, GEORGE (Prof.) Univ. of Chicago 1884–1897, Folder 36.

826 Georg Baur, New Haven, CT to Henry Fairfield Osborn, Princeton, NJ, 18 June 1889, VPA 1/8, General Correspondence Bas–Bel, BAUR, GEORGE (Prof.) Univ. of Chicago 1884–1897, Folder 36. The Article was written by British paleontologist Harry Govier Seeley (1839–1909) in the "Quarterly Journal of the Geological Society of London". Harry Grovier Seeley: Note on the Pelvis of Ornithopsis, in: Quarterly Journal of the Geological Society of London, vol. 45 (1889), pp. 391–397.

827 Albert Smith Bickmore (1839–1914) was one of the founders of the AMNH.

828 Georg Baur, New Haven, CT to Henry Fairfield Osborn, Princeton, NJ, 25 June 1889, VPA 1/8, General Correspondence Bas–Bel, BAUR, GEORGE (Prof.) Univ. of Chicago 1884–1897, Folder 36.

> I have received your letter of the 26th to day. After I have been in bed for a day I feel better. You must excuse the restlessness shown in my last letters. It is produced by that miserable malaria sticking in my bones. I thank you very much for your kind letter, and I hope to cheer up!

Despite the undoubtedly encouraging and kind nature of Osborn's letter, it must have informed Baur that the position at the museum had already gone to another candidate, for Baur replies:

> No doubt, there is no better man for the position, needed at the N.Y. Museum and I should be very glad for him, if he could get the place, besides Hatcher who, has done some splendid collecting, as you will see from a paper just published by Marsh, which will be of special interest to you.[829]

The rest of the letter contains professional anatomical advice to the colleague.

In July of the same year, being aware that the position at the New York Museum was no longer available, Baur wrote to Osborn, once again complaining about Marsh's work ethics and qualifications. Marsh had just published an article called "Discovery of Cretaceous Mammalia"[830] and Baur's harsh, one might say polemic, criticism probably influenced Osborn's review of Marsh's text, published in the "Naturalist" almost two years later, in July 1891.[831]

> What do you say about 'the Discovery of Cretaceous Mammalia'? I came to Museum yesterday afternoon after I had been away a few days on account of my health, and M. gave to me this remarkable piece of work; remarkable for its absolute ignorance, its untrue statements, remarkable for the fact, that it was written by him alone in every respect. I have not seen a line of it before it was in my hands in its complete form. It looks to me that he made about 6 or more genera from a Plagiaulacoid; that makes about a new genus out of each tooth in the upper and lower jaw. Wortman and Cope are Discoverers[?]. Not a single specimen of Marsh's was found in the rock, but all exactly in the same manner as Wortman found his, mixed with teeth of Dinosaurs, pieces of turtle shells, scales of Fishes, vertebrae etc. etc. The figures are good and I hope that Cope to

829 Georg Baur, New Haven, CT to Henry Fairfield Osborn, Princeton, NJ, 28 June 1889, VPA 1/8, General Correspondence Bas–Bel, BAUR, GEORGE (Prof.) Univ. of Chicago 1884–1897, Folder 36.

830 Othniel Charles Marsh: Discovery of Cretaceous Mammalia, in: The American Journal of Science, ser. 3, vol. 38, no. 223, (Jul. 1889), pp. 81–92.

831 Osborn, Henry Fairfield: A Review of the "Discovery of the Cretaceous Mammalia", in: The American Naturalist, vol. 25, no. 295 (Jul. 1891), pp. 595–611.

> who's Puerco Mammals is even not alluded, may find them useful. I have never seen any paper of M.'s which could show his noble character and his splendid scientific quality better than this one. I shall see him tomorrow and I shall ask him some questions, in regard to his specimens. The whole thing will appear of course in the Geol. Mag. and will be advertised over the whole world. I have written to Schlosser about it, and somebody besides Cope, ought to go for it here. Why not Scott?[832]

Baur's next surviving letter to Osborn is dated January 14, 1890, and is written in German. Note that this letter is not written on the papers of the Yale College Museums, as most of Baur's other letters had been up to this point. Baur told Osborn that he now finally felt "morally compelled" ("Ich sehe mich moralisch gezwungen") to take his leave from Marsh, that he would under no circumstances take Marsh's side in the, now more than ever, public conflict fought on the pages of "New York Herald" (see below). Baur wrote that his arguments against his employer were not of a personal nature but for the sake of the "dignity and honor of science, which had been trampled under Marsh's feet" ("um die Würde und Ehre der Wissenschaft, die von Marsh mit den Füssen getreten wurde"). He would send his family to Germany and join Hatcher on his fossil hunting expedition in the West. He would then return to Germany himself if he had found no employment in the US by late fall. Maybe Clark University would accept him. He further speculated that Marsh would only retaliate against Cope in the hope that the others mentioned in the article would remain silent, but that "would and must not happen. The affair has to be fought out now!" ("Dies wird, und darf nicht geschehen. Die Sache muss diesmal durchgekämpft werden!")[833] What follows is a series of letters detailing Baur's involvement in the "New York Herald's" articles, constituting

832 Georg Baur, New Haven, CT to Henry Fairfield Osborn, Princeton, NJ, 29 July 1889, VPA 1/8, General Correspondence Bas–Bel, BAUR, GEORGE (Prof.) Univ. of Chicago 1884–1897, Folder 36.

833 "[I]ch sehe mich moralisch gezwungen, meinen Abschied von Marsh zu nehmen. Ich Denke, Sie werden hierin mit mir übereinstimmen. Ich werde aber warten, bis ich von Ihnen gehört habe. Unter keiner Bedingung werde ich mich auf Marsh's Seite werfen; was ich zu sagen habe, ist durchaus unpersönlicher Natur, es handelt sich einfach um die Würde und Ehre der Wissenschaft, die von Marsh mit Füssen getreten wurde. Ich stehe also auf demselben Standpunct wie Sie, Scott und Meyer; mit der anderen Geschichte habe ich gar nichts zu thun. Meine Absicht ist meine Familie nach Deutschland zu senden. Ich selbst gehe mit Hatcher nach dem Westen, und sammle dieses Jahr für Woodward, Zittel, Gaudry etc. Im Spät-Herbst will ich nach Deutschland zurück, wenn ich bis dahin hier nichts gefunden habe. Vielleicht lässt sich doch etwas an Clark Univ. machen. Wie gesagt, warte ich bis ich von Ihnen höre. Ich glaube, dass Marsh's Erwiderung allein gegen Cope gerichtet sein wird, wahrscheinlich denkt er, dass dann die Übrigen, die mit hineingezogen worden sind, schweigen werden. Dies wird, und darf nicht geschehen. Die Sache muss diesmal durchgekämpft werden!" Georg Baur, New Haven, CT to Henry Fairfield Osborn, Princeton, NJ, 14 January 1890, VPA 1/8, General Correspondence Bas–Bel, BAUR, GEORGE (Prof.) Univ. of Chicago 1884–1897, Folder 36.

the most public phase of the "Bone Wars." The next letter, dated January 21, is written in English. Baur informs Osborn of his further dealings with the "Herald":

> I have been ill yesterday, but have written a paper to the Herald to day, and finished so far. I am not used to write such language, and the paper needs revision. I think it best to have it in Sunday's Herald; it will appear with a long report by Dr. O. Meyer, whom I saw in New York and who has been here to day. Could you and Prof. Scott publish something at the same time? All my statements are based on facts, and I can defend every one. I saw Prof. James D. Dana today, he expressed his regret, that I am going to leave and said that it was not right of Marsh, not to publish my whole statement.[834] He will know more about Marsh, when my paper is out; he will know that he was deceived by Marsh, who showed him an incomplete copy of Dinocerata as a complete one, and who wrote a review of his own work. Sidney Smith is sick,[835] but he is with me, Williston is in bed too, and is forbidden by the faculty to say more. I hope that the blow I am going to give Marsh will be a strong one, perhaps sufficient enough to open Powell's eyes. I am glad that the thing is out. I saw Mrs. Harger to day, who came to visit Mrs. Baur; she was very indignant; Mr. Harger left a diary with many important notes on Marsh. I probably shall go to New York tomorrow, to see the Herald-Reporter about the time of publishing. I shall telegraph you, when I go, perhaps you can come there, or Prof. Scott. I feel lonesome. I received your telegram yesterday night, but have not yet received your letter; (9. pm.) I hope that every thing will come out right.[836]

Baur's next letter, which he wrote on the following day, clarifies that he had not yet traveled to New York to meet Ballou,[837] but had received Osborn's letter. He was planning his departure from Marsh and New Haven:

---

834 Ballou: Marsh Hurls Azoic Facts at Cope, p. 11. For a detailed analysis of the article see below.

835 Sidney Irving Smith (1843–1926), professor of comparative anatomy at Yale.

836 Georg Baur, New Haven, CT to Henry Fairfield Osborn, New York, 21 January 1890, VPA 1/8, General Correspondence Bas–Bel, BAUR, GEORGE (Prof.) Univ. of Chicago 1884–1897, Folder 36.

837 "New York Herald" journalist William Hosea Ballou (1837–1957). As to the character of the journalist, Osborn wrote: "Ballou was adventurous and fond of a good fight. He lent a willing journalistic ear to professor Cope's lifelong troubles with Professor Marsh and recent difficulties with Director Powell, and when they reached the boiling point persuaded Cope that now was the time to strike." Osborn: Cope, p. 402. In 1908 Ballou wrote an article for the "Chautanquan," a real puff piece praising Cope as the most important scientist of the nineteenth century, who would "form most of the base of the American scientific pyramid" (p. 101). Ballou also considered himself to be "on most intimate terms" (p. 103) with Cope. William Hosea Ballou: Some Great American Scientists. VII. Edward Drinker Cope, in: The Chautauquan, vol. 50, no. 1 (Mar. 1908), pp. 100–117.

> Many thanks for your letter and contacts received this noon. The step I have taken is certainly an important one, and I hope that the truth will come out at last. I have nothing to do with the reporters, there was one here last Monday night, but I told him that, what I had to say I would say over my own signature. I am in a very trying position at present, I have to bring every thing in order in the Museum, all the specimens on which I have been working for the last week; I have to transfer all the books and lots of other things, which takes quite a time. Of course I must all do this, before I can say anything. There is no doubt, that Marsh wants me back, but how can I do this? I think I shall be through with my things in the Museum to morrow; after this I shall have my resignation granted. If this is done I shall go to New York, to bring my [illegible] to the Herald; I should be very glad if I could meet you there, that you could see the article, which I said before, is not a model of English language. I am hotheaded and nervous and need somebody's advice very much in this affair.

These statements are followed by a conspiratorial exchange in which Baur promised to send Osborn back his last letter and told him that he had archived every letter written by Osborn. It stands to reason that Osborn wanted to keep his involvement in Baur's "betrayal" to Marsh a secret: "In regard to your letters, I have every one ever written by you, preserved. Your last letter, I send back. I do not need to copy it." After some details on his intended communication with the "Herald," Baur once again employs Osborn's advocacy for a future position at Clark University: "Would you be kind enough to write to Stanley Hall of Clark Uni.[838] Whether there is soon possibility to get a place as docent or any other position for the next future."[839]

The next letter, dated January 26, starts with the promise to obscure Osborn's agitations against Marsh: "Included find list of papers not mentioned on your list. I send you all the papers of which I have copies left. Of some I have never received any." It remains unclear whether said papers were scientific papers or other articles written by Osborn as part of the newspaper battle.

Concerning the ongoing battle, he commented:

> I had a very good time in New York, I visited Speir again and read him my article; in fact I thought to publish it: Now Meyer's is out, and I think it is first class. What do you think about the last sentence?![840] I had a letter from Barbour

838 Granville Stanley Hall (1846–1924), the first president of Clark University.

839 Georg Baur, New Haven, CT to Henry Fairfield Osborn, New York, 22 January 1890, VPA 1/8, General Correspondence Bas–Bel, BAUR, GEORGE (Prof.) Univ. of Chicago 1884–1897, Folder 36.

840 Meyer called Marsh a "big man with a little head," in a twist on a polemic statement made by Marsh in a previous "Herald" article. William Hosea Ballou: Some More Nuts for Marsh to Crack, in: The New

> today, he wants to go for M. also. I have not seen anybody yet today. I shall have my resignation granted to-morrow or Tuesday. I think I shall give, what I have to say in the next Sunday Herald. I do not think it hurt me. Whitman wrote to me today, asking whether I had really given up my position, and whether I had something else. I shall write to him after this, and tell him my position. If I only could get a place there. (Clark) I think everything will come out all right; I have no idea what Marsh will do, and whether Dana will explain the note [illegible] the appearance of the Dinocerata. I think Yale will open her eyes, and see what Marsh really is.[841]

Fortunately for Baur, his job application with Whitman and Clark University was successful. Yale however never "opened her eyes" to Marsh's alleged treacheries, at least no consequences ever followed.

After all correspondences and elaboration, Baur did not publish his accusations against Marsh in the "Herald." In his next letter to Osborn (which is written in German) he explains that he had not published the article because Marsh had not as of yet responded to Meyer's "challenge" ("Meyer's Herausforderung") and that his article would mostly reaffirm the allegations made in Meyer's paper. He had messaged Marsh on Friday, telling him that he was expecting his job application to Whitman to go unheard and that he would try for some position at the Smithsonian Institution for the next two months. In the meantime, he would send his family to Germany.[842] Indeed, Baur did send his family to Germany and his employment at Clark University was rekindled by February 14.[843]

The next surviving letter of the Baur-Osborn correspondence was written on June 19, 1890, in Russel Springs, Kansas. Baur had followed through with his expedition plans and was now unearthing fossils: "I have splendid luck; found a wonderful Mo-

---

York Herald, 26 January 1890, p. 25.

841 Georg Baur, New Haven, CT to Henry Fairfield Osborn, New York, 26 January 1890, VPA 1/8, General Correspondence Bas-Bel, BAUR, GEORGE (Prof.) Univ. of Chicago 1884–1897, Folder 36.

842 "Sie werden gestern vergebens nach einem Artikel von mir im Herald gesucht haben. Ich hielt es nach reichlicher Überlegung für das Beste nichts zu sagen, wenn Marsh auf Meyer's Herausforderung schweigen sollte. Er hat geschwiegen u. somit wäre die Aufgabe des Herald in dieser Sache vorerst zum Abschluss gekommen. Was ich zu sagen hatte, war zum grössten Theil eine Bestätigung der Aussagen Meyers mit einigen weiteren Angaben u. näheren Ausführungen. Ich bin am Freitag mit Marsh fertig geworden. Habe vorerst noch nichts zu thun. Habe heute an [illegible] geschrieben und ihn gefragt, ob ich am Smithsonian für etwa 2 monate Arbeit finden könnte. Von Whitman habe ich noch nichts gehört, ich glaube kaum, dass dort was zu machen ist." Georg Baur, New Haven, CT to Henry Fairfield Osborn, New York, 3 February 1890, VPA 1/8, General Correspondence Bas-Bel, BAUR, GEORGE (Prof.) Univ. of Chicago 1884–1897, Folder 36.

843 Georg Baur, New Haven, CT to Henry Fairfield Osborn, New York, 14 February 1890, VPA 1/8, General Correspondence Bas-Bel, BAUR, GEORGE (Prof.) Univ. of Chicago 1884–1897, Folder 36.

sasaur. […] I feel very good and can work like one who has done this kind of work for years.” Still, the whole Marsh-affair occupied some of his mind: “I am glad you have left O. C. M. alone. Williston wrote me that you[?] will have some notes about the great man. I have written a short note on his latest Dinosaur paper for the Naturalist.”[844] Baur kept in touch with Cope as well: “Prof. Cope sent me the May Naturalist.”[845]

In July 1890 Baur had arrived in Worcester, Massachusetts, ready to begin his work at Clark University. He had rented a place to live and felt very much at home. Things were starting to look up, it seems. But Baur still had to lend money from Osborn, who this time obliged:

> I have found however a very nice tenement near the University, at the low rent of $ 15 per month. […] I am ashamed to say that you would do me a great favor by lending me $ 25 more. The first of October I shall receive my salary for October. […] You have no idea how glad I shall be, to find me again in a regular condition. I like Worcester very much and feel more at home after two weeks than at New Haven after six years.

The letter ends with a postscript written in German in which Baur again asks for Osborn’s forgiveness for being so blunt and insistent on his financial help, stating that he was terribly sorry for his impertinence.[846] Baur proceeded with his paleontological work and was now describing and preparing the aforementioned *mosasaurus* for transportation: “I have work enough this year to get about without any collection. I am now handling the Mosasaurus and I shall describe the head before I send it to Zittel.” The rest of the letter is filled with remarks about the latest European paleontological publications, proving that Baur was still part of the international scientific discourse.[847] A few days later Baur’s financial troubles seem to have been sorted out: “I have just received some money from Munich, which keeps me straight now until I get my regular salary. I do not need therefore your assistance.” The last sentence is a clear indicator that Baur still had some financial issues and Osborn had most likely finally

---

844 Most likely Baur meant: George Baur: Prof. Marsh on Hallopus and Other Dinosaurs, in: The American Naturalist, vol. 24, no. 282 (Jun. 1890), pp. 569–571.

845 Georg Baur, Russel Springs, KS to Henry Fairfield Osborn, New York, 19 June 1890, VPA 1/8, General Correspondence Bas–Bel, BAUR, GEORGE (Prof.) Univ. of Chicago 1884–1897, Folder 36.

846 “P.S. Bitte nehmen Sie mir eine wiederholte Bitterei nicht übel, es ist mir furchtbar unangenehm, aber die Verhältnisse haben es leider so mit sich gebracht.” Georg Baur, Worcester, MA to Henry Fairfield Osborn, New York, 8 September 1890, VPA 1/8, General Correspondence Bas–Bel, BAUR, GEORGE (Prof.) Univ. of Chicago 1884–1897, Folder 36.

847 Georg Baur, Worcester, MA to Henry Fairfield Osborn, New York, 14 September 1890, VPA 1/8, General Correspondence Bas–Bel, BAUR, GEORGE (Prof.) Univ. of Chicago 1884–1897, Folder 36.

offered his help. The rest of the letter illuminates Baur's view on the Darwinian theory of evolution and proves he was a Neo-Lamarcist:

> I have read Wallace's Darwinism again, but with the same result. I am unable to believe to [sic!] Darwinism. I have just sent proof of my paper on the Galapagos Lizards to the printers (Biol. Centralbl.) With the coolution [sic!] of these species Darwinism has nothing to do whatsoever. I can only repeat what I have said long ago; that Darwinism simply presents the species made by Lamarckism.[848]

It seems Baur had started over in Worcester, leaving his personal and financial troubles behind, for the next surviving letter to Osborn is of purely scientific nature.[849] Alas, a few days later the ghosts of the past seem to have returned to haunt him:

> At least I received news from Hatcher. What made him stay with M? I do not understand it. There is certainly some 'Teufelei' [devilry] behind it. I think my Galapagos-problem will also go to water. It is hardly to believe that it is impossible to raise a thousand Dollars for such an undertaking.[850]

In Baur's mind poor Hatcher was somehow forced by Marsh to stay in his employment and that the professor must have used some kind of wickedness to ensnare him.[851] Indeed, there are several letters written by Hatcher to Marsh that suggest that Hatcher had some serious financial grievances with his employer. He complained on various occasions that salaries and other founds promised by Marsh had not yet arrived. When his young son died in 1889 Hatcher wrote a letter to Marsh asking for a loan to pay for the funeral of the child. It is rather revealing of Hatchers financial situation that he needed to borrow money in the first place, and the letter also reveals how desperate

848 Georg Baur, Worcester, MA to Henry Fairfield Osborn, New York, 19 September 1890, VPA 1/8, General Correspondence Bas-Bel, BAUR, GEORGE (Prof.) Univ. of Chicago 1884–1897, Folder 36.

849 Georg Baur, Worcester, MA to Henry Fairfield Osborn, New York, 15 January 1891, VPA 1/8, General Correspondence Bas-Bel, BAUR, GEORGE (Prof.) Univ. of Chicago 1884–1897, Folder 36.

850 Georg Baur, Worcester, MA to Henry Fairfield Osborn, New York, 21 January 1891, VPA 1/8, General Correspondence Bas-Bel, BAUR, GEORGE (Prof.) Univ. of Chicago 1884–1897, Folder 36.

851 According to Schuchert and LeVene Hatcher had a contract with Marsh that had him employed until 1896, but when the USGS declined to pay for his services in 1892 due to budget cuts Marsh would have had to pay Hatcher out of his own pocket and the contract was dissolved in 1893. After that Hatcher was employed by Princeton and continued his fieldwork, besides being plagued by rheumatism. It seems that Hatcher had some difficulties with Marsh and harbored some bitterness towards him, but that bitterness subsided in later years and he is quoted to have said that Marsh was the best paleontologist of them all. Schuchert; LeVene: O. C. Marsh, pp. 222–225.

he must have been; Hatcher promises Marsh to pay him “whatever rate of interest” he would deem to be right.[852] Marsh lent him $ 250 for the funeral and graveyard lot.[853]

That Hatcher remained loyal to Marsh is evidenced in a letter from January 1890 in which he writes that Baur had asked him to resign from Marsh and join him in his plans to go into the “collecting business”:

> I had had a letter from Baur signifying that he would most likely resign & asking me definitely if I would go into the collecting business with him. Of course it is impossible for me at present to go into such a business with anyone without doing you an injustice & even if I were inclined to I could not now under my present financial conditions. I think I offended him in my reply as I have not heard from him since. However I will state definitely so that there may be no misunderstanding between you and I that untill [sic!] my time with you is up I shall enter into no such business with anyone unless it be by your consent & approval.[854]

Still, there might have been some hope for Baur that Hatcher would leave Marsh in 1890 for the relationship between Marsh and Hatcher worsened with time. In March Hatcher asked Marsh to support his application for a permanent position with a scientific institution, as Marsh had apparently promised to do several years before, and also asked him for a letter of recommendation. He told Marsh:

> It is with a great deal of reluctance that I have thought of closing my connections with you, for I have much to be thankful to you for. But during the past year matters have so shaped themselves that I no longer feel contented. Perhaps I am too sensitive but I feel sore over somethings, & if I am, it is my fault, & I will have to suffer for it. You no longer have a place for me in the museum winters, & at times I think you are only waiting for an opportune moment to get rid of me in the field.[855]

852 John Bell Hatcher, Long Pine, NE, to Othniel Charles Marsh, New Haven, CT, 21 April 1889, MS 343, Series I. Correspondence, Box 15, Folder 612.

853 See: John Bell Hatcher, Long Pine, NE, to Othniel Charles Marsh, New Haven, CT, 2 May 1889, MS 343, Series I. Correspondence, Box 15, Folder 613.

854 John Bell Hatcher, [?], NE, to Othniel Charles Marsh, New Haven, CT, 27 January 1890, MS 343, Series I. Correspondence, Box 15, Folder 615.

855 John Bell Hatcher, [?], WY, to Othniel Charles Marsh, New Haven, CT, 30 March 1890, MS 343, Series I. Correspondence, Box 15, Folder 615.

In the same letter Hatcher promised not to join forces with any of Marsh's enemies: "The position I have in view is in a Western institution & not with any of your enemies." The relationship deteriorated further, and Hatcher wrote to Marsh:

> Since you did not send me a recommend I suppose you had rather not give me one or think me unworthy of one. As for giving Yale the first chance & the National Museum the second I will say that I had rather be connected with either of those institutions than any other in the world for it is there that the result of six of the best years work of my life are stored. But when I work as I have I like to go up one round at least in the ladder instead of being pulled down two or three as you have done with me the past year.[856]

Hatcher goes on to complain that Marsh had bestowed more trust and career opportunities upon Beecher, another one of his collectors, than he had granted Hatcher. He closes his letter by threatening to quit unless Marsh would be willing to give Hatcher "full credit" for his work. Nevertheless, the situation did not improve but Hatcher did not quit. In May he had still not received a letter of recommendation and still threatened to terminate his employment:

> I would like to quit to be on perfectly friendly terms with you. But in view of the turn matters have taken the last few months I do not think I can ever be contented as I was before & then there is no chance for me ever to do anything of any account but work in the field as long as I am with you. Your letter of the 22nd gives me no encouragement, as I made you a fair offer in my previous letters & you do not even so much as refer to it. You say you rather give me too much credit instead of too little. I should like to know in what instance you have given me too much credit. [...] You cannot surely blame me for wanting to get a permanent position for I am getting to that age when I ought to make up my mind & settle down to something. I now have a chance of getting a position & I may never have another such a chance if I let this one go, for one of them I know is a good one.[857]

In the same letter Hatcher also tells Marsh that Baur had written to him and told him of his recent appointment at Clark University and his plans to collect fossils in Kansas. Likely the threat of Baur coming to Kansas and working with a disgruntled Hatcher

856 John Bell Hatcher, Long Pine, NB, to Othniel Charles Marsh, New Haven, CT, 16 April 1890, MS 343, Series I. Correspondence, Box 15, Folder 615.

857 John Bell Hatcher, Lusk, WY, to Othniel Charles Marsh, New Haven, CT, 1 May 1890, MS 343, Series I. Correspondence, Box 15, Folder 616.

against him took effect. Marsh swiftly promised not only to write the recommendation but also to see that Hatcher would receive a position with the Smithsonian or at Yale, and he furthermore raised Hatcher’s salary by $ 50 monthly.[858] A few days later Hatcher answered, called Marsh’s raise offer “a very liberal one,” and the matter was settled for the moment.[859] After Marsh’s operation under Hatcher’s supervision had endured the menacing Baur visit (see below), tensions rose anew. On August 1 Hatcher wrote to his boss that he had still not received the promised money, a long standing problem worsened by the fact that Hatcher now openly proclaimed he needed said money to pay off his indebtedness to Marsh, to close operations down and to quit. He ended the letter stating: “I am thoroughly disgusted with the way everything has gone the past year. You never pay any attention to my requests nor to anything I write to you.”[860] It appears that Marsh had sent a letter and the requested money to Hatcher but they arrived later than expected. Hatcher telegraphed his employer immediately and told him he had received the money on August 7.[861] Still, the tone present in Hatcher’s letter written on August 1, the harsh words chosen indicate the deteriorating state the Hatcher-Marsh relationship was in. In his next letter, written on August 7, Hatcher again told Marsh that he would “be able to secure a position [in a scientific institution] commencing October 1st.”[862] He further wrote that he would like to accept the position, for he wanted to settle down and not spend twelve months a year collecting fossils. On August 20 he wrote to Marsh that he had “fully decided to accept a permanent position this fall.”[863] He then elaborated on his reasons for doing so:

> I have endeavored during the last seven years by my work with you to show that I was worthy of a permanent position & I have several times brought the matter before you, but always to no purpose. Since my marriage three years ago; instead of considering my changed position & trying to make matters more agreeable to me, you have kept me in in the field constantly. A few days after arriving in New Haven last winter one of your bosom friends told me that you

858 Handwritten copy of: Othniel Charles Marsh, New Haven, CT to John Bell Hatcher, Lusk, WY, 9 May 1890, MS 343, Series I. Correspondence, Box 15, Folder 616.

859 John Bell Hatcher, Lusk, WY, to Othniel Charles Marsh, New Haven, CT, 24 May 1890, MS 343, Series I. Correspondence, Box 15, Folder 616.

860 John Bell Hatcher, Lusk, WY, to Othniel Charles Marsh, New Haven, CT, 1 August 1890, MS 343, Series I. Correspondence, Box 15, Folder 618.

861 Telegram of John Bell Hatcher, Lusk, WY, to Othniel Charles Marsh, New Haven, CT, 7 August 1890, MS 343, Series I. Correspondence, Box 15, Folder 618.

862 John Bell Hatcher, Lusk, WY, to Othniel Charles Marsh, New Haven, CT, 7 August 1890, MS 343, Series I. Correspondence, Box 15, Folder 618.

863 John Bell Hatcher, Lance Creek, WY, to Othniel Charles Marsh, New Haven, CT, 7 August 1890, MS 343, Series I. Correspondence, Box 15, Folder 618.

> considered me an element of discord in the museum & that you did not want me there. On the following day I told you that I would quit July 1st or as much sooner as you preferred & I at one commenced looking for another position. This summer you wrote me that you had several positions to offer me & I wrote back wanting to know what they were (for I had much rather accept a position under you). You have never told me of a single position, although I have waited over three months. I have tried a great many places & I find that positions of that kind I want are very scarce. Have had several offers at a starvation salary & a few very fair[,] I now have an offer of just such a position as I want about 4 months in the year collecting & the remainder of the time working on collection with an assistant & at the same salary that you are now giving me. It looks hard to have to leave the immense collection I have made under you the last seven years & commence all over again & it is hard; for I know that no person ever took more pride in their work than I have taken in mine since I commenced with you, but if I have to break off, the sooner the better. I wish you would come out here & bring my notes with you. For I want to settle up everything with you & leave in a perfectly friendly, fair & honorable way.

Those were clear words; Hatcher was very frustrated to be stalled for a better job by Marsh, and this must have been especially frustrating after not siding with Baur on his visit (see below). But it also speaks volumes concerning Marsh's background. The wealthy professor probably had little understanding as to why someone would need more money or a better position, and that in a timely manner. The life-long bachelor probably also had little sympathy for Hatcher, whose needs had changed when he had decided to start a family, who could no longer devote all his time and energy towards paleontology. But the letter at least prompted Marsh to pay Hatcher and his operation a visit in September of the same year. It seems that Hatcher and Marsh talked things over during the professor's stay in Wyoming; the next letter Hatcher wrote to his boss seems much more amiable then the one partially cited above,[864] and in November he wrote to Marsh that he would come to New Haven to look for a house and that he would remain with Marsh.[865] In December, after Hatcher had relocated to New Haven, he struck a new contract with Marsh. He promised to stay with Marsh for another five years, until July 1, 1896 (for the contract began in July 1891), and he was to be paid $ 1,800 annually. His position would be "Assistant in Geology," and he would work six months of the year in the field, five months "in the East," and had one month va-

864 John Bell Hatcher, Camp on Lance Creek, WY, to Othniel Charles Marsh, New Haven, CT, 16 October 1890, MS 343, Series I. Correspondence, Box 15, Folder 619.

865 John Bell Hatcher, Long Pine, NE, to Othniel Charles Marsh, New Haven, CT, 16 November 1890, MS 343, Series I. Correspondence, Box 15, Folder 619.

cation.[866] The crisis was over and Hatcher would remain loyal to Marsh, as evidenced in the next letter he wrote back in the field in February 1891, where he told Marsh that Osborn had been trying to hire one of the members of Hatcher's team and that Osborn had tried to hire Hatcher himself before.[867] Another proof of his loyalty came in February 1892 when Hatcher wrote to Marsh telling him that he had asked Baur to pay him back some money he owed him:

> I wrote him last Spring asking him to send me the money he owed me & he asked me to wait until Oct. & I have not heard from him since. I wrote him again about a month ago asking him to send it & he has not answered if he does not answer & show a disposition to pay I shall give the account to a lawyer to collect. When he borrowed the money he was going to pay it back right away.[868]

As a reward for his loyalty Hatcher finally received a letter of recommendation from Marsh in January 1893.[869] They had terminated their contract in December of the previous year and Hatcher signed a contract to do field work for Princeton University.

As mentioned above Baur sought contact with Hatcher in the summer of 1890 and paid him a visit in the field. Hatcher wrote to Marsh immediately that Baur had started collecting fossils in Kansas;[870] in June Baur had apparently not yet met "very great success" in his fossil-hunting endeavor.[871] One month later Hatcher informed Marsh that he had received a letter form Baur and that Baur had had some success in Kansas and was going to visit Hatcher before returning home. Hatcher quickly ensured Marsh that he was not worried that Baur and his team would find anything at Hatcher's dig site and that he had collected anything of interest there and thus Baur could do "no damage."[872] Later he informed Marsh that he expected Baur would arrive at July 19, and that he would receive him as a guest and would claim anything Baur should

---

866 John Bell Hatcher, New Haven, CT to Othniel Charles Marsh, New Haven, CT, 20 December 1890, MS 343, Series I. Correspondence, Box 15, Folder 619.

867 John Bell Hatcher, Lusk, WY to Othniel Charles Marsh, New Haven, CT, 2 February 1891, MS 343, Series I. Correspondence, Box 15, Folder 620.

868 John Bell Hatcher, Long Pine, NE to Othniel Charles Marsh, New Haven, CT, 12 February 1892, MS 343, Series I. Correspondence, Box 15, Folder 627.

869 See: Othniel Charles Marsh, New Haven, CT to John Bell Hatcher, New Haven, CT 10 January 1893, MS 343, Series I. Correspondence, Box 15, Folder 625.

870 See: John Bell Hatcher, Lusk, WY to Othniel Charles Marsh, New Haven, CT, 31 May 1890, MS 343, Series I. Correspondence, Box 15, Folder 616.

871 See: John Bell Hatcher, Lusk, WY to Othniel Charles Marsh, New Haven, CT, 10 June 1890, MS 343, Series I. Correspondence, Box 15, Folder 616.

872 John Bell Hatcher, Lusk, WY to Othniel Charles Marsh, New Haven, CT, 1 July 1890, MS 343, Series I. Correspondence, Box 15, Folder 616.

find in Marsh's name: "I shall take everything he finds, if he goes out, myself & ship it to you. Under no other conditions can he come."[873] Despite all these endurances Marsh immediately responded via telegram: "Very important not to take B. to camp. He is now open enemy of survey and me, and doing all he can against both. His visit will lead to serious trouble, and I hope you will telegraph him not to come."[874] It is worth mentioning that Hatcher did not follow orders, did not uninvite Baur, but telegraphed to Marsh that "no injury shall occur to you through Bs visit."[875] Two days later Hatcher gave Marsh an update on the Baur situation:

> Dr. Baur came yesterday, four days before I expected him. I have told him that I would receive him or any other guest & that if he went out to camp anything & everything that he should find would be turned over to me, & he said certainly. But he will only be at camp one or two days, & then I will go down to long Pine with him for a day or two.[876]

Three day later Hatcher described Baur's visit in more detail, starting with an apology:

> I am very sorry that Baur's visit has caused you so much uneasiness, for I am sure you have no cause to be at all uneasy. I told him before I took him out to camp that I received him as a friend & a gentleman & that I should expect him to make no use of anything he saw while with me in anyway, & I am sure he will not. If you think Baur can use me to your injury in anyway [sic!] you do me an injustice. I have always protected myself & you out here & I am not at all alarmed yet, nor would I be if Baur, Scott, Cope & all should invade my territory. Baur will stay here a few days with me & then goes on to New York. He is through collecting for this season. He was at camp only one day & was greatly pleased with his visit.[877]

No wonder Marsh's and Hatcher's relationship soured after this clear insubordination (see above), yet still Hatcher stuck with Marsh, did not side with any of his ene-

---

873 John Bell Hatcher, Lusk, WY to Othniel Charles Marsh, New Haven, CT, 9 July 1890, MS 343, Series I. Correspondence, Box 15, Folder 616.

874 Telegram Othniel Charles Marsh, New Haven, CT to John Bell Hatcher, Lusk, WY, 13 July 1890, MS 343, Series I. Correspondence, Box 15, Folder 616.

875 Telegram John Bell Hatcher, Lusk, WY to Othniel Charles Marsh, New Haven, CT, 14 July 1890, MS 343, Series I. Correspondence, Box 15, Folder 616.

876 John Bell Hatcher, Lusk, WY to Othniel Charles Marsh, New Haven, CT, 16 July 1890, MS 343, Series I. Correspondence, Box 15, Folder 616.

877 John Bell Hatcher, Long Pine, NE to Othniel Charles Marsh, New Haven, CT, 19 July 1890, MS 343, Series I. Correspondence, Box 15, Folder 616.

mies. Note that in August 1894 Baur went to see Hatcher in the field in South Dakota. He stayed there until September 27 and returned with 3000 pounds of fossils, as he wrote to his friend Franz Uri Boas (1858–1942), another German-born scientist, who had been employed by Clark University until recently.[878]

In an article discussing Hatcher’s career and his quarrel with Marsh Rainger concludes concerning the precarious working conditions of Hatcher, Baur, and Marsh’s other assistants:

> Marsh thwarted the aspirations of George Baur, a German paleontologist who joined his staff in 1884, blocking Baur from taking a professional appointment at Harvard. In effect, Marsh sought to prevent his assistants from structuring their own careers in vertebrate paleontology. The situation of Hatcher and others was a trying one, not least because they had little recourse against Marsh. A wealthy, independent man, Marsh was able to run his own show at the Peabody Museum and the Geological Survey. The museum, although located on Yale’s campus, was separate from Yale College, the Sheffield Scientific School, and the Medical School. With Marsh as its nominal director and with its own board of trustees, the museum was an independent institution under Marsh’s direction. Marsh thus operated as he wished, and not only Hatcher but other collectors [...] complained that Marsh failed to pay them properly, to provide them with adequate instructions, or to acknowledge their labors on his behalf. Moreover, Marsh’s assistants Williston, Otto Meyer, and others claimed that they did much of the research for Marsh’s publications without receiving credit or the opportunity to publish. [...] Marsh had little interest in providing opportunities to others whom he basically considered paid employees.[879]

Back to Baur: it seems that the 1890 season of collecting was a success (“habe sehr guten Erfolg”). In the second paragraph he wrote that he had not seen any Native Americans nor their remains yet, but he would be on the lookout for them (“Indianer habe ich zwar keine gesehen, ebensowenig irgendwelche Reste derselben. Ich werde übrigens meine Augen offen halten.”) The German scientist became a real American scientist on the “frontier.” But Baur did not only write about Native Americans to

878 “[A]m 20. August fuhren Stejneger, Mr. Fem und ich den Bad Lands in South Dakota, wo wir mit Hatcher zusammentrafen. Wir blieben bis zum 27 September und hatten sehr guten Erfolg; schickten 3000 pfd Fossilien hierher.” Georg Baur, Chicago to Franz Boas, location unknown, 23 November 1894, American Philosophical Society Library, Franz Boas Papers (Mss B B61).

879 Ronald Rainger: Collectors and Entrepreneurs. Hatcher, Wortmann, and the Structure of American Vertebrate Paleontology Circa 1900, in: Earth Sciences History, vol. 9, no. 1 (1990), pp. 14–21. Quote on page 17.

evoke adventurous "frontier life," but mainly because Boas was interested in Native American remains and artifacts for his anthropological studies.[880]

Though it seemed his luck had finally changed, Baur had trouble financing his proposed expedition to the Galapagos Islands. Haunted by the memories of his financial and personal struggles of the recent past, Baur nonetheless managed to finance the expedition after all since the next surviving letter to Osborn was sent from Guayaquil, Ecuador. Baur very cheerfully detailed the journey so far, including the vast-seeming and very useful German-network (he had met the son of a friend of his father's, and had spent a very agreeable evening with him at the "German Club" of Guayaquil).[881] Note that Osborn had lent Baur $ 200 for his visit to the Galapagos Islands.[882] Beginning in 1893 the correspondence with Osborn is less frequent. The next letter Baur sent to Osborn was written in Chicago, after the transition to the University of Chicago:

> Things are going slow here, and it will take another year before we are fully settled. The Museum is going up now, but our building is not furnished yet, though the plans are ready. Hatcher is looking for a position and I hope to get him here, if possible. There are many things here which are not very satisfactory, but we have to wait what the future will bring.

As part of the transatlantic paleontological network he kept up with the European publications and had written a brief review of Zittel's latest publication:

> I have received Zittel's Mammals I a few days ago; I do not like it at all. There is not a new point in the whole book; a flat compilation without good critic; for the geologist however it may be good enough.[883]

Nearly one year later Baur was still trying to cope with his new situation in Chicago:

> You ask, what I am doing especially now. Not very much, I must say; the fact is, that since I am here, I have written less than in any previous year. But you must not think that I have been idle. For my lectures in Comparative Osteology I had to go over the whole field 'de novo'; many points being very unsatisfactory, and

880 Georg Baur, New York to Franz Boas, Worcester, MA 1 May 1891, American Philosophical Society Library, Franz Boas Papers (Mss B B61).

881 Georg Baur, Chicago to Henry Fairfield Osborn, New York, 29 January 1893, VPA 1/8, General Correspondence Bas–Bel, BAUR, GEORGE (Prof.) Univ. of Chicago 1884–1897, Folder 36.

882 Baur stated so in his letter to Boas, see: Baur to Boas, 1 May 1891.

883 Georg Baur, Chicago to Henry Fairfield Osborn, New York, 21 January 1891, VPA 1/8, General Correspondence Bas–Bel, BAUR, GEORGE (Prof.) Univ. of Chicago 1884–1897, Folder 36.

> the textbooks mostly useless. [...] Chicago, as a city to live in, is horrible. You must some day establish a Department of Reptiles in the N. Y. Museum; and get me out from here.[884]

In another letter Baur again proposed that the American Museum should establish a department of reptiles, setting the tone for a never ending repetition of that request: "In case your Museum intends to establish some day a Department of Reptiles and Batrachians, including beside the living forms, the whole Osteology and Paleontology, you may think of me." In the same letter he mentions that his family had grown, a new daughter had been born: "Last week a little girl arrived in my family, both she and Mrs. Baur get along splendidly."[885]

For two years the correspondence consists of scientific discussions and staffing issues at the university and the museums of Chicago. Still, Baur had not given up on his dream of Osborn establishing a department for reptile-studies in New York: "I have not yet given up the idea, that you will open some day a Department of Reptiles in the widest sense."[886]

In April 1896 Baur wrote to Osborn concerning a more personal matter, undoubtedly reacting to news of Cope's failing health: "these are sad news about Cope! After your last letter it looked already very hopeless. What will be done with all his material?"[887] After a vacation in Munich, Baur returned to Chicago and again corresponded with Osborn, who tried to send some family photos to the German professor. Besides a scientific exchange about taxonomic details, Baur still inquired about the reptile-department:

> I am sorry that the photographs you sent to Hess-Str 32,[888] did not arrive. My sister sent me the sad news some time ago. Now you better send me one here.

884 Georg Baur, Chicago to Henry Fairfield Osborn, New York, 5 February 1894, VPA 1/8, General Correspondence Bas–Bel, BAUR, GEORGE (Prof.) Univ. of Chicago 1884–1897, Folder 36.

885 This letter of Baur's, recorded in the Osborn correspondence, is dated November 23, with the year illegible. Someone has dated it with a pencil "1897?" but it is more likely older. It is written on the paper of the University of Chicago, so it had to be written after 1892 since Baur had only been employed there after 1892. The logo however is in the same style as on his other letters written in 1894 and 1895. The font of the logo changes in 1896 and stays that way in 1897. In addition, Baur mentioned with no word his sick leave due to overwork which had plagued him in 1897. Georg Baur, Chicago to Henry Fairfield Osborn, New York, 23 November 1894 [?], VPA 1/8, General Correspondence Bas–Bel, BAUR, GEORGE (Prof.) Univ. of Chicago 1884–1897, Folder 36.

886 Georg Baur, Chicago to Henry Fairfield Osborn, New York, 13 May 1895, VPA 1/8, General Correspondence Bas–Bel, BAUR, GEORGE (Prof.) Univ. of Chicago 1884–1897, Folder 36.

887 Georg Baur, Chicago to Henry Fairfield Osborn, New York, 13 April 1896, VPA 1/8, General Correspondence Bas–Bel, BAUR, GEORGE (Prof.) Univ. of Chicago 1884–1897, Folder 36.

888 Baur's address in Munich.

> [...] What are the prospects for a Reptile-Department at the American Museum? If you start it, do not forget me.[889]

A carbon copy of Osborn's response to Baur's abovementioned letter, written on November 28, is preserved: "I will send you the family group to Chicago, as you request. [...] We have an opening of our Department next Monday afternoon, and I am beginning to think about reptiles. It would greatly strengthen us if we had you here to stir up the reptiles, and I shall not forget you."[890]

In a letter dated March 16 Cope's health is the matter of discussion once again. Osborn, being in close contact with the mortally ill Cope, informed Baur of his treatment:

> You will be much grieved to hear that Cope is seriously ill. [I] have the gravest anxiety about him. He has an enlarged prostate, and instead of resorting promptly to surgical treatment he has been using morphine, and in his last letter writes me, formalin. I can hardly believe the latter, but he tells me in sober earnest, and of course he will not live long under this treatment. I have urged him most strongly to have an immediate operation by an expert. I think you will be interested to hear this, because in your correspondence with him it is a good thing to know that the man must be very ill. I have not written him in this vein, of course but I have writtencheering [sic!] him up. I am trying to go on to see him today.[891]

The next surviving letter of the correspondence is a copy of Osborn's response to a (lost) letter of Baur's, written on April 13. Again, the letter is mostly about Cope; he had died the previous day:

> Cope's death is indeed very sad. It leaves a terrible gap in our science, which we must do our best to fill in various ways. I did everything I could to prolong his life, but he shrank from an operation. He made me one of his executors, and his fossil vertebrate collection will be offered for sale. They are talking of a movement in Philadelphia to purchase it, and I suppose we are honor bound to offer it

889 Georg Baur, Chicago to Henry Fairfield Osborn, New York, 22 November 1896, VPA 1/8, General Correspondence Bas–Bel, BAUR, GEORGE (Prof.) Univ. of Chicago 1884–1897, Folder 36.

890 Copy of a letter, written by Henry Fairfield Osborn, New York to Georg Baur, Chicago, 28 November 1896, VPA 1/8, General Correspondence Bas–Bel, BAUR, GEORGE (Prof.) Univ. of Chicago 1884–1897, Folder 36.

891 Copy of a letter, written by Henry Fairfield Osborn, New York to Georg Baur, Chicago, 16 March 1897, VPA 1/8, General Correspondence Bas–Bel, BAUR, GEORGE (Prof.) Univ. of Chicago 1884–1897, Folder 36.

> there first. I telegraphed you this morning asking you to send me your estimate of Cope's most important contributions to Batrachia and Reptilia. I am to review his life in 'Science' and I desire to make it as adequate as possible.[892]

Furthermore, Osborn had nothing but praise for his dead friend and mentor and again had to put off Baur's ambitions to purchase the late Cope's collection, preferably for the American Museum and as the basis for the proposed reptile department:

> I was glad to get your letter and enclosures of April 20th. I did not receive it in time to make much use of it for the Amphibia, but I put in your manuscript bodily for the Reptilia. I am perfectly amazed in this review at the extent and value of Professor Cope's work. Do you agree with me, that he ranks with Cuvier, and is in some respects a superior of Owen and Huxley.? [sic!] I am placed in a delicate position as regards the collection, because Cope has established a professorship at the Academy of Natural Sciences, to be founded by the sale of his collection. I know that it was his preference the collection should go to Philadelphia, and as one of his executors I am morally bound to respect this wish. We propose to give the Academy of Natural Sciences the option upon its purchase at $ 50,000, which will expire at the end of three month. I shall then be free to negotiate for it myself, but it will not be an easy matter since we have just recovered from the strain of paying for the mammalia, I am sure I wish we could get it here, and you with it. We could make the grandest department in the world.[893]

Now Baur, sensing opportunity, made preparations to succeed Cope in Philadelphia, asking Osborn for his help in that matter:

> If Cope's collection remain[s] in Philadelphia, I should like [to] take the Professorship at the Philadelphia Academy. The question is, whether I could not at the same time take Cope's place at the University. I should be very much obliged to you if you would assist me in this matter. I think with the opportunities in Philadelphia I could do much more better work than here. I prefer the East very much to the West, besides we are here entirely isolated, completely

892 Copy of a letter, written by Henry Fairfield Osborn, New York to Georg Baur, Chicago, 17 April 1897, VPA 1/8, General Correspondence Bas–Bel, BAUR, GEORGE (Prof.) Univ. of Chicago 1884–1897, Folder 36.

893 Copy of a letter, written by Henry Fairfield Osborn, New York to Georg Baur, Chicago, 30 April 1897, VPA 1/8, General Correspondence Bas–Bel, BAUR, GEORGE (Prof.) Univ. of Chicago 1884–1897, Folder 36.

> removed from scientific intercourse. Could you tell me to whom I have to apply in Philadelphia about the University Position. I had written to Frazer,[894] and and [sic!] received an answer from him, from Stockholm. He is on his way to the [illegible] Congress. He writes: 'I have no doubt that Prof. Cope would have been glad to know that the Professorship would come to you and if I can in any way aid in accomplishing his wishes I shall be glad to do so.' I have now occupied my new quarters in the Zoological building. Great improvement, but I have no Assistant![895]

The ever ambitious, ever industrious Baur kept in contact with Osborn for the rest of his short life. When his health was again starting to fail, this time marking the beginning of the end, he took a vacation to Lake Oconomowoc to heal his seemingly damaged nerves from stress, related to him overworking: "Since two weeks I am here at the beautiful Oconomowoc Lake, to rest from the work I have done during the last 9 months. I was very sorry I could not come to Detroit and Toronto, but it was not possible."

The very next sentences reveal Baur's industrious nature, which might have contributed to his untimely death at the age of 39, for even during his vacation home he was thinking about and delegating work:

> I have written to Mrs. Baur to send you abstract [sic!] of paper on The Pelycosauria and the Origin of Mammals. [...] My reviews of Seeley's papers have now been printed and you will find them in the 2 or 3 last numbers of the Neues Jahrb. F. Geol. Min. Pal. I have no copies. – This seems to be a slow proceeding in Philadelphia; what do you think about it? [...] As soon as a [sic!] I am back in Chicago; I shall write a paper that the Carboniferous Anthracosauria are true Reptiles, ancestral to the Permian Pareiasauria.[896]

A copy of the last letter Osborn wrote to his friend is archived with the Osborn Correspondence:

> I am very sorry to learn from your friend, Dr. Boaz, that you have not been well lately, and have been obliged temporarily to let up from your work. I write

894 Persifor Frazer (1851–1899), mineralogist and chemist, professor at the University of Pennsylvania and an editor for the "Naturalist."

895 Georg Baur, Chicago to Henry Fairfield Osborn, New York, 18 July 1897, VPA 1/8, General Correspondence Bas–Bel, BAUR, GEORGE (Prof.) Univ. of Chicago 1884–1897, Folder 36.

896 Georg Baur, Oconomowoc, WI to Henry Fairfield Osborn, New York, 17 October 1897, VPA 1/8, General Correspondence Bas–Bel, BAUR, GEORGE (Prof.) Univ. of Chicago 1884–1897, Folder 36.

> especially to cheer you up, and send my friendly greetings and best wishes for your speedy recovery, and at the same time you will allow me, as an old friend, to give you a word of advice. I sincerely hope that you will take a thoroughly good rest, strictly follow the advice of the doctor and ward off any possibility of a serious breakdown from overwork. You have had a great deal to worry you, and I imagine that you have been working tremendously hard. I can easily imagine, although I have never experienced it myself, how hard it would have been temporarily to suspend work, but it seems to me that if the doctor advises it, as Boaz informs me he does, that you must acquiesce, and for a time at least take a thorough change and rest. Your work is so important in the present state of paleontology, especially since Cope has gone, that you must not run even the slightest risk of a breakdown. By taking good care of yourself now when you have had a warning, get a complete recovery, and then when you renew your work but on a little less steam.[897]

But Baur never did recover; he moved back to Germany to take a long vacation but died in an asylum in 1898.

6.2.4 Osborn, Cope, and the German Assistants

In his Cope biography Osborn quoted various letters from the Cope Correspondence. In one letter cited Cope wrote about the unhappiness of Marsh's assistants with whom he had made contact. The letter is dated October 27, 1885, and Cope described how four men who had recently left Marsh's employment had confirmed that Marsh was scientifically incompetent and even "more of a pretender" than Cope had suspected; and that Marsh was merely a "scientifico-political adventurer," but not a real scientist.[898] The four men could bear witness to Marsh's fraudulent conduct and were "anxious to publish in both Europe and America." Then Cope wrote about Baur, deeming him "as important as the four," and attested that he had "a good deal of a backbone" but was also indebted to Marsh, whom he owed $650. Cope was looking for someone to lend Baur the money but could not do it himself. He only promised to try to raise $200 and proceeded to ask Osborn if he could lend Baur the rest of the money. In a letter dated November 28 Cope lamented that Baur could not be helped in the foreseeable future and that Baur could not air his disagreements with Marsh while indebted

897 Copy of a letter, written by Henry Fairfield Osborn, New York to Georg Baur, Chicago, 6 December 1897, VPA 1/8, General Correspondence Bas–Bel, BAUR, GEORGE (Prof.) Univ. of Chicago 1884–1897, Folder 36.

898 Osborn: Cope, p. 380.

to him.[899] Therefore it became more important to secure Schlosser's support against Marsh. Meyer, it seems, was already wholly on board:

> As Baur is thus apparently out of the project, it becomes more important to have Schlosser's signature. Where the number is so small every one counts. I will ask Meyer to send him a copy of the document so that he can wire his signature.[900]

In the same letter Cope wrote that they should start their public attack as soon as possible, and that a "Newspaper man of New York" had told him that the attack would make it to the press as well.

In a letter dated August 6, 1886, Cope wrote: "I spent a day in New Haven, recently, with Baur and Williston. Baur had a new daughter while I was there."[901] In a letter dated 21 October, 1886, Cope stated that he had secured Schlosser's support against Marsh.[902] On November 21, 1887, Cope told Osborn that he was still gathering witness accounts against Marsh and was looking for someone other than a direct rival or employee of Marsh, and that the accounts of Williston, Meyer, and himself were not enough.[903]

Few contracts between Marsh and his assistants survived. Schuchert and LeVene state that Harger, Marsh's first permanent assistant, was paid $ 35 a month and that his salary was increased until he earned $ 1500 a year in 1883. Williston was guaranteed $ 40 in his first contract with Marsh, signed in 1876. In 1879 he signed another three-year contract and was to be paid $ 2500. In 1882 a five-year contract assured Williston (who had received a M.D. degree in the meantime) $ 1500 per year. This contract was terminated in 1885 after Williston had received a Ph.D. for his research on flies and started practicing medicine and teaching anatomy.[904]

6.2.5 Preliminary Conclusion

When Baur was asked by Marsh via Zittel to work at the laboratories of the Peabody Museum in 1883, the proposition must have seemed like a once-in-a-lifetime opportunity. Marsh had an outstanding international scientific reputation. The fossils of the

899 Osborn: Cope, p. 381. In the next cited letter, dated February 14, 1886, Cope again wrote that Baur could not criticize Marsh while he was still in his employment. Osborn: Cope, p. 382.

900 Osborn: Cope, p. 381.

901 Osborn: Cope, p. 384.

902 Osborn: Cope, p. 385.

903 Osborn: Cope, p. 388.

904 Schuchert; LeVene: O. C. Marsh, pp. 295–302.

American West, particularly the spectacular dinosaur skeletons in which Baur had a special professional interest, were also of world renown. Together with his colleague (and presumably friend) Max Schlosser he ventured to New Haven, and shortly after Baur married his wife Auguste. Therefore, the shock must have hit all the harder when Baur realized his employment at Yale was somewhat of a dead end. While Baur published more scientific papers than Marsh's other assistants (at least while they were employed by Marsh), he felt that his talent was squandered because his own research and publications were being hindered. Furthermore, due to his financial obligations to Marsh, Baur could not leave New Haven and grew increasingly desperate and frustrated, maybe even depressed, as shown in his letters to Osborn, and one very accusatory article published in the "American Scientist" in 1890 (see below). Baur not only lost any faith he had had in Marsh but grew to despise his employer and criticized his scientific ability as well as his integrity. Either Baur found an ally against Marsh in Cope and Osborn, or Cope took advantage of Baur during the "Bone Wars." In either case, most of their communication (Marsh might have called it plotting) was conducted through Osborn and Scott. When Baur finally escaped from New Haven he did some field work in Kansas and remained a member of the US-American scientific community for the rest of his life. Baur was never truly content with his academic career and always struggled with funding for new projects. Nonetheless, he was a hard worker and might have even worked himself to death, suffering from burnout and depression. Schuchert and LeVene call Baur "brilliant", and he certainly was very productive and made an impression on the small US-American paleontological community, to which he contributed through his studies and publications.

6.3 Max Schlosser

Max Schlosser was born on February 5, 1854, in Munich, where he spent most of his life. Beginning in 1873 he studied natural sciences, under Zittel amongst others. In 1880 he received his doctorate by writing a dissertation about the Jurassic fauna of the shell bearing limestone Kelheim (Kelheimer Muschelkalk). Together with Baur and due to Zittel's recommendation (see above), he left for New Haven in 1884, but returned to Munich in 1885. Baur told Osborn that Schlosser was fed up with Marsh and his exploitative work habits. In Munich Schlosser remained in Zittel's service, became a curator in 1890 and a conservator and custodian in 1900. He retired in 1924 but continued his work at the fossil collection of the Bavarian state. He died in 1932.[905]

905 Helmut Mayr: Schlosser, Max, in: Neue Deutsche Biographie 23 (2007), pp. 107–108 (online version, URL: https://www.deutsche-biographie.de/gnd117330353.html#ndbcontent, as consulted online on

No letters of Max Schlosser are preserved in the Yale University's Marsh papers. During his year in New Haven Schlosser kept in touch with the German paleontological establishment, as evidenced by a letter dated September 26, 1884, in which he asked Carus to publish a correction in the "Zoologischer Anzeiger."[906] Carus, one of the founders of the journal, obliged and had the errata printed in issue 182.[907] In another letter, dated January "2th" 1885, Schlosser asked for some copies of issue 182. Again, Carus obliged: a note at the bottom of the petitionary letter, signed by Carus, instructs for five copies of issue 182 to be sent to Schlosser's return address.[908]

Schuchert and LeVene assess Schlosser's stay at Yale as follows:

> Max Schlosser (1854–1932), although recommended as an able, conscientious, dependable, and industrious worker, stayed in New Haven even less time than Meyer, returning to the Alte Akademie in Munich early in 1885. In contrast to Meyer however, he remained an ardent student of fossil mammals throughout his life, and became their leading interpreter in Germany. [...] Gibb and Westbrook remembered Schlosser as a very difficult man to live with, suspicious, exceedingly nervous, outspoken, and excitable. One day while walking with Baur past the old Treasury Building he was run into some students just dismissed from the building. He at once challenged the lot of them, and raised his cane, threatened to whip them all. Baur, however, got him away unharmed. Schlosser did not like American ways, and Westbrook agreed that he never fitted himself into the Marsh laboratory. His return to Germany was due in part to difficulties with Marsh, and in part to illness.[909]

There are, however, some letters written to Osborn. The relationship between Osborn and Schlosser is more professional and less personal than the one between Osborn and Baur. All letters Schlosser sent to Osborn were written in German. Schlosser's handwriting was very clean and neat at the beginning but grew increasingly shaky over time. Osborn's letters, as far as can be deduced from the few surviving carbon copies of his typewritten letters, were written in English. As stated in a letter to Baur (see above), Osborn had met with Schlosser in Munich in 1885, presumably in Zittel's

---

January 19, 2018).

906 Max Schlosser, New Haven, CT to Julius Victor Carus, Leipzig, 26 September 1884, Staatsbibliothek zu Berlin, La 1880 (13), Schlosser, Max.

907 Max Schlosser: Nachträge und Berichtigungen zu: die Nager des Europäischen Tertiärs. Palaeontographica 31. Band, in: Zoologischer Anzeiger, vol. 7, no. 182 (Dec. 1884), p. 639.

908 Max Schlosser, New Haven, CT to Julius Victor Carus, Leipzig, 2 January 1885, Staatsbibliothek zu Berlin, La 1880 (13), Schlosser, Max.

909 Schuchert; LeVene: O. C. Marsh, p. 303.

laboratory. There is no evidence that they spoke about Schlosser's recently terminated employment in New Haven and the German's dissatisfaction with Marsh and his payment practices. The first surviving letter of the Schlosser-Osborn correspondence is dated February 17, 1888; it is a reply to a letter Osborn must have sent to Schlosser some time before, for he opened the letter with an apology for his belated reply. He further responded to Osborn's promise to send him a copy of an abstract about North American fossil mammals. Osborn must also have promised to send the teeth and other fragments of the mammals to Munich. Schlosser wrote that he would send some specimens and casts of the Munich Museum to Princeton.[910] These exchanges are the topic of the next letter, complemented by the information that Scott had safely arrived in Munich and would be staying there for fourteen days.[911]

The next surviving letter is dated Mach 29, 1891. Unfortunately, there are no letters that shed light on Schlosser's reaction to the "New York Herald's" articles and his involvement in the campaign against Marsh. In this letter, however, he congratulated Osborn on his criticism of Marsh's "cretaceous mammalia" (see above) and stated that the article might help to remediate the public image of "Mr. Marsh" ("Hoffentlich wird Ihre Abhandlung dazu beitragen das Urtheil über Herrn Marsh zu modificiren") and that Zittel's handbook also would not favor Marsh ("Auch v. Zittels Handbuch wird demselben wenig günstig werden.").[912] The actual passages of the handbook concerning Marsh and his American discoveries were nonetheless quite flattering.[913]

Once Osborn had assumed the presidency of the New York Academy of Sciences he made Schlosser a corresponding member, as can be concluded from a very appreciative letter written by Schlosser in 1899. After he formally thanked Osborn for this honor, Schlosser asserted himself to have been one of the first Europeans who had recognized the "truth" and "high value" of Cope's theories, which were then "masterfully" built upon by Osborn.[914]

---

910 Max Schlosser, Munich to Henry Fairfield Osborn, Princeton, NJ, 17 February 1888, American Museum of Natural History, VPA 1/84, General Correspondence Schau-Sci, SCHLOSSER, Dr. Max Paleo. Mus., Munich 1888–1925, Folder 50.

911 Max Schlosser, Munich to Henry Fairfield Osborn, Princeton, NJ, 8 July 1888, VPA 1/84, General Correspondence Schau-Sci, SCHLOSSER, Dr. Max Paleo. Mus., Munich 1888–1925, Folder 50.

912 Max Schlosser, Munich to Henry Fairfield Osborn, New York, 29 March 1891, VPA 1/84, General Correspondence Schau-Sci, SCHLOSSER, Dr. Max Paleo. Mus., Munich 1888–1925, Folder 50.

913 "Im Jahre 1878 erhielt man zuerst Kenntnis von den Großartigen Funden im Westen, welche O. C. Marsh 20 Jahre lang das Material zu einer großen Anzahl, meist prächtig ausgestatteter Abhandlungen boten und zu einer grundlegenden Reform der Dinosauriersystematik Veranlassung wurden. Die Schriften von O. C. Marsh über die fossilen Reptilien und namentlich Dinosaurier bezeichnen eine wichtige Epoche in der Entwicklung der Paläontologie der Wirbeltiere." Zittel: Geschichte der Geologie und Paläontologie, p. 832.

914 "Wenn ich mir wirklich ein Verdienst zuschreiben darf, so besteht es nur darin, dass ich wohl al seiner jener europäischer Paelaeontologen gelten kann, welche zuerst die Richtigkeit und den hohen

In 1899 Schlosser's opinion of Marsh had not improved. He had published an article about *leptodon graecus*,[915] in which he attacked some presumptions made about the mammal's genus by Fraas.[916] Schlosser carefully pointed out that his remarks were directed at Fraas' mistakes and in no way a personal attack against Osborn, who had been wrongfully cited by Fraas. Schlosser assured his friend that he knew that he harbored a great "love for truth" and therefore would not misconstrue a factual correction into a personal attack, and furthermore that Osborn had become a victim of the "Fraas-dynasty."[917] Schlosser then remarks that Fraas in his dishonesty and incompetence was on his way to become a "second Marsh" if no one would stop him, and Schlosser intended to "clip the claws" of this beast in due time.[918] A few sentences later he further remarks that he had almost denounced the corresponding membership of the New York Academy of Sciences, for which he had thanked Osborn so enthusiastically in his earlier letter cited above, when he realized that Fraas had also been nominated a corresponding member.[919] Schlosser was very careful not to insult or affront Osborn with his harsh words; yet his contempt for Fraas is very bluntly stated and likened to Schlosser's scorn for Marsh, to whom Schlosser compares Fraas.

In another letter, written in October 1899, Schlosser inquires what happened to the late Marsh's library, whether it had been sold, and if he could buy his own reports to

---

Werth jener Theorien erkannt haben, welche unser unvergesslicher Freund und College E.D. Cope aufgestellt hat und welche dann von Ihnen in so meisterhafter Weise weiter ausgebaut wurden." Max Schlosser, Munich to Henry Fairfield Osborn, New York, 11 April 1899, VPA 1/84, General Correspondence Schau–Sci, SCHLOSSER, Dr. Max Paleo. Mus., Munich 1888–1925, Folder 50.

915 Leptodon graecus is a synonym for pliohyrax graecus, a Pliocene mammal discovered by Gaudry in 1862.

916 Max Schlosser: Über neue Funde von Leptodon Graecus Gaudry und die Systematische Stellung dieses Säugethieres, in: Zoologischer Anzeiger, vol. 22, no. 597 (Sep. 18, 1899), pp. 378–380; and continuded in: Max Schlosser: Über neue Funde von Leptodon Graecus Gaudry und die Systematische Stellung dieses Säugethieres, in: Zoologischer Anzeiger, vol. 22, no. 598 (Oct. 2, 1899), pp. 385–387.

917 "[I]ch Weiss sehr wohl, dass Sie die Wahrheit über Alles lieben und daher auch etwaige Correcturen nicht als persönliche Beleidigung auffassen werden. Zudem geht die Sache auch ursprünglich sicher nicht von Ihnen aus, sondern Sie sind aller Wahrscheinlichkeit nach das Opfer der in der Dynsastie-Fraas erblichen unwissenheit geworden." Max Schlosser, Munich to Henry Fairfield Osborn, New York, 15 October 1899, VPA 1/84, General Correspondence Schau–Sci, SCHLOSSER, Dr. Max Paleo. Mus., Munich 1888–1925, Folder 50.

918 "[Fraas] ist auf dem besten Wege dazu, in all und jeder Beziehung ein zweiter Marsh zu werden, doch werde ich schon Sorge tragen, dass ihm die Krallen noch rechtzeitig beschnitten werden." Schlosser to Osborn 15 October 1899.

919 "Als ich leider erst sehr viel später erfuhr, dass ausser Jäckel und mir auch E. Fraas correspondirendes Mitglied der New Yorker Academie geworden sei, war ich fest entschlossen, mein Diplom zurück zu senden." Schlosser to Osborn 15 October 1899.

Marsh this way.[920] Said reports are not part of the Marsh papers and it is unknown if they survived or had even been bought by Schlosser.

Schlosser stated his opinion about “national” science very poignantly in a letter dated February 13, 1905. He wrote that the newly founded scientific museum in Munich was now to be named “Deutsches Museum” (“German Museum”) to appease the emperor. According to Schlosser, the original title was to be “Museum von Meisterwerken der Naturwissenschaft und Technik” (“Museum for Masterpieces of Science and Technology”). The new name was nonsensical in his opinion because science and technology were inherently and fundamentally international, and there could be no national, independent science. Furthermore, a name like this would discourage international attention for the museum.[921] The Republic of Letters was alive and well. Considering Osborn’s prominent position at the *American* Museum of Natural History in New York, Schlosser’s statement might be read as thinly veiled criticism of Osborn’s museum.

In 1921 Schlosser asked Osborn for financial aid for the publication of a book on the study of the Bavarian Alps during the Eocene and Oligocene. He stated that the publication costs in Germany had skyrocketed and sincere and important projects like this book could not be founded.[922] Osborn pleaded with William Diller Matthew (1871–1930), the curator of the American Museum, on Schlosser’s behalf. Matthews denied the request, giving a quite lengthy and arrogant lecture on the current political state of Germany:

> I have the feeling that what we need to encourage in Europe is not so much publication of results as getting together the evidence. The general tendency there, especially in Germany, is to write vast amounts of voluminous and learned research upon insufficient and fragmentary material. […] It is also necessary to say that the ‘frightful expense’ of publication in Germany is due simply and solely to the wicked and suicidal policy of the present German government in spending two or three times its income and making up the deficit by debasing

920 Max Schlosser, Munich to Henry Fairfield Osborn, New York, 31 October 1899, VPA 1/84, General Correspondence Schau–Sci, SCHLOSSER, Dr. Max Paleo. Mus., Munich 1888–1925, Folder 50.

921 “Vor Allem muss ich bemerken, dass der offizielle Titel jetzt nicht mehr Museum von Meiserwerken der Naturwissenschaft und Technik ist, sondern aus Wohldienerei gegen den Kaiser in ‘Deutsches Museum’ geändert wurde. Da es aber naturgemäss nur eine internationale Naturwissenschaft und Technik gibt, aber keine deutsche oder chinesische etc. so spricht sich dieses Museum eigentlich selbst jede Existenzberechtigung ab, und für ein ‘deutsches’ Museum sich zu erwärmen besteht für amerikanische Forscher wohl kaum ein Anlass.” Max Schlosser, Munich to Henry Fairfield Osborn, New York, 13 February 1905, VPA 1/84, General Correspondence Schau–Sci, SCHLOSSER, Dr. Max Paleo. Mus., Munich 1888–1925, Folder 50.

922 Max Schlosser, Munich to Henry Fairfield Osborn, New York, 13 October 1921, VPA 1/84, General Correspondence Schau–Sci, SCHLOSSER, Dr. Max Paleo. Mus., Munich 1888–1925, Folder 50.

> the currency. [...] The real grievance of the German scientist lies against his extravagant and reckless government. And until this is reformed nothing that American scientists can do will be of much real aid, nothing that American finance can do will drag them back from the pit that they have digged [sic!] for themselves, and are still busily engaged in deepening. With the example of Russia before them it is strange to me that the intelligent German people do not turn their backs on the illusions of socialism and semi-socialism, and get back to the old but sound principles of earning your living, keeping within your income, and reducing the government to its essential functions of police, education, and a few other really necessary activities. I would have more sympathy with the intelligentsia of Europe if I could see them getting together and organizing to drive home this view point, which, I doubt not, many of them see well enough, and trying to educate the rest of the people into seeing it as well.[923]

In 1924, now seventy years of age, Schlosser wrote the first letter (at least the first surviving in the archives of the AMNH) to Osborn which contained more personal details: after complaining that the publication of his monography on the Alps had still not been financed, he shared that he had had hernia-surgery three months before writing the letter, that an old housekeeper, who had served his family for seventy years, had just died, and that he had trouble with his aching feet.[924]

Still, Schlosser's relationship with Osborn was very impersonal and professional (as far as such a thing can be concluded from their letters). Their correspondence mostly covers scientific details in articles, the exchange of fossil specimens, and the occasional opinion concerning another scientist's work.[925]

In conclusion, Schlosser was discouraged by Marsh's scientific conduct and lax payment practices, left after just one year at the museum laboratories, and returned to Munich. His first-hand encounter with US-American paleontology was but a short episode in an otherwise solid career in paleontology. More than scientific enlightenment, he probably received a lesson in human nature at Yale.

---

923 William Diller Matthew, New York to Henry Fairfield Osborn, New York, 3 November 1921, VPA 1/84, General Correspondence Schau–Sci, SCHLOSSER, Dr. Max Paleo. Mus., Munich 1888–1925, Folder 50.

924 "Auch sonst habe ich mancherlei zu beklagen. Vor 3 Monaten wurde ich wegen Hernienbruchs operirt, was freilich sehr gut gelang. Dann starb meine alte Haushälterin, die siebzig Jahre in meiner Familie diente, und jetzt laborire ich an einem Fussleiden, das mir ziemliche Sorge verursacht." Max Schlosser, Munich to Henry Fairfield Osborn, New York, 16 May 1924, VPA 1/84, General Correspondence Schau–Sci, SCHLOSSER, Dr. Max Paleo. Mus., Munich 1888–1925, Folder 50.

925 See also: Max Schlosser, Munich to Henry Fairfield Osborn, New York, 5 August 1912, American Museum of Natural History, VPA 12, Correspondence loans and exchanges, Museums and Universities Aboard 1896–1934, Box 5, Folder 3.

6.4 Otto Meyer

Of Marsh's German assistants, Otto Meyer is the most elusive. No biography or recorded date of death exists. Some information about his childhood and youth can be found in the records of Meyer's alma mater, the University of Leipzig, and at the beginning of his dissertation. Later dates can be learned through his publications and correspondences, but this data is hardly extensive.

The only biographical account of Meyer's youth and family background is a half-page at the end of his dissertation and the certificate of enrollment at the University of Leipzig. In his dissertation Meyer stated that he was born in Coswig in Anhalt on June 5, 1856. His father was a "Dr. B. Meyer," a local physician. In 1866 he attended school in Dessau and began an apprenticeship in a drugstore because he wanted to become a chemist. One year later he moved to Berlin and attained his "Abitur" (the German university-entrance diploma) in 1875. He studied chemistry for two semesters in Berlin and for a third in Marburg. Then he quit chemistry and began his study of natural science in Leipzig, where he mainly studied geology and mineralogy under Professors Zirkel and Carus.[926] His dissertation titled "Untersuchungen über die Gesteine des Gotthardtunnels" (Investigations about the stones of the Gotthard Tunnel) was published in 1878.[927] The certificate of enrollment at the University of Leipzig of 1875 confirms Meyer's statements and adds information about his religion, stating that Meyer was Jewish.[928]

Meyer was not hired on Marsh's initiative or directly through Zittel but took the initiative himself and wrote a job application letter. On December 3, 1883, he wrote that he had heard from Henry Villard, who had invited Zittel to the Yellowstone National Park (see above), that Marsh had asked Zittel to look for young German scientists he could hire:

> I had to-day a conversation with Mr. Henry Villard, Presid. of the Northern Pacif. Rail. [?], in which he told me, that you asked Professor Zittel of München, when he was here, if he could send you a few young german geologists.[929]

---

926 Ferdinand Zirkel (1838–1912), professor of mineralogy and geology at the University of Leipzig.

927 Otto Meyer: Untersuchungen über die Gesteine des Gotthardtunnels, Berlin 1878.

928 N.N.: Eintrag in die Matrikel (Rektor M 29), Universität Leipzig. For further information on Meyer's graduation see: N.N.: Eintrag in das Doktorbuch der Philosophischen Fakultät (Phil. Fak. B 128 b), Universität Leipzig.

929 Otto Meyer, New York to Othniel Charles Marsh, New Haven, CT, 3 December 1883, MS 343, Series I. Correspondence, Box 23, Folder 929.

He then proceeded to introduce himself, cite his credentials, and state his willingness to work for Marsh:

> I do not know anything else about it but I allow me to write you the following about me. I am 27 years of age; studied natural sciences, principally Geology in Berlin, Marburg and Leipzig, Germany, passed the 'Oberlehrerexamen in Naturwissenschaften' of the first degree and published the following essays: 1) Untersuchungen über die Gesteine des Gotthardtunnels. Zeitschrift d. deutschen geologisch. Gesellschaft 1878 p.1–24[.] 2) Einiges über die mineralische Natur des Dolomits, - the same journal 1879 p. 445–452[.] 3) Palaeontologische Notizen aus dem Mainzer Tertiär. Jahresbericht der Senckenbergischen naturforschenden Gesellschaft 1879/80 p. 311–321 plate VI. 4) Aetzversuche an Kalkspath. – Neue[s] Jahrbuch für Mineralogie, Geologie, Palaeontologie vol I p. 74–78 plate II [.] 5) Untersuchungen aus dem Märkischen Rupelthon. Jahresbericht d Senckenb. Naturforsch. Gesellsch. Zu Frankfurt 1/M 1883/84 with 1 plate[.] Of the last publication, which is not yet issued, I had the proof the other day, have the plate on hand and expect the copies in a month. You see from these essays, that I have two specialties, mikroscopical [sic!] lithology and tertiary invertebrates. I am [a]member of the 'deutsche geologische Gesellschaft' and have the second, if not the first collection of german Oligocene invertebrates. This collection is standing in the American Museum of Natural History, here, where I was working in the last two months, studying mostly Claiborne Eocene, comparing it with our german species and exchanging doublettes.

Meyer proceeded to write about his former business experiences, but unfortunately never states of what nature his business in New York was.

> Before this time I was here ten months with success in business. At present I am very doubtful, what to begin. I am almost inclined to enter business again but I should prefer a science position. If you are able and willing to engage me, please write to me about. I never worked in vertebrate fossils but I have general zoological knowledge and the right to teach Zooglogie [sic!] in all classes of a 'Realsgymnasium.' You may inquire about me at Mr. Oswald Ottendorfer, Editor of the New York Staatszeitung [and] Mr. Henry Villard and Prof. Dr. Felix Adler New York.

Despite this not being the strongest of job application letters (Meyer was not even willing to fully commit to sciences and stated his unfamiliarity with vertebrates), Marsh employed him as an assistant. Possibly Marsh was interested in Meyer's collection of German invertebrate fossils, or perhaps he was impressed by Meyer's teaching qualifications, especially since the German educational system was held in the highest regard (see chapter 8. 3.). Maybe Meyer's connections with the press,[930] with Cornell University,[931] or finance via Villard induced Marsh's decision? Meyer, it seems, was very well connected with the German community of New York. Schuchert and LeVene quote Westbrook, who wrote that Meyer gave Marsh very little real help in the laboratory. Meyer then resigned in 1886 and went into business.[932]

The only other documents concerning Meyer included in the Marsh papers are his contract of employment and a certificate that stated that Marsh had advanced Meyer $ 255 of his salary.[933] The contract stated:

"One & ½ year to July 1st 1885 7 hours per day or 40 per week $50, per month or $ 600, per annum one month vacation. Payment once a quarter."[934] The contract was amended some time later, for it is stated on the same document: "July 1st 1884. Raised salary to $1000, per annum, if U.S.G.S. appropriates the money, as during the present year."

Note that the contract is dated December 7, 1883, a mere four days after Meyer had written his application. Marsh therefore most likely employed Meyer right away without further inquiry into his qualifications, sending him the contract with the return letter.

On February 19, 1886, Baur wrote to Osborn that Meyer (he spelled his name "Mayer") had been employed by Aldrich in Cincinnati.[935] Meyer was part of the Alabama Geological Survey in 1886.[936] His contributions to the survey were published in the same year.[937] He had met Baur in New York in 1889 or 1890, and visited him on January 21, 1890, in New Haven.[938]

---

930 Via the mentioned Ottendorfer, who in turn was well connected to the political establishment, unsuccessfully running for mayor of New York in 1874.

931 Via Adler, who taught "Hebrew and Oriental literature" at Cornell, had good ties with the Jewish community of New York, and had founded the New York Society of Ethical Culture in 1876.

932 Schuchert; LeVene: O. C. Marsh, p. 302.

933 Otto Meyer, New Haven, CT to Othniel Charles Marsh, New Haven, CT, 2 April 1885, MS 343, Series I. Correspondence, Box 23, Folder 929.

934 Otto Meyer, New Haven, CT to Othniel Charles Marsh, New Haven, CT, 7 Dec 1883, MS 343, Series I. Correspondence, Box 23, Folder 929.

935 Georg Baur, New Haven, CT to Henry Fairfield Osborn, Munich, 19 February 1886, VPA 1/8, General Correspondence Bas–Bel, BAUR, GEORGE (Prof.) Univ. of Chicago 1884–1897, Folder 36.

936 Thomas Waverly Palmer: A Register of the Officers and Students of the University of Alabama, 1831–1901, Tuscaloosa, AL 1901, p. 30.

937 Otto Meyer: Contributions to the Eocene Paleontology of Alabama and Mississippi, in: Geological Survey of Alabama – Bulletin, no. 1 (1886), pp. 61–85.

938 Georg Baur, New Haven, CT to Henry Fairfield Osborn, New York, 21 January 1890, VPA 1/8, General Correspondence Bas–Bel, BAUR, GEORGE (Prof.) Univ. of Chicago 1884–1897, Folder 36.

The "Bone Wars" reached their climax on the pages of the "New York Herald" in 1890 (see below). One of the seven "Herald" articles was practically a letter of grievance written by Meyer and listed some of the same criticisms Baur harbored towards his employer.[939] In an earlier article the journalist Ballou had judged: "Dr. Otto Meyer, of New York, is a German, and is an excellent comparative anatomist and has published several important papers."[940] Eugene Woldemar Hilgard (1833–1916), another German immigrant and professor of agricultural chemistry at the University of California, Berkeley, would not have agreed. He wrote a harsh and quite polemic review of some of Meyer's publications. Contrarily to Ballou's high praise, he stated that Meyer's

> second paper showed the extremely limited extent of his own observations, and his failure to even read, much less study, the literature of the subject, from which he quoted only disjointed sentences, selected to suit his ideas. The three articles in the October number of the journal,[941] [the "American Journal of Science and Arts"] from three observers whose observations he calmly sets aside as unworthy of confidence beside his own superior lights, expressed their astonishment at the cool assumption, grounded on such a slender basis, that pervades Dr. Meyer's methods and assertions; [...] He simply ignores facts pointedly stated, that completely overturn his whole scheme; [...] Fortunately, the geological area which he attests to turn wrong side up is now again under examination by competent observers, who have no hobby to ride, and whose results, I have reason to hope, will be made public before many months. In the mean time I commend Dr. Meyer's methods to the attention of ambitious young geologists as a conspicuous example of 'how not to do it.'[942]

This example once again illustrates how small the scientific community was for Hilgard was the cousin of Henry Villard, who was born as Ferdinand Heinrich Gustav Hilgard in the Palatinate in Germany. It is possible that Meyer took Hilgard's harsh

---

939 Ballou: Some More Nuts for Marsh to Crack, p. 25.

940 Ballou, William Hosea: Scientist Cope Fires Back at Marsh, in: The New York Herald, 20 January 1890, p. 3.

941 Otto Meyer: The Genealogy and the Age of the Species in the Southern Old-Tertiary, in: The American Journal of Science, ser. 3, vol. 29, no. 174 (Jun. 1885), pp. 457–468; Otto Meyer: The Genealogy and the Age of the Species in the Southern Old-Tertiary, in: The American Journal of Science, ser. 3, vol. 30, no. 175 (Jul. 1885), pp. 60–72; Otto Meyer: Successional Relations of the Species in the French Old-Tertiary, in: The American Journal of Science, ser. 3, vol. 30, no. 176 (Aug. 1885), pp. 151–153. For an exhaustive 730-page review of Meyer's and Aldrich's work see: Katherine van Winkle Palmer: The Claibornian Scaphopoda. Gastropoda and Dibranchiate Cephalopoda of the Southern United States, in: Bulletins of American Paleontology, vol. 7, no. 32 (Dec. 1937).

942 Eugene Woldemar Hilgard: Dr. Meyer and the South-Western Tertiary, in: Science, new ser., vol. 7, no. 152 (Jan. 1, 1886), p. 11.

criticism to heart because he never published new scientific papers after 1886.[943] It is most likely that he returned to his undefined business. If this proves to be true, one Hilgard brought him in contact with Marsh and the other turned him away from science altogether.

6.5 The Duel in the New York Herald

On what Schuchert and LeVene call a "memorable but depressing Sunday,"[944] January 12, 1890, Cope's carefully collected allegations against Marsh went public. Ballou published the first of seven articles detailing the decade long feud; it was titled "SCIENTISTS WAGE BITTER WARFARE,"[945] spanned one and a half pages, and included portraits of Cope, Marsh, and Powell. After a series of spectacular subtitles, designed to underline the importance of the piece for the uninformed reader (for example: "Heavy Blows Dealt in Attack and Defence [sic!]" and "Will Congress Investigate"), Ballou listed Cope's grievances and (mis-)quoted his various sources:

> Serious allegations are made against the present administration of the United States Geological Survey in the columns of the HERALD to-day. Professor E. D. Cope, of the University of Pennsylvania, supported by several of the most eminent scientists of the country, arraigns Director John W. Powell and his chief assistant, Professor O. C. Marsh, of Yale College, president of the National Academy of Sciences, and for a number of years past vertebrate palaeontologist of the survey, on charges of plagiarism and of gross ignorance and incompetence in the performance of the important public duties intrusted [sic!] to their care. The charges are supported by an abundance of documentary evidence. For some time past a volcano has been slumbering under the Geological Survey, and of late there have been indications that the time for an eruption was not far distant. Now it has arrived and the long pent up forces have gained their freedom with a rush and a roar which, if it does not indeed carry the present management of the survey to official destruction, will certainly disturb the entire scientific world of America and bring in its train a series of charges and

943 Meyer's last papers on record: Otto Meyer: Some Remarks on the Present State of Our Knowledge of the North American Eastern Tertiary, in: The American Geologist, vol. 2, no. 2 (Aug. 1888), pp. 88–94. Otto Meyer: Birds with Teeth, in: Popular Science Monthly, vol. 36, no. 3 (Jan. 1890), pp. 382–389.

944 Schuchert; LeVene: O. C. Marsh, p. 290.

945 William Hosea Ballou: Scientists Wage Bitter Warfare, in: The New York Herald, 12 January 1890, pp. 10–11.

> counter charges, recriminations and reproaches which will ring from one end of the land to the other. The primary cause for all this commotion is the dissatisfaction with which Professor Cope and some of his professional associates and co-workers in the field of vertebrate palaeontological research have long regarded the conduct of the survey under Director Powell and Professor Marsh. The beginning of the trouble dates back several years to about the time when Professor Cope served his active connection with the old Hayden survey. The movement has gone on gathering strength ever since.[946]

The charges were that Powell ran the USGS "on machine political methods"[947] and had removed it from all official control, that Marsh and Powell had conspired and used undue tactics to get Marsh elected as president of the NAS, and that Marsh had, with Powell's help, misappropriated "immense government collections." These collections were now stored at Yale and were inextricably intermingled with the college's collections. Additionally, Marsh had made the Yale-collections "inaccessible to visiting scientists and nearly every one else." Marsh was also accused of having published the original work of his assistants under his own name, or of having committed outright plagiarism in publications financed by the government. Powell was charged with knowingly accepting Marsh's fraudulent practices and accused of committing various plagiarisms himself. Under Powell's supervision the USGS had been made an "asylum for Congressmen's sons" in an effort to secure "favorable actions from Congress," journalists were said to have been bribed to squall all criticism of these dubious proceedings and "secure newspaper support." Because of the status of the accused and the severity of the charges, it was stated that Ballou had sent manuscripts of his accusatory article to both gentlemen. Powell's reply followed the opening statements cited and summarized above. Powell stated that "the charges made have been circulating in secret for many years, [...] but this is the first time that they have come in tangible form and in such a manner that their authors could be clearly identified." Powell would "prepare a simple statement" that would constitute "a complete refutation to all fair minded men." Marsh was said to have replied via telegram; he intended to prepare a counterpoint of his own to the allegations, to be published at a later date. The allegations of "incompetency, ignorance and plagiarism" were said to be backed by prominent scientists, namely Cope, Williston, Scott, Osborn, Meyer, Schlosser, "Barber [sic!]," Hatcher, Endlich, Winchell, Smith, Merriam, Warner, "Fraser" [sic!],

946 Ballou: Scientists Wage Bitter Warfare, p. 10.

947 Further into the article Ballou stated that "The methods by which Major Powell is said to have created a gigantic politico-scientific monopoly next in importance to Tammany Hall are worthy of the best machine bosses."

Wilson, Wheeler, Flower, Gaudry, Rutimeyer, and Zittel.[948] It was said that Marsh and Powell had conspired to gain unrestricted control of the USGS and the NAS by electing members of the USGS to the NAS and giving members of the NAS positions in the USGS, and by doing so they had secured the gratitude and support of numerous members of both institutions, thereby attaining a quasi-monopoly on US-American geology and paleontology:

> Marsh has, so far as possible, given Major Powell the prestige of Yale College, and that has formed no small backing for Powell's schemes. In consideration, therefore, Powell has allowed Marsh to store vast government collections at Yale College, where they are locked away from the people and no one is allowed to see them, not even visiting scientists. [...] for the purpose of increasing these collections. Professor Marsh has been allowed $60,000 per year by Powell and a salary of $4,000.

The correspondence of Baur and Osborn confirms that Marsh did indeed "lock away" the Yale fossil collections, guarding them jealously from potential scientific competitors. In addition, Marsh was accused of having

> sent out large field collecting parties into the Territories, which were instructed to gather everything of use to the survey and break all specimens which it was unnecessary to bring East, in order that other scientists should not have them, or, rather, should not be able to anticipate the work of the survey. Beside breaking specimens and destroying the possibility of any competition in the acquirement of knowledge, Professor Marsh is now charged with retaining the salaries of members of his field parties.

Judging from Baur's letters, Marsh indeed had a less than stellar track record when it came to paying his employees. Baur also lamented the fact that Marsh had published

---

948 Barber, should be E. H. Barbour. The others have either already been mentioned in this study or are: Frederick Miller Endlich (1851–1899), a German geologist who had immigrated to America and worked for the Smithsonian Institution. Eugene Allen Smith (1841–1927), professor of geology at the University of Alabama who had received a doctorate at the University of Heidelberg in 1868. Clinton Hart Merriam (1855–1942) was a naturalist who specialized in the study of mammals and was nonetheless one of the founders of the American Ornithologists' Union. Adoniram Judson Warner (1834–1910), who was a Civil War general and a member of Congress from 1879 to 1881. "Fraser" is: Persifor Frazer. Thomas Wilson (1832–1902) was the curator of archeology at the Smithsonian. William Henry Flower was a comparative anatomist from England, a champion for the theory of evolution, and director of the Natural History Museum in London. Karl Ludwig Rütimeyer (1825–1895), Swiss paleontologist, professor of zoology and comparative anatomy at the University of Basel.

the work of his assistants as if it were his own without giving any credit to the original authors. This constitutes the next point of contention stated by the article:

> But the most asinuating [sic!] charge against Professor Marsh is that all of the work purporting to be his, as published by the government, is not his own, but in part that of his employés [sic!], the remainder being a collection of plagiarisms.

Under a subheading, reading "PROFESSOR COPE SPEAKS HIS MIND," Ballou recounted his interview with the complainant:

> I saw Professor Cope at his residence, No. 2.102 Pine street, Philadelphia, a few days ago. He was engaged in naming new species of fossil animals, but he cheerfully put aside his work when the nature of my call was made known to him. He is a man not over forty-five years of age and of distinguished appearance. 'What is the origin, Professor, of this war against the Geological Survey?' I asked. 'It may be found in the outrageous order I have received from the Secretary of the Interior to turn over my collections to the National Museum at Washington. I have no more than a bushel of specimens belonging to the government, and to those it is welcome. The fact is, I sent out my own exploring parties and secured my collection at an expense of about $80,000 of my own money, to say nothing of the value of the time I have expended upon them.'
>
> 'Who is the author of the order?'
>
> 'Why to be sure, who but Major Powell. The object of the absurd order to place my collection in the National Museum is to gain control of them, so that my work may be postponed until it has been done by Professor Marsh, of Yale College, and this in spite of the fact that the preliminary work has been already published by me, and that the truth is sure to come out at some future time.'
>
> 'What is the nature of Professor Marsh's methods?'
>
> 'The collections made by Professor Marsh as the vertebrate palaeontologist of the geological Survey, which he has been making during the past ten years at an expense of some $60,000 per year to the government, are all stored at Yale College, with no assured record as to what belongs to the government and what to the college. Professor Marsh has shown that he never was competent to do work of this kind. Unable to properly classify and name the fossils his explorers secured, he employed American and foreign assistants who did the work for

> him and to which he has signed his name. Of those who originally entered his employ one has died and three have left him, unable to stand the vassalage of their positions. Fearing that some rival scientists might secure and be the first to name the specimens in the great deposits of fossil animals in the Territories, Marsh’s expeditions were ordered to smash all duplicates which they could not send East, or of which the survey had a sufficiency for its own uses.’

Indeed, Marsh’s oversight to clearly mark which fossils had been acquired using government funds and to distinguish those from his own collection would come back to haunt him when he was forced to give up a sizeable part of his fossils eventually.[949] As to the three employees who could no longer stand their “vassalage,” Cope was most likely referring to Schlosser, Meyer and Williston. The one who had died was Harger. Baur was still in Marsh’s service at New Haven and would leave only after the publication of the interview. Baur fully agreed with Cope’s judgment that Marsh was unable and unwilling to do the scientific work himself, as seen above in his correspondence with Osborn. He wrote to Osborn that Schlosser had left because of his disenfranchisement by Marsh and the latter’s loose payment practices. The third candidate mentioned by Cope is most likely Williston (Hatcher was, beside all dissatisfaction, still working with Marsh).

The next subheading of the article reads: “IS PROFESSOR MARSH A PLAGIARIST?” Ballou continued his interview:

> ’What is this charge of plagiarism against Professor Marsh?’

> ‘First, his alleged discovery of the evolution of the horse. It will be remembered that this paper attracted great public attention some years ago as showing the divided hoofs of the ancestors of the horse. This work was mostly plagiarized from Professor Kowalevsky, of Moscow, who complained bitterly when he heard of Professor Marsh’s theft of his important life work.’

Vladimir Onufrievich Kovalevsky (1842–1883) had studied law in London (where he met Darwin) and Saint Petersburg; later he became more and more interested in geology and paleontology, and visited the fossil collections in Heidelberg, Jena, Munich, and Paris. For a while he stayed in Jena and studied under Ernst Haeckel (1834–1919), the champion of the theory of evolution in Germany. He studied the evolution of the horse. After some major financial missteps, Kovalevsky shot himself in 1883.

949 The rest of his collection became the core collection of the Peabody Museum when Marsh gifted it to the museum in 1898. Schuchert; LeVene: O. C. Marsh, pp. 313–333.

'Second, his alleged work on toothed birds (*Odontornithes*) was written by his assistants,[950] one of whom was Professor Williston. Third, his work on the *Dinocerata*, or horned mammals, was done chiefly by his assistants. The generalizations were dictated by George Baur, who repealed what he knew from my own work on the subject, Marsh changing the names of divisions of classification. This attempted theft from my work is making a laugh all over Europe.'

Indeed, Baur had complained about practically dictating the fruits of his own scientific labors concerning the *dinocerata* to Marsh in a letter to Osborn dated 26 September 1886 (see above). If this information was passed on by Osborn or if Baur had brought his grievance directly to Cope cannot be ascertained due to the incomplete nature of the Cope correspondence.

'Fourth, Professor Marsh's work on the saurian of the West was written by Williston, and fifth, his paper on the mammals of the Laramie formation was really written by himself, and is the most remarkable collection of errors and ignorance of anatomy and the literature on the subject ever displayed. Sixth, his papers on the horned saurians of the Laramie is a pretended discovery which I fully described thirteen years since.'

What follows is a list of some of Marsh's alleged scientific inadequacies, evoking the Cope-Marsh feud fought out on the pages of the "Naturalist" some 17 years before (see chapter 4). Then Cope vented his anger concerning Powell, his practices and, above all, his scientific incompetence and dishonesty, and summarized the Powell-Marsh relationship as follows:

Major Powell must be held responsible for the blunders of his agent, Professor Marsh. He has used Marsh as a tool and Marsh has used him as a tool. They together have used the National Academy of Sciences as a tool for their mutual purposes. [...] As to the naturalists in the Academy they have been generally subsidized by the survey. Scientific men usually are very simple of thought, unsuspecting of the machinations of others, and thus have been the victims of a pair of political scientists, more political than scientific.

Concerning Marsh's assistants, Cope is quoted to have said:

950 See: Othniel Charles Marsh: Odontornithes. A Monograph on the Extinct Toothed Birds of North America, with Thirty-Four Plates and Forty Woodcuts, Washington, DC 1880.

> He [Marsh] also used the men in his own office until they were compelled to escape from him in order to retain their self-respect. He used their investigations as his own. Among these men were S. W. Williston, Ph. D., M. D.; Otto Meyer, Ph. D.; M. Schlosser, Ph. D.; E. H. Barber, B. A., and J. B. Hatcher. Ph. D

A letter, sent to Cope by Williston, is used to illustrate and specify some of the accusations:

> I wait with patience the light that will surely be shed over Professor Marsh and his work. Is it possible for a man whom all his colleagues call a liar to retain a general reputation for veracity! * * *I do not worry about his ultimate position in science. He will find his level, possibly fall below it. There is one thing I have always felt was a burning disgrace—that such a man should be chosen to the highest position in science as the president of the National Academy of Science, while men of the deepest erudition and unspotted reputation are passed by unnoticed. Professor Marsh did once indirectly request me to destroy Kansas fossils rather than let them fall into your hands. It is necessary for me to say that I only despised him for it.

Another accusatory and strongly worded letter of Williston's is quoted next:

> The assertion of Professor Marsh that he devotes his entire time to the preparation of his reports is so supremely absurd, or rather so supremely untrue, that it can only produce an audible smite from his most devoted admirers. I have known him intimately for ten years. During most of the time while in his employ I never knew him to do two consecutive, honest days' work in science, nor am I exaggerating when I say that he has not averaged more than one hour's work per day. He is absent from the Museum fully half of the time, and when in New Haven he rarely appears at the museum till two o'clock or later and stays but an hour or two, devoting his time chiefly to the most absurd details and old maid crotchets. The larger part of the papers published since my connection with him in 1878 have been either the work or the actual language of his assistants. At least I can positively assert that papers have been published on Dinosaurs which were chiefly written by me. * * * Professor Marsh's reputation for veracity among his colleagues is very slight. In fact he has none. * * * Those who know him best say—and I concur in the opinion—that he has never been known to tell the truth when a falsehood would serve the purpose as well. These are strong statements to make of one holding such a position as he does, but I state them the more freely from the fact that everybody here (Yale College) concurs in them. He has no friends here save those who do not know him well.

Later, near the end of page 10, Ballou quoted Williston as having said that he had written the condemning letters some years before.[951] Ballou heavily implied that Williston and others had been pressured by Marsh to retract their statements.

An equally condemning letter, written by Meyer, is quoted next:

> I consider it necessary that a scientist, especially a leading one, ought to be well posted in constantly appearing literature relating to his specialty. Professor Marsh neglects to do so in a surprising manner, at least in later years. Frequently, indeed, the writings of contemporaries are not removed from the wrappers in which he receives them. Thus it sometimes happens that he is unaware of the existence of papers treating of subjects in which he is especially interested. Professor Marsh performs very little, I might say almost no scientific work, at least in later years. The main part of his work is done by assistants. It is not allowed to assistants of Professor Marsh to publish, under their own names, any material concerning the government collections of fossils, even that which Professor Marsh has no intention to work upon.

Ballou continued: "Professor Meyer further charges that Professor Marsh's restorations of fossils are unscientific; that he makes statements as scientific, with prodigal disregard of the facts; that he deliberately and systematically plagiarizes, and that his work is wholly unreliable."

Note that Ballou called Meyer "Professor Meyer," indicating that he held a teaching position at that time; this is most likely an error on Ballou's part. The correspondence between Powell, Secretary of the Interior John W. Noble (1831–1912), and Cope, mainly concerning a planned monography of Cope's, his fossil collection, and the possible transfer of the collection to Washington D.C., is detailed in three cited letters, printed in full in the article. After that an interview with Scott is printed. Among other accusations against Marsh, he is quoted to have said:

> I can testify that Professor Marsh is wholly incompetent as a palaeontologist. His last book is a pure plagiarism. It has but one idea in it, and that was stolen from Professor Cope. Europeans are as much disgusted with Professor Marsh as we are in this country, and to think that he has been permitted to disgust them by the use of over $500,000 of government funds!

951 "Professor Williston was mysteriously moved and wrote to the HERALD that his letters concerning Professor Marsh were mostly written some years ago, under exasperating circumstances, before he had become connected with Yale College."

Next Endlich is quoted, chiming in with criticisms directed at Powell and the operation of the USGS, then Frazer confirms Endlich's grievances and raises some of his own. Then Powell's reply to Ballou is printed in full, as advertised at the beginning of the article. Osborn wrote that Ballou had sent the proposed "Herald" article to Marsh and Powell in order to "guard himself against libel and to lay a fresh stock of sensational material for the *Herald*."[952] Powell refuted Cope's accusations of corruption, favoritism for, and conspiracy with, Marsh. According to Powell, Marsh had fulfilled his duties with "eminent ability" and "exhibited in his work great industry." Concerning the supposed enmity between Marsh and his (former) employees, Powell made the following statement:

> It is charged that Professor Marsh treats his assistants unfairly and that they are his enemies. This I know to be untrue of all those who are employed by the Geological Survey, for they work with zeal, fidelity and loyalty and speak of him in the highest terms. In past years, before being connected with the United States Geological Survey and to some extent since that time, Professor Marsh has employed men in the field and in the laboratory in a subordinate capacity and paid their wages himself. Any disagreement which he may have had with these men is a private affair, in which the director of the survey and the public have no interest. Some years ago professor Cope caused to be published a statement that one of Professor Marsh's official volumes had been written by an assistant, who was himself its real author. At Professor Marsh's request I investigated this matter and found that two or three of the Professor's assistants had indeed written a part of the volume, as clerks, from the dictation and notes of Professor Marsh – a method of writing adopted by most public men, and especially by scientific man engaged in large work.

It is highly dubious that Baur, who at that time was still employed by Marsh, was amongst those who, according to Powell, worked "with zeal, fidelity and loyalty" and spoke of Marsh "in the highest terms." Baur had not publicly attacked Marsh at this time and had, upon some reflection, decided not to contribute to Ballou's article.[953]

Concerning the allegations that Marsh had misappropriated government funds for the Yale fossil collection, Powell noted that some of the specimens collected with public money were studied and worked upon at the laboratories of the Peabody Museum "without expense to the government." Furthermore, the two collections were "scrupulously labelled and segregated by themselves, so that if Professor Marsh and his assistants should suddenly die, together with the members of the Geological Sur-

952 Osborn: Cope, p. 403.

953 See the letter from Baur to Osborn dated 14 January 1890, quoted above.

vey, a stranger could go to the laboratories of the college and readily identify every specimen belonging to the government." This was an overstatement, for indeed the differentiation of the two collections would cause some headache later on. Powell then shifted the blame for the shoddy payment practices from Marsh to the offices of the USGS: "Professor Marsh is not responsible for the payment of the men in his division. That function is performed by the disbursing officer in Washington, and all payments have been regularly and properly made." It could be argued that Baur and the other assistants were employees of the USGS because Marsh's contract with his employees stated that they were only to be employed if the USGS would provide Marsh with the respective amount of money (see above). This again indicates that it is impossible to differentiate between Marsh's functions at Yale and as the chief paleontologist of the USGS. As with the fossil collections, the money funds and responsibilities were inextricably interwoven. A good portion of the rest of the statement refuted Cope's allegations of squandering, mismanagement, and the suppression of Cope's own scientific work, but does not relate to Marsh and his assistants. Powell even fired back, painting Cope as an envious paranoiac, and stated:

> Professor Cope's mental and moral characteristics unfit him for any position of trust and responsibility. In addition to his great vanity, which leads him into vicious species work, he is inordinately jealous and suspicious of every other worker, and these two traits combined give him that hysterical temper and gift of voluble denunciation rarely found in persons of his sex. In fact his general ravings about scientific men, members of the National Academy, professors in colleges and geologists in general, whom he believes are all in league against himself, make it impossible for him to associate on terms of co-operation with other men engaged in kindred work.

He implied that Cope had to be the sole instigator of the unfair allegations, the master puppeteering and spurring on the other claimants cited in the article: "He [Cope] is the only one of the coterie who has scientific standing. The others are simply his tools and act on his inspiration." Roughly a year later Marsh would echo this statement in a letter to Leidy when inquiring whether Osborn had also fallen to the machinations of his "enemy" Cope.[954] At the end of the article Ballou listed a few statements, telegrams forwarded by Marsh. This concluding portion is titled "THEY ALL DENY" with the subhead "DENIALS WHICH SOUND FUNNY IN THE LIGHT OF PREVIOUS EVENTS." Osborn had written to Marsh that he had not "seen or authorized any article whatsoever," Williston had supposedly written:

954 See the letter written by Marsh to Leidy on 24 January 1891, quoted above.

> I have not authorized Professor Cope or any other Person to make an attack on your character or your work, and I wish to add that in all your published scientific work you have treated Mr. Harger and myself with entire fairness.

Baur gave a similar statement but emphasized that his statement had been given voluntarily. Finally, Scott assured to never have authorized the usage of his name and claimed that he had not "sent any written statement to any one bearing upon the subject."

Ballou's next article, published the very next day, was titled "VOLLEY FOR VOLLEY IN GREAT SCIENTIFIC WAR."[955] A subheading promises "MORE LETTERS AGAINST PROFESSOR MARSH," the next subheadings informs that

> Incompetency, Ignorance, Plagiarism Still the MAIN Offences in the Bitter Budget Charges Against Yale's Famous Geologist. MEN OF SCIENCE AGOG. Some Shocked, All stirred Up by the Sensational Disclosures in the Herald, and Many Unable to Believe the Accusations. LONG SMOULDERING EMBERS OF HATRED. Friends of Powell and Marsh Declare that the Pennsylvania Professor Is Inspired by Disappointed Ambition, Jealousy, Envy and Other Unworthy Motives.

The article was basically a continuation of the one published the day before; one of its first sentences promised further mudslinging and sensationalized the feud: "Professor Cope, at his home in Philadelphia, returns to the attack, [...] and he does it with a freedom of speech and apparently reckless disregard of consequences, that is sure to draw fire again from the enemy."

Cope and his supporters were called "the leaders of the rebellion against the sway of the Washingtonian and Yale professor." Ballou stated that he was writing about "warlike scientists," and that the whole affair was a "very pretty fight." Ballou's efforts to make the quarrel interesting to the general public would shape the way the "Bone Wars" would be remembered to this day. The next subheading reads "PROFESSOR COPE STRIKES BACK." Ballou described how he met Cope in his house at Pine Street and interviewed the Professor right after he had read the "Herald" article, thus making Cope's response appear spontaneous, genuine and up to date. Cope mostly refuted some of Powell's remarks made in the previous "Herald" article, which are not relevant for this study. He then stated:

---

955 William Hosea Ballou: Volley for Volley in Great Scientific War, in: The New York Herald, 13 January 1890, p. 3.

> There is another thing I wish to say concerning my statement regarding the failure of Professor Marsh to pay certain of his employés [sic!]. That does not refer to employés [sic!] of the Geological Survey, but to Professor Marsh's own employés [sic!]. My information on this subject I regard as perfectly reliable

Cope most likely referred to private correspondence with Baur, Meyer, and Williston, who could be considered employees of the USGS, as stated above. At least in Baur's case it can be shown that he wrote about Marsh's lackluster payment practice to Osborn, and maybe he did share his gripes directly with Cope as well. Cope then recounted his early encounters and collaboration with Marsh, culminating in Marsh's enticement of the Haddonfield quarries' foremen, undoubtedly a betrayal in Cope's eyes:

> During my early acquaintance with Professor Marsh I ottered [sic!] him every facility to examine my own collection, and gave him all information regarding localities of fossils that I knew of. I took him through New Jersey and showed him the localities. Soon after, in endeavoring to obtain fossils from those localities, I found everything closed to me and pledged to Marsh for money considerations.

Next Cope asserted that, despite his alleged incompetency, Marsh's money and social skills had brought him his current position at the top of American science; and that, for some reason, the same thing would not have happened in Germany, implying that corruption in science was an inertly American problem: "Such a man as Marsh could not attain any position in science in Germany, or if he should succeed, he would not be tolerated for a moment." Cope was apparently concerned about the international reputation of American science, especially in regard to Germany, and feared that Marsh's despicable actions[956] and incompetency might damage the scientific reputation of the nation:

> Concerning Professor Marsh's reputation in Europe among the cultivators of his especial department, Schlosser, of Munich, in an important quarto work, recently issued in Vienna, says: —'Marsh's work is so superficial that it must be received with the greatest caution. He is accustomed to neglect entirely the

956 In the same article Cope alleged that Marsh would order some fossils to be destroyed rather than let them fall in the hands of his enemies: "distinguished professor in a New England college informed me that Professor Marsh has indulged his specimen smashing proclivities in New England as well as in the West. He obtained a monopoly of all specimens of the celebrated reptile and bird tracks obtained at some localities in the Connecticut Valley, and all specimens not selected by Marsh were ordered to be broken up and thrown away."

> materials and publications of other authors.' In a recent letter from a distinguished English scientist, whose name I am not at liberty to give, occurs the sentence, 'Marsh has made a nice mess of his cretaceous mammals.' Referring to Marsh's disregard of the work of others, another English scientist, holding an important public position, remarked with reference to Marsh's methods in a recent letter:—'However wrong it may be, and no doubt is, yet, as you say, there is a strong temptation, when you get an entire beast out, to ignore prior discoveries of parts and go in for a new name of one's own.' And this is Professor Marsh's uniform custom, both as regards European and American work.

According to Cope, Marsh's alleged tendencies to steal the intellectual credit from his assistants had the consequence that English journals had stopped publishing Marsh's discoveries; this was yet another example for how Marsh allegedly harmed the international recognition of American science:

> Dr. Meyer was at considerable pains to convince Professor Marsh that certain bones from a Rocky Mountain locality demonstrated the existence of a new form of bateachia [sic!], at a period previously unknown to include such a form of life. When Professor Marsh understood the matter, he read a paper before the meeting of the British Association for the Advancement of Science, held at Aberdeen, announcing the discovery as his own. This sort of thing has now become so notorious that in place of Professor Marsh's reported discoveries being at once republished in England, as formerly, his announcements for the last several months have remained unnoticed.

He then stated that he saw the disclosure of Marsh's shortcomings as his "duty" to his "country's reputation," and that he hoped to have demonstrated "that no man can, by the use of money only, palm himself off successfully as a representative of the science of America."

Later Ballou wrote about Scott's great dissatisfaction with the "Herald" article of the previous day that Scott had felt misrepresented by the paper and claimed he had been tricked into giving a statement, which he had understood to be off the record:

> Professor Scott was very angry yesterday. He said that the man who furnished the story to the HERALD had acted most dishonorably toward him, [...] 'I have been grossly misrepresented.' he said. 'Some time ago I got a letter from one W. H. Baltou [sic!], whom I met two years ago at Professor Cope's, and who pretended to be an amateur in science. He broached this subject, which I had heard much about, though I had no personal information. I did not answer the letter. Later on he called on me in the guise of a scientific man who was

> interested in the subject and said he was preparing an article on the subject and wanted a statement from me. 'I told him I had none to make; in the first place because I thought it unbecoming in men of science to wash their dirty linen in public. I told him I knew nothing except by hearsay. Then I talked to him as one gentleman to another. It seems he was not one. He took no notes. I did not dream that he was acting the part of a reporter, and he has most grossly misrepresented what I said to him.

Note the critical nature of Scott's comment about the journalistic integrity of Ballou, whose name is misspelled as "Baltou." This either reveals the rushed nature of the article and the shoddy quality control at the newspaper, or implies that the article was a collaboration of more than one journalist and not the work of Ballou alone (who presumably would not have misspelled his own name). Neither Scott nor anyone else of Ballou's quoted sources retracted the accusations against Marsh, but they objected vehemently to the methods with which the journalist had attained them and stressed that they had never given their consent to be quoted publicly.[957] Indeed, Scott added that there was a case to be made against the scientific legitimacy of Marsh (and that this was a matter of "much public importance"), but that Ballou had "weakened" this cause by "attaching falsehood and misrepresentation" to the otherwise legitimate points of contempt.

Next Ballou interviewed Osborn, who stated:

> 'Yes, I have read the HERALD article,' he said 'with very great interest. I am glad that the matter has at last come out. It will clear the atmosphere. The truth will be sifted out from the falsehood, and great good will be accomplished. I don't think I want to say another word.[']

Osborn managed to stay neutral in public and stated that he would write to both Powell and Marsh in the matter at hand. Unfortunately, Osborn's letter is not preserved in the Marsh papers. Osborn is then quoted saying that he had seen the fossils at Yale and that they were correctly labeled, so that the Yale collection could be told apart from the government-owned specimens. When asked if that was also true for the vast amounts of fossils hidden from the eyes of the public and from visiting scholars, stored away in

957 The statement made by Scott is very representative of the feelings of most of Ballou's sources: "Now, I don't wish you to understand," said Professor Scott, "that I come forward as the defender of Professor Marsh in any way. I simply wish to make it clear that my participation in this dirty business has been purely an involuntary one." Later in the article Williston is quoted to have said that the accusatory letter sent to Cope had never been intended for publication and was of a wholly private nature.

the cellars of the Peabody Museum, Osborn supposedly stopped speaking and after a while replied: "I don't care to say."

Ballou had also called Meyer, who did not agree to an interview about the Marsh-Cope controversy. He said he had written a letter explaining his opinion of Marsh to Cope in 1886 but stated that this letter should not be published. While he had not agreed to be interviewed about the feud between the professors, Meyer's own misgivings with Marsh were published in the article (probably with Meyer's consent, for there is no later refutation of the statements):

> 'I had been two years with Professor Marsh.' said Dr. Meyer. 'I am no longer interested in geology. I read the HERALD'S article this morning and it seems to me Professor Cope has made the mistake in overstating his case, for example, in saying that the government material in New Haven cannot be separated from the college material. The collections can be separated. It seems, when you come to look at it, that the two properties are mixed. Every article is marked, however, in the books, with corresponding marks on each article, and in that way you can determine which belongs to the government and which to the college. Nevertheless there is a basis for the charges which I see in the HERALD. I was disgusted with Professor Marsh, and that is why I left him.' 'Well,' I asked, 'why were you disgusted with him?' 'One of my reasons,' replied Dr. Meyer, 'was that I was opposed to his methods.' Dr. Meyer further said that it was true Professor Marsh did not do much work. The main portion of his work had been done by his assistants; not only the clerical work, as Professor Marsh put it, but also the investigations and discoveries. His principal assistant, said the Doctor, was Mr. Oscar Harger, who was more than fifteen years carefully collecting and investigating, but never published a line over his own name. All was put down to the credit of Professor Marsh.

One of the few Yale professors willing to make a public statement concerning the controversy was William Henry Brewer (1828–1910), professor for agriculture. He is quoted to have said that Marsh was above the petty accusations and that "men in Professor Marsh's position were often the object of the attacks of jealous and envious men." A series of interview snippets with patrons of a gentlemen club situated in Washington D.C. follows; the interviewees mostly defend Marsh, Powell, and the USGS: they claim that Cope's accusations were born out of jealousy and that he probably wanted to supersede Marsh in his official function at the USGS. For Professor Nathaniel Southgate Shaler (1841–1906) of Haverford College (and also a member of the USGS), who knew about the long-lasting squabble, Cope's accusations were nothing new:

> 'It's a very old feud, existing for, I was going to say, twenty years. It really seems as long ago that I first heard of it. As I recollect, the quarrel between Professors Cope and Powell grew out of some discoveries of fossils by Professor Marsh, of Yale, which were claimed by Professor Cope. In the wrangle which ensued Major Powell took sides with Professor Marsh and consequently suffered the displeasure of Professor Cope. [...] The story of Professor Cope's feud against the survey is so old and has been rehearsed so many times that scientific men are sick of it. You can say for me and the Harvard men interested in geological matters that little weight should be attached to Professor Cope's statements, as they grow out of disappointment, envy and other causes.'

The next day Ballou published a third article about the rivalry; it was titled "WIDENING THAT GEOLOGICAL CHASM."[958] A subheading read: "Like Kilkenny Cats, if the Squabble Much Longer Continues There Won't Be Much Left of Either Combatant."[959] Again, the competing scientists were denominated as combatants in an effort to capture the interest of readers, who otherwise had no interest in paleontology, working conditions at Yale, or the workings of the USGS. This article is considerably shorter than the others; encompassing just two columns (the "Herald" printed six columns per page). It seems there was not much news to report, so the article basically consists of comments on the previous two instalments, described above, without adding much.

However, a letter is reproduced that Professor Karl Ludwig Rütimeyer (1825–1895) of Basel, Switzerland, had sent to Osborn. It did not contribute much to the current dispute but sang high praises to the state that US-American paleontology had achieved as a scientific discipline in the last three decades:

> I am sorry to hear of the difficulties in the way of the publication of Professor E. D. Cope's Book [...] on Tertiary Vertebrate of North America. In view of the extent of this work such difficulties could readily be understood were it supposed that the publication depended on private means or on a scientific society of limited resources. Now, in Europe we consider this work as a na-

---

958 William Hosea Ballou: Widening that Geological Chasm, in: The New York Herald, 14 January 1890, p. 4.

959 While the true origin of the term "Kilkenny cats" seems to be untraceable, it was used since at least the seventeenth century, describing two cats fighting each other so ferociously that both perished. In some versions of the story the cats destroyed each other's bodies at such a rate that only the tails remained identifiable. Considering that, indeed, both Marsh and Cope ruined their considerable family fortunes in the pursuit of the "Bone Wars" and greatly damaged their scientific reputations by quarreling in the "Herald" and on the pages of the "Naturalist," the metaphor seems justified. Then again, both men are today considered to be among the most accomplished US-American paleontologists of all time.

> tional one, in the fullest sense of the word. Since the publication of that foundation of mammalian palaeontological science, Baron Cuvier's researches on the fossil animals discovered in France, certainly no similar compact bulk of well ascertained facts respecting the history of the highest vertebrate class has been accumulated and delivered to the scientific public as that contained in the first half of Professor Cope's work. Everywhere, therefore, the scientific public impatiently expects the Completion of this enterprise, which is considered by every one to be worthy of the great country from which it has issued.

As seen in chapter 2. 5., being likened to Cuvier seems to be the highest praise possible for a paleontologist. Because of Cope's and Marsh's scientific reputation, their differences – and the resulting acceleration of paleontological publishing – could have been understood to be of international importance. This impression was affirmed by Gaudry in another letter to Osborn, included in the article:

> The sympathy we have in France for the American people causes us to follow with pleasure its rapid progress in science, as in everything else. Especially the discoveries made within a few years in North America are of such importance to palaeontology in general that the learned men of every country cannot fail to be interested in them.

Finally, Zittel was reported to have chimed in in the same manner:

> The publications of Professor Cope are exceedingly highly valued in Germany, and I assure you that the second part of the Tertiary Vertebrata will be awaited with anxiety. The government of the United States has always sustained in an admirable manner the advancement of science, and especially the sciences of geology and palaeontology, and has thus called into existence museums and works which old Europe envies.

After a short discussion of the financial situations and backgrounds of the feuding paleontologists, Ballou concludes that

> Moreover both he [Cope] and Professor Marsh are so well in trenched and buttressed in the Geological Survey that it seems almost a Quixotic undertaking for Professor Cope to attempt to dislodge them. Nevertheless, although his adversaries are rich, powerful and have influential friends in almost every department of the government, the plucky Pennsylvania professor springs as gallantly to the fray as ever any knight of old.

These words invoke another transfiguration of the squabble into the sphere of martial competition and constitute a romanization by invoking images of shining armor, noble steeds, and the excitement of the joust.

Next, a friend of Cope's was quoted anonymously; he was said to have stated that Cope had "turned his whole attention to these attacks," and that he had wanted to sell the "Naturalist" but had kept his shares to "keep up a regular attack on Professors Marsh and Powell." Near the end of the article Ballou described how he had met Marsh in the Peabody Museum and how Marsh was preparing his reply, soon to be published in the "Herald." One statement of Marsh's is noteworthy due to its hard rhetoric, for he stated that Cope was "either insane or insanely jealous." Ballou's subheadings continued the trend of martial metaphors, as he wrote that Cope was still "WARLIKE" and Marsh would be "LOADING FOR BEAR."

On Sunday, January 19, a fourth article about the Marsh-Cope feud was published in the "Herald," titled "MARSH HURLS AZOIC FACTS AT COPE."[960] The subheadings continued the warlike rhetoric,[961] promising that Marsh would "pick up the gauntlet" and defend his "scientific reputation" in a "royal battle." Another subheading read: "WAR CARRIED INTO AFRICA." Now it was Marsh's turn to refute the allegations and fire back. After a very brief summary of the events so far, Marsh's rebottle was published in full length, filling the best part of the six-column page.

First Marsh called Cope's allegations "slanders" which Cope had been "repeating for years" while he had "devoted some of his best years" to the preparation of the public accusations, which Marsh sarcastically called Cope's "crowning work" and a "crusade." He continued:

> To meet these charges one and all is an easy task, but not a pleasant one, as I shall have to use plain words and say many things which I should otherwise wish to leave unsaid. If my language may seem severe it should be remembered that for ten years I have suffered those attacks in silence.

Marsh then denied the alleged favoritism and cronyism which, according to Cope, ran rampant within the NAS and the USGS; all his appointments in relations to government work had been made "after the fullest consultation with all the members of the council of the academy." He asserted that the government-paid specimens of the fossil collection could very well be told apart from his own collection, and added: "A part of

960 Ballou: Marsh Hurls Azoic Facts at Cope, p. 11.

961 Cope also used this kind of phrasing, saw himself as a soldier or knight in a battle to save the honor of science, as evidenced in his correspondence with Osborn, see for example: "when a wrong is to be righted, the press is the best & most Christian medium of doing it. It replaces the old time shot gun & bludgeon & is a great improvement." See: Osborn: Cope, p. 411.

these specimens have already been sent to Washington, and others will follow from time to time as their investigation is completed." To the allegation that he would jealously hide his collection from visiting scientists, Marsh replied: "Visiting scientists of good moral character are always welcome, but I have learned caution by experience." Obviously, Cope and his allies were, in Marsh's judgement, not "of good moral character." Marsh described how Cope, accompanied by Silliman, had entered the museum after hours (and without Marsh's knowledge) and had inspected rooms that Marsh considered private and that were full of specimens and lithographic plates not yet fully studied and published. Cope supposedly gave his word not to publish on what he had seen at the Peabody Museum, but did it anyway because he was not a "man of honor;" he also added "many falsehoods" in the publication of what he saw. Marsh later added: "[That] the museums of Berlin and Paris likewise suffered at Professor Cope's hands is no secret." He then listed further museums that Cope had allegedly intellectually robbed.

Marsh then denied that he had squandered government money and added that the allegations that he had ordered the destruction of fossils were lies and slander, conveniently timed before important elections were held at the NAS. Marsh also denied not having paid members of his expedition parties and added that Cope had merely been following Marsh's teams into the West for years, sometimes bribing and enticing his employees. Marsh then wrote that many of his former employees had kept amiable ties with him and even boasted about his friendship with Buffalo Bill (see chapter 5.3.2.2.).

Concerning his assistants, he wrote:

> My assistants at New Haven have been comparatively few in number, perhaps fifty in all during the past twenty years. Nearly all of these have been faithful to their duties and faithful friends to me. Some of these were men of rare ability, and I could only retain their services for a limited time, as wider fields were before them and I rejoiced at their success. [...] Assistants, who remained with me for a shorter time and then passed on to other scenes, were nearly all employed in mechanical and clerical work alone, as most of them were not sufficiently versed in scientific work to make their services of special value to me. Of these some left of their own accord. The work of others was not satisfactory and I could not retain them. [...] For the latter Professor Cope promptly expressed great sympathy and friendship, and most of his ammunition in the present attack is derived from them alone. In his fertile imagination men whom I had employed simply to clean fossils or measure them became at once profound anatomists, whose opinions on the most difficult problems of paleontology were conclusive. A touch of Professor Cope's magic wand, and again the same men became authors, who kindly wrote the works that I had for years

> in preparation. As the spirit of his own vanity pervaded them they imitated closely their master, and proved his apt pupils in prodigality of accusation and economy of truth. Their ranks were joined by a few other young partisans of this disappointed leader, and we have their combined efforts in the present attack. Little men with big heads, unscrupulous in warfare, are not confined to Africa, and Stanley will recognize them here when he returns to America. Of such dwarfs we have unfortunately a few in science, and some of them have fallen ready victims to the wiles of Professor Cope's flattery and promises of friendship. How reliable his friendship is many, both dwarfs and larger man, have learned to their cost.

So according to Marsh, Meyer, Schlosser, Williston, and the others had just done mechanical and clerical work, were "little men with big heads" and "dwarfs," who were ensnared by their treacherous leader Cope. The remaining assistants, who were currently employed by him, Marsh considered faithful and his friends: "With such assistants, and all others who are faithful in their work, discreet in conduct and truthful in speech, I have no serious difficulties, and all such I believe are my friends." Baur at this time was still employed by Marsh and most certainly did not consider himself a friend of the professor and would have described his position and employment as very difficult indeed.

Marsh then denied ever having committed plagiarism and stated that had indeed always given due credit and recognition to his coworkers who had advised him:

> Dr. Baur and Dr. Schlosser were likewise my assistants during the printing of the volume ["Dinocerata"] and for what assistance they gave me I endeavored to return them full credit in my preface. Dr. Williston was with me at the time and began the preparation of the bibliography, which was finished by other hands. He also acted as my amanuensis for a while, and I have given him due credit for all his work, as he himself testifies in the letter, of which this is a copy:–

What follows is a reproduction of a letter Williston had sent to Marsh; in it he quotes a written statement given by Baur in January 1890: "I hereby certify I did not dictate any of the generalizations of Professor Marsh's volume one the *Dinocerata*." This of course stands in stark contrast to the letter Baur himself had written to Osborn and in which he explicitly stated to have dictated parts of the "Dinocerata" to Marsh.[962]

962 "Den Schlusstheil der Dinoceraten den ich ihm so ziemlich dictirt habe." Baur to Osborn 23 October, 1886.

Marsh then wrote that he had not plagiarized the genealogy of the horse from Kowalevsky, and even wrote the following about the character of the late Russian scientist:

> [Cope and Kovalevsky are] twin brothers in work and methods. I have already alluded to Professor Cope's depredations on the museums of the scientific world. Kowalevsky's were of a similar character. [...]Kowalevsky was at last stricken with remorse and ended his unfortunate career by blowing out his own brains. Cope still lives, unrepentant.

He went on to deny that Williston had written about the dinosaurs of the West for him, denied other accusations of plagiarism, and wrote about the important ways in which he contributed to opening the West to science.

Marsh then proceeded to write about some "ancient history," as he called it, describing how he had met Cope for the first time in Berlin and how they became friends, and what had finally ended their friendship. As the prime reason for the outbreak of open hostilities he refers to the story of Copes *elasmosaurus* (see chapter 4. 1.):

> The skeleton itself was arranged at the Museum of the Philadelphia Academy of Sciences, according to this restoration, and when Professor Cope showed it to me and explained its peculiarities I noticed that the articulation of the vertebrate were reversed and suggested to him gently that he had the whole thing wrong end foremost. His indignation was great, and he asserted in strong language that he had studied the animal for many months and ought at least to know one end from the other. It seems he did not, [...] his wounded vanity received a shock from which it has never recovered, and he has since been my bitter enemy.

On the next day, January 20, Ballou continued the series with an article titled "SCIENTIST COPE FIRES BACK AT MARSH."[963] The bulk of the article is basically Cope's comeback to Marsh's response. After Cope's initial charges were repeated once again, they were illustrated with some new examples of Marsh's duplicity in exhausting detail; engaging journalism this was not. Interestingly, Cope mentions the supposed reaction of the German scientific community to Marsh's long running frauds and incompetence: "Professor Marsh's course is simply fatuous, for the truth comes out at last, and indeed already Professor Marsh's work is receiving the just verdict of scientific Germany." But alas: "Major Powell and the American public do not read German scientific books and they are deceived just the same." After likening himself to Martin

963 Ballou, William Hosea: Scientist Cope Fires Back at Marsh.

Luther,[964] Cope continued to refute Marsh's countercharges point-by-point. Later, Cope brought up the context of US-American science being young and the subject of a relatively small community; scoundrels like Marsh would hurt it a good deal more than it would a greater community like that of Britain or Germany:

> In America, where the number of men devoted to pure science is relatively small, the standard of excellence of scientific work is constantly liable to be lowered by the entrance into the field of man who are more or less incompetent and sometimes by mere seekers after notoriety."

Following a subheading that reads: "DR. BAUER'S [sic!] INDIGNATION. HE RESIGNS AS PROFESSOR MARSH'S ASSISTANT IN CUT TERMS." Ballou had received a telegram from Baur, in which he told him that he had read Marsh's reply and would terminate his employment with the Yale professor immediately. He added in reference to his statement printed on the day before:

> The following is the copy of the statement given by me to Professor Marsh January 13:—'I hereby certify that I did not dictate any of the generalizations of Professor Marsh's volume on the *dinocerata*. My assistance was purely advisory. We discussed all the points in full, especially on classification, and as his assistant I gave my advice, considerable part of which was adopted by Professor Marsh. With the rest of the volume I have nothing to do except some points on the tarsus, for which special credit was given in [the] volume. [...] I may add that this statement was written down in the presence of Professor Marsh. Originally the word 'dictate' was in quotation marks. On the 16th Professor Marsh importuned me twice to remove them. I objected. On the 17th the Professor came again with the same request. I was tired of it and told him if he thought it would be of so very great importance for him I would remove them. He brought the statement and I removed the marks. Of this statement only the first sentence was published by Professor Marsh in the HERALD.'

The article ends with a rather silly poem titled: "PALAEOZOIC POETRY. THE UNFORTUNATE PTERODACTYL WINGS ITS FLIGHT THROUGH PROSODY."

In an effort to keep his story relevant, or at least alive, Ballou published a very short article on January 22. The article was simply titled "SCOTT WRITES TO MARSH. PRINCETON'S SCIENTIST TELLS THE YALE MAN THAT HE DISAPPROVES OF HIS

---

964 "If I am insane in this matter then the Bar of New York was insane when it demanded the impeachment of judge Barnard, or Luther was insane when he publicly denounced the official dishonesty of the emissary of the pope."

METHODS."[965] Scott wrote that he felt misunderstood and had been drawn into the controversy. To clarify his position, he provided the "Herald" with a copy of a letter he had sent to Marsh on January 17. In it Scott stated that his "personal relations" to Marsh had "always been civil," and that "my participation in the affair was not only involuntary, but against my protest," and that it had never been his wish to participate in a public attack on Marsh. But still he disapproved of Marsh's work, his methods, and his administration of his office.

The final "Herald" article concerning the "Bone Wars" was published on January 26 on page 25 and occupied the better part of two columns. It was titled "SOME MORE NUTS FOR MARSH TO CRACK." And the supplementary subheads read: "His Former Assistant, Otto Meyer, Ph. D., Makes Several Extraordinary Statements Implying Dishonest Methods." and "BIG MEN WITH LITTLE HEADS."[966] The article is a letter Meyer wrote to the "Herald" to further the allegations against his former employer and to clarify his own position in the matter. Some instances of Marsh's alleged incompetence and failure to keep up with the scientific literature are described in great detail. Later Meyer inferred that Marsh had stolen or at least purchased a stolen fossilized tooth from a German museum:

> He [Marsh] pointed out to me once in the drawers of his private room a certain type specimen of a German Jurassic mammal, which consists of a tooth. As far as my information through friends and through the literature goes, this specimen once disappeared mysteriously from a German museum. Does Professor Marsh wish to state how he came into the possession of it?

Still, Meyer regretted that the controversy had come to light in this very public way. He remarked that his criticism was of a purely professional nature, and that he had never given Ballou the authority to publish any of his statements:

> I have now, as always before, no grievance of a personal nature against him [Marsh]. He has always treated me fairly and I have always given him credit for his numerous real merits for paleontology, and shall continue doing so. Few people in the United Sates, for instance, will spend a fortune for science and show a zeal and persistence in collecting, & c., as Professor Marsh has done. I wish, furthermore, to state that nobody had been authorized to use my name in a newspaper attack on Professor Marsh: that there have been some inaccuracies in the HERALD in what I am purported to have written and said, for which I assume no responsibility: that I am sorry that this matter has come

965 William Hosea Ballou: Scott Writes to Marsh, in: The New York Herald, 22 January 1890, p. 5.
966 Ballou: Some More Nuts for Marsh to Crack.

> to a newspaper controversy, and that to publish these statements is a task disagreeable to me in many ways.

He stressed his scientific credibility and the fact that he had helped Marsh not only with mechanical work but also in scientific matters. He concluded his statement with a twist on Marsh's jibe about the "little men with big heads" in the article of January 19:[967] "As for the 'little men with big heads,' I presume that all true scientists have more regard for a little man with a big head than for a big man with a little head."

This most public chapter of the "Bone Wars" ended with this "Herald" article. It appears that the general public was not too interested in the bickering of the paleontologists and the whole affair was nothing new to the scientific community, who had had ample opportunity to witness the conflict in the pages of the "Naturalist" and in private correspondences. The "Herald" articles were reprinted and rehashed in other newspapers,[968] among the most influential ones were the "New York Times," the "Chicago Tribune," and the "Philadelphia Inquirer." The "Times" called the original article "a long, rambling statement," [969] and a "screed." Furthermore, a "Times" reporter had investigated the charges made in the article and found them to be "unworthy of consideration." The whole affair was summarized in an interview by three scientists, among them Alexander Winchell (1824–1891) of Ann Abor University. It was said that Marsh had hired assistants to help him with the more tedious aspects of the paleontological work. He still had "ordered, instituted, carried on, and supervised" the discoveries and should be "honored" for that. Cope, on the other hand, was said to have been embittered by his own financial mishaps and Marsh's scientific successes. Supposedly, Cope was jealous of Marsh's ties to the government via the USGS since he himself had "long been anxious to become attached to the Government Survey." Winchell was then quoted saying that he indeed had some grievances with Marsh and that he had "told Prof. Marsh that he ought to be more liberal with the young men who were employed by him and ought to allow them the credit and honor of their discoveries." But that failure would not amount to plagiarism. Concerning Cope, he stated that "Prof. Cope is a genius in his way, and has all the erratic peculiarities of genius." Finally, the

---

967 Ballou: Marsh Hurls Azoic Facts at Cope.

968 See for example: New Haven Evening Register, 13 January 1890, p. 1; Springfield Republican, 13 January 1890, p. 5; Philadelphia Inquirer, 14 January 1890, p. 2; The Row Among the Geologists, in: Boston Evening Transcript, 16 January 1890, p. 6; Idaho Daily Statesman, 25 January 1890, p. 2; the Tombstone Epitaph Prospector commented on the events in a most bizarre and metaphorical way: "Yale students are said to be observing the fur fly in the bout between Professor Cope and Professor Marsh with emotions akin to those with which the woman cheered on the death struggle between her husband and the bear.", Tombstone Epitaph Prospector, 1 February 1890, p. 4.

969 N.N.: An Old Grievance Aired, In: New York Times, 13 January 1890, p. 8.

“Times” judged “the whole attack was declared to be inspired by a spirit unworthy of the persons engaged in it.”[970]

The “Chicago Tribune” chimed in with an article titled “WAR AMONG THE SCIENTISTS.”[971] It was a simple rehash of the “Herald” article and did not add anything of substance, only the last few sentences warrant quotation to illustrate the smug attitude of most journalists towards the entire affair:

> All savantdom is agog and ready for the fray, and any day may witness a general onset and crash between the consolidated columns of geology and paleontology, with all the other ologies careering about as sharpshooters and outriding guerrillas. That such passions can disturb the ordinarily peaceful scientific breast will seem astonishing, but alas, the votaries of the ologies and the artists of the fossils, the glacial periods, and prehistoric men are only human after all. We hope, however, they may go slow until the influenza is over, and remember that overheating always invites attacks from ‘la grippe.’

The “Philadelphia Inquirer” published an article on the same day; it kept with the theme of martial rhetoric and was titled “Scientist at War.”[972] The “Inquirer” stated that

> the charges have been matter of common rumor among scientific men in this city for over a year past, but the newspapers have declined to discuss them, owing to the eminence of men concerned, and the rather indefinite nature of the evidence adduced against them.

The “Herald” article is then summarized in some detail. The “Inquirer” also saw Cope’s ambitions to restructure the management of the USGS, including a high-ranking position for himself, as the prime motivation for the attack. A new perspective was added in the form of an interview with Leidy, who was also a prominent member of Philadelphia’s scientific elite. Leidy is quoted to have said that it was very “unfortunate that this thing should have gotten into the newspapers,” and that it would “simply cause strife and vexation among the scientific men of the country.” As to the origin of the dispute, Leidy identified Cope’s personality: “While Cope is a man of great ability,

---

970 Other newspapers also saw the futility of the conflict, deeming it unworthy, see for example: “Professor Marsh of Yale and Professor Cope of Philadelphia continue their controversy, which has little interest for the general public and is discreditable for American science.”, Worcester Daily Spy, 22 January 1890, p. 4.

971 War Among the Scientists, in: Chicago Tribune, 16 January 1890, p. 4.

972 Scientist at War, in: Philadelphia Inquirer, 13 January 1890, p. 2.

he has a faculty of getting into hot water with his associates." Leidy had clashed with Cope himself and witnessed the "faculty of getting into hot water" firsthand (see chapter 4. 4.). Regarding the charges of plagiarism, Leidy defended Marsh and specified that:

> all his descriptions are original and all his reports are excellent and highly thought of by scientific men generally. He was perfectly justified in allowing his assistants to relieve him of some of the work. [...] I have known Prof. Marsh for many years and I have always found him to be an upright and honorable man.

After the mainstream newspapers had lost interest in the story,[973] Cope used his position as an editor of the "Naturalist" to bring the "Herald" articles to the pages of said magazine.[974] He recapped that the conflict which had long smoldered in secret, had reached the general public, and brought the charge of Marsh's practice of stealing the scientific work of his assistants to the "Naturalist." Maybe this was to prepare the readers for the articles of Baur and Barbour, which were published in the "Naturalist" in subsequent issues:

First there was Baur's "A Review of the Charges Against the Paleontological Department of the U. S. Geological Survey, and of the Defence [sic!] made by Prof. O. C. Marsh," published in March 1890.[975] Baur recapped each of the major points of contention listed by Cope and subsequently refuted by Marsh, and in turn added his criticisms. First, he admitted that indeed all fossils stored at Yale were labeled and government specimens could not be mixed up with those that belonged to the Peabody Museum. But he added that the labeling was "entirely in the hand of Prof. Marsh, without any control from the geological Survey." And the question remained who really paid for the acquisition of the fossils: "Can Prof. Marsh pay his collectors this month out of his own pocket, and the following out of the pocket of the government?" The second point Baur addressed was that of the availability of Yale specimens to visiting scientists. He stated that Marsh had been lying even to scientists of unquestionable "good moral character," that he had been putting them off by telling them that the specimens were stored in boxes and were inaccessible, when in truth they lay about

973 The "Aberdeen Daily News" put it most poignantly: "they continued their fossil throwing until the general public at last was fatigued.", Aberdeen Daily News, 18 September 1890, p. 2, see also: The Scientists New President, in: Columbus Daily Enquirer, 22 September 1890, p. 6.

974 Edward Drinker Cope; John Sterling Kingsley: EDITORIAL, in: American Naturalist, vol. 24, no. 278 (Feb. 1890), pp. 158–160.

975 Georg Baur: A Review of the Charges Against the Paleontological Department of the U. S. Geological Survey, and of the Defence made by Prof. O. C. Marsh, in: American Naturalist, vol. 24, no. 279 (Mar. 1890), pp. 298–304.

on a table in the same room but had been covered with cloth on the professor's orders the day before. The third point Baur addressed was probably closest to his heart and of the most relevance for this study. It was the point that "the greater part of Prof. Marsh's published work has been done by his assistants." Baur claimed that huge parts of the "Odontornithes" had probably been written by Harger. He went on to describe Marsh's work practice:

> Prof. Marsh asks them [his assistants] questions, the answers of which he either immediately puts down in black and white, or he makes out a list of questions to be worked out by his assistants, for instance: 'What are the principal characters of the skull of the Sauropoda?' or, 'What are the relations between the different groups of Dinosaurs?' and so on. The assistant, if not yet fully familiar with these questions, begins to work; he goes over the whole literature, a thing rarely done by the Professor, and studies the specimens in the collection. After this is done, the Professor receives the note of the assistant, or he asks questions, writing down the answers he receives. In this way he accumulates a great quantity of notes, written in his own handwriting, or in that of the assistants. By comparing and using these notes it is easy for him to dictate a paper to any person who can write. This person, of course, when asked, can testify that the work was dictated by Prof. Marsh, without telling a falsehood.

Baur then detailed his own involvement in writing the "Dinocerata":

> On two Sundays I spent a number of hours at Prof. Marsh's house, to 'go over his conclusions.' Questions were asked and answered, new points were brought up by me and adopted, and when it came to the classification of Ungulata, I gave my opinion, which was mainly based on Prof. Cope's work, introducing small changes only. I gave the classification with Prof. Cope's names, as I informed him; but these were all changed by Prof. Marsh. There is no doubt Prof. Marsh had never studied Prof. Cope's papers on this subject, since he not only did not know the names of the orders, but he even asked how to spell them. That the descriptive part of the Dinocerata was mainly the work of Mr. O. Harger, I know. He made both descriptions and measurements of different bones, which were used by Prof. Marsh when he wrote his text, or dictated it.

Baur went on to demonstrate Marsh's supposed scientific incompetence and sloppiness by describing an instance during which Marsh had published a faulty description: he had mistaken a badly weathered dinosaur vertebra for the jaw of a rather huge Jurassic mammal. Marsh had recognized his mistake later, but never bothered to correct it. The fourth point Baur addressed is Marsh's alleged plagiarism. Even though he believed that Marsh had never read Kovalevsky's work, Baur insisted that Marsh must have read about the Russian's paper in the "Palaeontographica." Furthermore, in 1878 Marsh had supposedly published discoveries of the British geologist T.W. Hulke,[976] which had already been published by the original author years before. He added:

> I have devoted considerable time to the study of the evolution of the skeleton of the ostrich. Among others, I made a discovery which was of especial importance, as it throws new light on the question of the relation between birds and dinosaurs. I told Prof. Marsh about this discovery, and did not publish it. When Prof. Marsh wrote his paper on *Ornithomimus* he simply claimed the discovery as his own, not mentioning me at all. This I saw when he gave me the proof-sheets of the papers. It was after a discussion of nearly two hours that Prof. Marsh agreed to give me credit for it (in a place where it could be easily overlooked) in the explanation of the figures.

The fifth point of contention was that Meyer had made the statement that fossil specimens had been restored in a very unscientific manner under Marsh's direction. He had ordered the use of colored plaster, making it very hard to tell which part of a reconstruction showed the original fossil and which was made of plaster, rendering them nearly useless for studies. Drawings of the specimens were equally useless because it was not made apparent which parts of the skeleton had been found and which had been reconstructed by deduction; though Baur admitted that this practice had stopped in 1885 and that missing parts were being "shaded in" now. He added that Marsh had written the review of the "Dinocerata" himself. Baur apparently forgot about the sixth point and continued with his seventh point, defusing an accusation made by Meyer that Marsh had come into possession of a fossil tooth that had been stolen from the Stuttgart Museum. Point eight is just a stab at Marsh:

> Prof. Cope thinks 'that an investigation as to who has delivered Prof. Marsh's lectures at Yale College during the past years will yield some interesting results.' To this I have to say, that such an investigation is not necessary; *Prof. Marsh does not lecture at Yale at all.*

976 Baur probably had John Whitaker Hulke (1830–1895) in mind, who indeed was at one time president of the Geological Society of London and had published the article mentioned by Baur (though not on page 334, as Baur suggested). See: John Whitaker Hulke: Appendix to "Note on a Modified Form of Dinosaurian Ilium, Hitherto Reputed Scapula", in: The Quarterly Journal of the Geological Society of London, vol. 32 (1876), pp. 364–366.

This of course was true because he was not employed by the college as a lecturer, and, at that time, received no salary from Yale.

In his ninth and final point Baur quoted Scott’s letter to Marsh, published in the “Herald” on January 22:

> I feel constrained to say that I disapprove of your work, your methods and your administration of the office which you hold. This disapproval does not rest on what I have heard from others, nor upon any personal considerations, but upon my own experiences and my studies in the field to which both you and I are devoted. If called upon to testify in any investigation, this is the line to which, however reluctantly, I shall be compelled to adhere.

Babour continued the attack in the next issue of the “Naturalist,” titled “Notes on the Paleontological Laboratory of the United States Geological Survey under Professor Marsh.”[977] In the very first paragraph he wrote about Marsh’s practice to camouflage the plaster used to reconstruct the dinosaur skeletons, and put it quite poetically:

> If there is any truth under the sun then judgment must fall on the scientist who walks the halls of the Yale Museum armed with a wet sponge. Why a wet sponge? You say. Perhaps it was to wipe the dust from some noble fossil? Far from it! [B]ut rather to wash the purity of a truth out of the blackness of a falsehood.[978]

The next paragraph hints at the international ramifications of this dishonest practice, which could supposedly hurt American science in general:

> To those in foreign lands, especially in Germany, who have marveled at the exceptional beauty and perfect preservation of Prof. Marsh’s specimens, let it be said that although you cannot apply the sponge test to his faultless, fractureless [sic!] plates you can to the specimens from which they were drawn.[979]

Barbour then described in some detail how an ideal mixture of paint and materials had been found through trial and error and how perfect the results were. Even he, who had participated in some of the reconstructions, could not tell where the bones ended

977 Erwin Hinckley Barbour: Notes on the Paleontological Laboratory of the United States Geological Survey under Professor Marsh, in: American Naturalist, vol. 24, no. 280 (Apr. 1890), pp. 388–400.

978 Barbour: Notes on the Paleontological Laboratory, p. 388.

979 Barbour: Notes on the Paleontological Laboratory, p. 388.

and the colored plaster began.[980] The merge of government and private finances came into play on the next page, as Barbour illustrated how the separation of those two was next to impossible:

> 'Life was too short' to give a Government specimen more than a lick and a promise, but quite long enough to devote months and years of Government time and money in beautifying his own private collections. [...] This abuse of public trust led us to frequent and spirited disagreements and our relations became exceedingly strained, and still more so when I refused to add to the crime of misappropriating Government time [to] that of deceiving in the restorations.[981]

Barbour then proceeded to describe how Marsh had misappropriated government money and how he had exaggerated the size of some of his fossils in his descriptions and documented the deceptions with illustrations. Along the way he also pointed out some cases of Marsh's awkward mistakes, stemming from his alleged paleontological incompetence.[982] Among the many points of contention there was the issue of how Marsh had treated his assistants:

> Not only does he avoid helping his assistants to better positions in geological fields, but he often hinders them by trampling over their good names when gone. We assistants watched the evolution of a falsehood from his lips, from the day when he said, 'that man has resigned' to the month when he said 'I had to let him go; he was a bad lot,' until still later he 'dismissed him because he was unreliable and light-fingered.' [...] by his ever-recurring, never-ending expressions of hatred and distrust, Professor Marsh methodically tries to fill to saturation the minds of his young assistants with prejudice against his contemporary in paleontology (Professor Cope). These are but allusions to his hindrances, put in the way of others in his attempts to monopolize paleontology in the East and West.[983]

Later he added:

> One important assistant, on private pay, not independent at the time (drawing a small salary, not half his just desserts), was asked as a favor to be listed on the

980 Barbour: Notes on the Paleontological Laboratory, p. 389.

981 Barbour: Notes on the Paleontological Laboratory, p. 390.

982 Barbour: Notes on the Paleontological Laboratory, pp. 391–396.

983 Barbour: Notes on the Paleontological Laboratory, pp. 396–397.

> Government pay-roll, to which he readily agreed as a matter of accommodation, but only to find, the next quarter, that his salary had been cut down two hundred dollars. These facts, and many that are necessarily suppressed for the nonce, in consideration of the present members of his force, coupled with his insincerity in scientific work, will help to explain why the *personnel* of his force undergoes such constant and rapid change. High-spirited young men, college graduates, cannot and will not tolerate such associations and environments.[984]

On the last pages of his writings Barbour exemplified how Marsh had squandered government money on unnecessarily grandiose lithographic plates and detailed the professor's habit to pay his employees late and only quarterly, not monthly.[985] In addition, some of the plaster Marsh's assistants used was made from ground-up greenbacks, redeemed by the Treasury. Thus, an argument can be made that government money was indeed interwoven with the fossils.[986]

As to Marsh's payment ethics, or the lack thereof, Schuchert and LeVene write:

> He was a hard bargainer, it is true, but it does not appear that the wages he paid were much if any below the scale of the day, which was certainly not high, even for University professors. However, whatever wage he paid was bound to seem niggardly, coming from a man of such obvious wealth. In irregularity of payment, on the other hand, he was a decided sinner. [...] Never having depended on a salary himself, he seems to have had no vision of how all-important it is, especially when one's income is small, to have it come to hand on a given date.[987]

Concerning the alleged plagiarism and his tendencies to deny his assistants the due recognition in his publications, Schuchert and LeVene write that the practice was commonplace. But the fact that some of his assistants (namely Baur and Harger) felt they were superior to their employer had contributed to their outrage. In regard to the actual day-to-day work, they write that each assistant was assigned a series of fossil bones that was to be studied within the context of its nearest affinities and of course in the context of the latest research published. Then, after some time, he would discuss their findings and take notes on the various fossils. His final conclusions he would seldom share with his assistants but publish them without giving the research assistant any credit. Schuchert and LeVene write that Marsh's work practice must

984 Barbour: Notes on the Paleontological Laboratory, p. 397.

985 Barbour: Notes on the Paleontological Laboratory, pp. 397–400.

986 Schuchert; LeVene: O. C. Marsh, p. 297.

987 Schuchert; LeVene: O. C. Marsh, p. 307.

have been clear to all his assistants; that it must have been evident to them that Marsh alone was to publish about the fossil vertebrates of the museum. Because Marsh had not employed any assistants before 1872 but still had managed to publish 51 notes and papers (210 printed pages in all), he cannot have been as scientifically incompetent as alleged by Cope and the others. Still, due to Marsh's habit of micromanaging every aspect of the paleontological work done at the museum, his assistants were left with many unoccupied hours during which they discussed the shortcomings of their employer, furthering their dissatisfaction. Schuchert and LeVene cite Westbrook, who attested that Marsh was a very hard worker at first, completely absorbed by his love for paleontology. But his increasing wealth and the administrative obligations in the USGS and the NAS cost him a lot of time and mental energy; he became a procrastinator. To make matters worse, he became overwhelmed by the sheer number of fossils collected after 1882, necessitating the employment of more assistants in the first place. At the same time, he became preoccupied with networking and securing money for the completion of the Peabody Museum. The final factor driving a wedge between Marsh and his assistants must have been outside influence. Cope made contact with Baur,[988] and thanks to the help of Osborn the grievances of the disgruntled assistants found their way, in one form or the other, to the pages of the "Herald" and the "Naturalist" in 1890.[989]

The articles had a damaging effect on Marsh's career in the long run. After two years of drought, pressures mounted on the USGS to concentrate on plans of reservoirs and irrigation projects. Hilary A. Herbert, a Congressman from Alabama, proposed to cut all funding for paleontology by the USGS. Marsh's epic monography on "Birds with Teeth" became synonymous with a waste of government funding; public money should further the material well-being of the citizens and not be spent on seemingly purely academic endeavors. Powell and Marsh had to resign from the USGS:[990] "Appropriation cut off. Please send your resignation at once."[991] But soon after Powell appointed Marsh "Honorary Paleontologist" of the USGS:

> With the state of the appropriations it is impossible to continue your work on a salary, but I sincerely hope that, by reason of great interest in the science, you

988 "Cope might have been attracted to Baur by the similarity of their scientific fields, and by their like philosophic trend of thought; but it is not to assume that their conversation, once they did meet, was confined to Vertebrate Paleontology in the abstract. Such disaffection as was already in the minds of the men mentioned above was exactly the sort of culture to grow the seed that Cope was only too anxious to sow, be his motive what you will.", Schuchert; LeVene: O. C. Marsh, p. 311.

989 Schuchert; LeVene: O. C. Marsh, pp. 308–312.

990 Lanham: The Bone Hunters, pp. 260, 261.

991 John Wesley Powell, Washington, DC to Othniel Charles Marsh, New Haven, CT, 20 July 1892, MS 343, Series I. Correspondence, Box 26, Folder 1096.

> will find time to supervise the preparation of the monographs now under way and to continue in the care and custody of the collections now in your charge. With this end in view you are hereby appointed honorary Paleontologist of the Geological Survey, in charge of Vertebrate Paleontology.[992]

6.6 Chapter Conclusion

Other than Baur finally resigning, there were no immediate consequences for Cope, Marsh, or Powell when the battle in the "Herald" was over. Nonetheless, this chapter of the "Bone Wars" had some long-term consequences for paleontological research in the US: For years, some congressmen for the western states had doubts about the USGS's leadership and mission. They were concerned with the practical applications of this public-funded research, namely the completion of a geological map of the United States, which could be used for the practical purpose of locating yet untapped water sources for irrigation. 1890 was a drought year; the political pressure on Powell and his USGS rose and the western congressmen sought to transform the USGS into a more economic and practical operation. Hilary Abner Herbert (1834–1919), a congressman from Alabama, led the charge against the USGS and particularly its department of paleontology. Alexander Agassiz had told him that paleontology was a science that did not need to be funded by the government and could be managed and conducted more efficiently by private individuals. Marsh became the focal point of Herbert's attack and his very lavish "Odontornithes" (expensively bound, with gilded edges and numerous illustrations) the evidence for the heedless dissipation of government money. This was somewhat uncalled for because Marsh had funded the publication of this extravagant edition himself and before he had joined the USGS. Still, "Odontornithes" remained a rallying point for the enemies of the paleontological department in Congress, and in July 1892, after Osborn had joined the fray against Marsh and Powell, and not without controversial debate, the Senate cut the budget of the USGS drastically. On July 20, Powell asked for Marsh's resignation via telegram. With the USGS income gone and his life-long Peabody-funds not yielding nearly as much as they had some twenty years before, Marsh had to ask Yale University for a regular salary.[993] The negative publicity Marsh and Powell had received on the pages of the "Herald" had certainly contributed to the hostile attitudes of the congressmen. Ballou surely believed it; in 1908 he wrote:

---

992 John Wesley Powell, Washington, DC to Othniel Charles Marsh, New Haven, CT, 5 August 1892, MS 343, Series I. Correspondence, Box 26, Folder 1096.

993 Schuchert; LeVene: O. C. Marsh, pp. 313–322.

> He [Cope] could not get justice in Washington, nor his works printed, nor his pay from the government. I undertook to straighten things out for him at which he was greatly amused. 'Why, little boy,' he said, 'you don't want to get in between. They will make dust of you so fine they can't even see it on their feet.' 'I'll show you,' said I, 'that the press rules in this country.' Thereupon followed my fifty-two-column exposé running eight days in the *New York Herald*, of one of the worst rings that ever fastened on the government, smashing the ring completely.[994]

Moreover, the manner in which the decade-old conflict between Cope and Marsh was portrayed and sensationalized by Ballou has influenced and affected the way it is remembered to this day. Terms like "Bone Wars" and "Fossil Feud" bear the same martial rhetoric. On the other hand, that animosity is the main reason Cope and Marsh are still somewhat famous today while most of the other paleontologists of that time have been forgotten by the public.

Reingold writes a very comprehensive statement about the relationship between Marsh and his assistants:

> Marsh had one fatal flaw – his relations with his subordinates at the Peabody Museum at Yale. Marsh did not teach (until 1896 when he was financially embarrassed, Marsh did not receive a salary) and had no students. The assistants at the museum, many of them good scientists in their own right, were regarded merely as hired hands who worked at small bits of his research, so that they could not either rival Marsh or leak information to Cope. The museum was a research factory for the glory of Marsh, not an institution of colleagues engaged in research.[995]

The "Bone Wars" were a transformative force in the international paleontological network. Cope, Osborn, and Scott fanned the flames of the discontent Baur, Schlosser, and Meyer had harbored toward Marsh. While Meyer quit science altogether and Schlosser continued his career in Germany, Baur remained in the US and transplanted some of the German scientific traditions to his new home. All of this was made possible because Marsh had visited Europe in the 1860s, kept in touch with the German scientific community, and employed a member of this community, Zittel, to look for German "Privatdozenten" who were willing to work in the US and study the great fossils of the American West.

---

994 Ballou: Some Great American Scientists, p. 104.

995 Reingold: Science in Nineteenth-Century America, p. 237.

As illustrated in chapter 4, the “Bone Wars” were a formative force in the development of US paleontology. They spurred a plethora of acquisitions and subsequent descriptions of fossils in the United States. These fossils then introduced dinosaurs, and to a lesser degree paleontology itself, to US-American popular culture, making dinosaurs both “American” and a source of “Americanness,” of US-American identity. The series of newspaper articles published in the New York Herald in 1890 ushered in the final and most palpable chapter in the conflict. Finally, Marsh’s assistants and employees, especially Baur, Meyer, and Williston could air their grievances publicly, which in turn tells a lot about the relationship they had with their former employer and about the working conditions at Marsh’s paleontological laboratory. Now the newspaper-reading public audience realized how bitterly, and in parts bizarrely, the decade-long feud had been waged; previously only the scientifically literate public had had insight into the “Bone Wars.” Furthermore, now the conflict itself became an inspiration for fictional and non-fictional works of art, making the “Bone Wars” a part of US popular culture.

7

"To Doubt Evolution To-Day is to Doubt Science" – Evolution and Paleontology

This chapter highlights the importance of paleontology in the context of Darwin's theory of evolution. US-American fossils contributed significant evidence for Darwin's theory, then still hotly disputed within the scientific community:

> Americans contributed to the Darwinian theory in two interrelated ways. The first was often unintentional: flora, fauna, and rocks collected by American scientists sometimes provided data for Darwin. Geologists, a very active scientific group in nineteenth-century America, were particularly important because of the fossil remains they brought to light. The second contribution was made by a small number of Americans who became correspondents of Darwin's, providing him with information and criticism.[996]

Marsh contributed in both ways.

Although the field of paleontology remained mostly closed to women until the twentieth century, it is worth mentioning that the success of Darwinian evolutionary theory inspired a wave of feminism in the second half of the nineteenth century. Darwin's "The Descent of Man"[997] familiarized many a reader with the concept of sexual selection and its potential consequences for human evolution. The importance of choosing a mate suggested that women had had more agency throughout the history of the human race then previously assumed. Kimberly Hamlin, who "analyzes American responses to evolutionary theory through the lens of gender,"[998] provides a fascinating analysis of the history of feminism and women's rights in nineteenth-century America. She describes how evolutionary theory replaced regressive Christian ideas of female inferiority,[999] and how in the 1870s women began to seek scientific education and to participate in science as an emancipatory act.[1000]

The first subchapter will detail in briefness how paleontology delivered evidence for Darwin's theory of evolution like no other scientific discipline could during the nineteenth century.

The second part of this chapter will detail Marsh's place in an international discourse discussing Darwin's theory of evolution. Marsh's correspondence with proponents of the theory, mainly Huxley, Haeckel, and Darwin, will be the foundation of this subchapter.

---

996 Reingold: Science in Nineteenth-Century America, pp. 162–163.

997 Charles Darwin: The Descent of Man, and Selection in Relation to Sex, London 1871.

998 Hamlin, Kimberly A.: From Eve to Evolution. Darwin, Science, and Women's Rights in Gilded Age America. Chicago 2014, DOI:10.7208/chicago/9780226134758.001.0001, p. 2.

999 Hamlin: From Eve to Evolution, pp. 25–56.

1000 Hamlin: From Eve to Evolution, pp. 57–78.

7.1 Paleontology and Theories of Evolution

"If we must locate our confidence about evolution in evidence for history – in part directed from the fossil record, but usually indirectly by inference from modern organisms – by what rules of reason, or canons of evidence, shall history then be established?"[1001]

Up until the second half of the nineteenth century the main goal of higher education was not to acquire knowledge, but to be morally educated and to discipline the mind. For a long time, religion very much influenced how higher education was structured. Religious moral imperatives were taught primarily, scientific learning only secondarily. To publicly doubt the "biblical truth" was, for many colleges and universities, reason enough to expel a student. To further the moral education of the students, institutions of higher education interfered with almost all aspects of their lives. Natural sciences were only allowed to be taught if they did not contradict Christian beliefs and the Bible. Nonetheless, simply advocating for Darwin's theory of evolution was almost never a reason to expel a student or fire a professor. Yet, around the year 1880, the system of moral education had been all but abandoned (see chapter 8. 3.).[1002] Note that the most resounding criticism of Darwinian evolution is – and always has been – inspired by religious beliefs. Michael Ruse describes very neatly how for centuries and throughout many different cultures and geographical regions the history of life on Earth has been imagined as being designed by a god, or gods. Accordingly, Darwinian evolution encountered heavy resistance.[1003] Religious resistance to evolutionary theory is always vehement when the origin of human beings is discussed, or as Ruse puts it: "If Darwinian evolutionary theory did not extend its grasp to cover us humans, no one would ever say anything nasty about it."[1004] While Darwin tried to circumvent the question of human evolution, Huxley – being an agnostic out in the open – never had the same reservations.[1005] In his 1859-opus "On the Origin of Species by Means of Natural Selection, or the Preservation of Favoured Races in the Struggle for Life"[1006] he never directly addressed human evolution, but did so with "The Descent of Man" almost two decades later.

1001 Stephen Jay Gould: The Structure of Evolutionary Theory, Cambridge, MA 2002, p. 101.

1002 Veysey: The Emergence of the American University, pp. 21–56.

1003 Michael Ruse: Darwin and Design. Does Evolution Have a Purpose? Cambridge, MA 2003, pp. 11–67.

1004 Michael Ruse: Darwinism and Its Discontents, Cambridge, MA 2006, p. 166.

1005 Ruse: Darwin and Design, pp. 103–106. Also see: Peter J. Bowler; Iwan Rhys Morus: Making Modern Science. A Historical Survey, Chicago 2005, pp. 154–157.

1006 Charles Darwin: On the Origin of Species by Means of Natural Selection, or the Preservation of Favoured Races in the Struggle for Life, London 1859.

The idea that the history of life on earth constituted a progression, that "lower" lifeforms had to give way to "higher" organisms, and that the progression of lifeforms had a ultimate (clear? definite? ultimate?) goal and found its completion in the emergence of humans first came up in Early Modern times. The idea of progression became especially prominent during the early nineteenth century, with the emergence of theories of evolution and extinction. Fossils had always played an important role in the discourse, even in early debates at the beginning of the century. They were used as evidence by both sides, one arguing that the history of life was ordered by progression, the other arguing against it. Another, more general point of contention was whether evolution by natural means had occurred at all, or whether all life had been divinely created. Both sides used fossils as proof for their claims. Even though the debate concerning evolution vs. creationism was far from over during the first half of the century, progression was seldom denied. The debate focused more and more on the cause of progression.[1007]

Lyell, one of the most influential geologists of the period (see chapter 2. 5.), argued against progression and said that there was no real evidence to be found in the fossil record. It was simply too incomplete to answer the question. This shows how important a comprehensive fossil collection was as evidence for scientific theories. Not only quality but also quantity was of great importance to argue for or against natural processes that took millennia and many generations to occur. Lyell was not swayed to accept progression even when the fossil record improved during the course of the century. In the 1860s he gradually accepted the Darwinian theory of evolution, for which fossil evidence was now undeniable. But he always remained skeptical concerning human evolution, thinking that humans needed to fill a special place in this order of creation, again showing how external, in this case religious, circumstances influence the pursuit of science.[1008] This external influence is even more apparent when looking at the reception of human evolution. To most observers, "progression" in the history of life meant a progression towards higher life and most scientists believed that human life was the highest possible lifeform of them all, that all progression had been leading towards the emergence of the human.[1009] This demonstrates another external, societal influence on scientific thinking, for the all-important hierarchies that ordered Victorian societies now found their expression in the way scientists imagined nature to be ordered. The fact that their own society was ordered into different distinct social classes influenced scientists to order nature accordingly, to order it into different distinct classes that followed a hierarchy.

1007 Peter J. Bowler: Fossils and Progress. Paleontology and the Idea of Progressive Evolution in the Nineteenth Century, New York 1976, pp. 1–65.

1008 Bowler: Fossils and Progress, pp. 67–77.

1009 Bowler: Fossils and Progress, pp. 93–115.

Darwin offered a natural explanation for evolutionary processes. Even if evolution was progress, said progress no longer had to be guided or directed by a divine plan. The unearthing and analysis of different fossils could further scientific debates and provide valuable evidence, as Bowler points out:

> [I]t seems not unreasonable to assume that the evidence itself forced the majority of naturalists to re-evaluate the old ideas about design, progression, and the linear approach to taxonomy. Although it was not enough to demolish belief in the divine control of nature, it *was* enough to make naturalists revise and qualify their views on the nature of that control.[1010]

In 1855 an incomplete bird skeleton was found in the Altmühl Valley in Bavaria. Soon thereafter German paleontologist Hermann von Meyer (1801–1869) misidentified it as being a *pterosaur.*[1011] When the fossil imprint of a feather was unearthed in the vicinity of Solnhofen, Bavaria, fossilized proof for Darwin's theory of evolution. Von Mayer identified the fossil as the first example of a bird, dating back to the Jurassic. He christened the specimen *archaeopteryx lithographica*. One year later a more complete skeleton of *archaeopteryx* was discovered in Solnhofen, and another one in 1862, which was then sold to London, where Richard Owen described the specimen. Owen supposed *archaeopteryx* to be a bird, and Thomas Henry Huxley discovered the similarities between *archaeopteryx* and *compsognathus*, a small Dinosaur. To Huxely *archaeopteryx* was a "missing link," an evolutionary steppingstone between reptiles and birds.[1012] *Archaeopteryx* is still somewhat of a paleontological super star, with just twelve known specimens. Most specimens have commonly known nicknames, draw a lot of public attention in the collections they are a part of, and whenever a new specimen is discovered, it is cause for much publicity. This happened in 2016, when a new specimen was exhibited at the Dinosaur Museum Altmühltal.[1013] Some of this special attention

---

1010 Bowler: Fossils and Progress, p. 112.

1011 In 1970 the specimen was identified as belonging to the genus archaeopteryx, and became known as the Haarlem specimen, for it was, and is still, kept at the Tyler Museum in Haarlem, Netherlands. A 2017 re-evaluation of the fossil concluded that the specimen actually belonged to a yet another genus: ostromia. See: Christian Foth; Oliver W.M. Rauhut: Re-evaluation of the Haarlem Archaeopteryx and the radiation of maniraptoran theropod dinosaurs, in: BMC Evolutionary Biology, vol. 17, art. no. 236 (Dec. 2017), https://doi.org/10.1186/s12862-017-1076-y.

1012 Buffetaut: A Short History of Vertebrate Palaeontology, pp. 110–111. For more information on the importance of archaeopteryx for the Darwinian theory of evolution see: Bowler; Morus: Making Modern Science, pp. 151–153.

1013 N.N.: Urvogel Archaeopteryx und andere Fossilienstars, https://www.naturpark-altmuehltal.de/fossilienstars/, as consulted online on December 03, 2019.

derives from *archaeopteryx*’ status as the “missing link,” and its significance for the theory of evolution.

Adam Goldstein judges that most US-scientists quickly adapted Darwin’s theory (Louis Agassiz was the most prominent exception), but in the following decades Lamarckian evolutionary theory was re-invoked in the form of Neo-Lamarckism, with Cope being one of the most influential proponents of this alternative theory.[1014] Reingold even attests that

> Agassiz’ presence in America converted the controversy over evolution in the United States from what might have been a provincial sideshow to a major incident in one of the great intellectual controversies of the nineteenth century.[1015]

Stephen Jay Gould writes that US-American contributions to the theory of evolution earned US-American science respect in Europe for the first time:

> For a nation still coming of age as a scientific power, and still bearing a reputation, at least in natural history, as supplier of data for the theory-mills of a more sophisticated Europe, the rise of an American movement, centered in a novel theoretical perspective, and generating both attention and respect in Europe, marked an important gain in maturity.[1016]

This goes hand-in-hand with the appreciation US-American paleontology received for its most complete and spectacular fossils, which in turn were the product of the uniquely American lands and “frontier experience” (see chapter 5).

7.2 Marsh and the Darwinian Theory of Evolution

“To doubt evolution to-day is to doubt science, and science is only another name for truth. Taking, then, evolution as the key to the mysteries of past life on the earth, I invite your attention to the subject I have chosen.”[1017]

1014 Adam A. Goldstein: Darwinism, in: Georgia M. Montgomery; Mark A. Largent (eds.): A Companion to the History of American Science, Chichester 2016, pp. 306–319, DOI:10.1002/9781119072218. Also see: Peter J. Bowler: Evolution. The History of an Idea, 3rd compl. rev. and exp. ed., Berkeley, CA 2003, pp. 186–187.

1015 Reingold: Science in Nineteenth-Century America, p. 181.

1016 Gould: The Structure of Evolutionary Theory, p. 366.

1017 From an address Marsh gave to the AAAS in 1877. Quoted after: Schuchert; LeVene: O. C. Marsh, pp. 240–241.

LeVene and Schuchert write the following about Marsh's contribution to the spread of the theory of evolution in the US:

> One of America's earliest and most ardent exponents of the Darwinian theory of natural selection, his startling discovery of birds possessing teeth and other reptilian characteristics bridged a gap in the evolutionary series, fulfilling a prophecy made by Huxley; and his carefully collected series of fossil horses demonstrated not only that the development of the horse had taken place mainly in this country, and not in Europe, as was formerly held, but that the line of descent so traced was an unanswerable argument in favor of the Darwinian hypothesis. These contributions of his were acknowledged by Darwin to be the best support that the evolution theory had received since the publication of The Origin of Species in 1859.[1018]

LeVene and Schuchert state that Marsh was familiar with creationist theories of evolution thanks to his education at Andover Academy, and had read Agassiz' book on the subject.[1019] They write that it cannot be said with any certainty when Marsh was convinced by Darwin's theory, whether it happened before, or, at the latest, during his journey through Europe (see chapter 3. 2.).[1020] As to the influence of the European journey on Marsh's view on the theory of evolution, they write:

> On his first visit to England, in 1862, he made the acquaintance of a number of the leading workers in science, notably Lyell, whose *Principles of Geology* had had so strong an influence on Darwin. It is not certain whether his new acquaintances of this year included Thomas Huxley, the brilliant zoologist who had become the leading exponent of the Darwinian ideas, but their meeting could not have been long delayed, because he spoke of Huxley in after years as one who had been 'guide, philosopher, and friend, almost from the time I made the choice of science as my life work.' We also know that as early as 1865 he had been at the country home of Darwin, and his library has a copy of the fourth English edition of the *Origin of Species*, dated June, 1866.[1021]

1018 Schuchert; LeVene: O. C. Marsh, pp. 1–2.

1019 Louis Agassiz; Augustus Addison Gould: Principles of Zoology. Touching the Structure, Development, Distribution, and Natural Arrangement of the Races of Animals, Living and Extinct, with Numerous Illustrations. Part I, Comparative Physiology, for the Use of Schools and Colleges, Boston 1851.

1020 Schuchert; LeVene: O. C. Marsh, pp. 226–230.

1021 Schuchert; LeVene: O. C. Marsh, p. 230.

Marsh was an “early adapter” of Darwin’s theory of evolution. The notes he took during lectures in Berlin contain numerus references to Darwin. In Peter’s first lecture (see chapter 3. 2.) on October 29, 1863, he noted that

> Darwin’s theory is ingenious but has no observations to support it. Of the lower animals 1000 generations can be observed, and in no case have important alterations been observed. The higher animals cannot be so observed as they are much longer lived. But fossils are good evidence.[1022]

Marsh already knew that fossils would provide the best available evidence for Darwin’s theory. This recognition foreshadows Marsh’s later role in the discussion and promotion of Darwinian evolution in the United States.

In lecture no. 13 on November 12, 1863, Marsh noted to much the same effect: “Darwin’s theory is very probably true, but has not yet been proved in the case of a simple species.”[1023] Directly beneath this statement Marsh noted that Huxley’s views on evolution would not take on, a conviction Marsh probably revised subsequently, at the latest when he began a close working relationship and, following, a friendship with Huxley (see below): “Huxley’s views will not hold, as the differences between men & apes is far too great to have been brought about in any length of time.”[1024] A later note, taken during lecture no. 39, on December 17, 1863, is even more revealing, for it takes note of the importance of the possible evolution of the horse in North America, the lineage of the horse in America becoming one of Marsh’s later contributions to proving Darwin’s theory of evolution: “Teeth of horses have been found in America (?) This is a very important discovery if true. (See Owen Pal. P. 398.) (See Darwin Voyage Beagle vol III p. 150)”[1025]

Grinnell attests that Marsh upheld the theory of evolution and was in contact with other champions of the idea:

> Prof. Marsh is a firm believer in evolution, and enjoys the personal acquaintance and friendship of Darwin, Huxley, Wallace, Spencer, and other prominent advocates of this doctrine. He is at present in England with his scientific friends, but will return in time for the St. Louis meeting of the Association for the Advancement of Science.[1026]

1022 MS 343, reel 24, frame 412.

1023 MS 343, reel 24, frame 434.

1024 MS 343, reel 24, frame 434.

1025 MS 343, reel 24, frame 524.

1026 Grinnell: Sketch of Professor O. C. Marsh, p. 615.

Georg von Bunsen, who had visited Marsh in 1893, stated that Marsh did not believe in the theory of evolution. He had revisited the "Odontornithes" and wrote:

> For a Nicht-Fachmann[1027] like myself, it was simply impossible to follow your descriptions or understand the bearings of each observation. Yet I gathered that satisfaction from personal acquaintance with the hesperornis & the ichthyornis as well as with our archaeopteryx which people have from the friendship of the Great & Good. Also I have begun to guess the deeper reasons why you & Virchow & others refuse to accept the theory of evolution.[1028]

This comes as a great surprise since Bunsen believed that Marsh refused the theory of evolution, while all other evidence points to the conclusion that he, in fact, did not reject said theory.

Marsh's extensive fossil collections from the American West became practical proof for the theory of evolution. In 1868 Marsh began his exploration of the West and published on the ancestry of the modern horse (Marsh's reconstruction of the evolutionary tree of the horse greatly impressed Huxley during his stay in the US in 1876). He undertook the next expedition in 1870 and crossed, according to Grinnell, the Rocky Mountains more than 77 times. Marsh's position as chief paleontologist of the USGS allowed him to very swiftly acquire an impressive fossil collection: "The scientific value of this entire collection Marsh believed to be far greater than that of any other collection of fossils made by any other geological survey in any part of the world".[1029] The 1872/73 discovery of the birds with teeth became the best proof for the evolutionary link between Cretaceous birds and reptiles, for the teeth were unmistakably reptilian features. Huxley studied the link between reptiles and birds, mainly focusing on *archaeopteryx* and *compsognathus*, the first dinosaur ever described as having had feathers by Huxley in a lecture in 1876.[1030] Bowler and Morus see Huxley's US-tour in a wider context of popular – and in many cases international – scientific lectures: "Traveling to the United States in 1876, he was the latest in a long line of British popular scientific lecturers to tour North America in this fashion. The geologist Charles Lyell lectured across the states in the 1840s."[1031]

But even more so than the toothed birds, it was Marsh's collection of extinct fossilized horses that brought him international recognition in the field of evolutionary

---

1027 Nicht-Fachmann = non-expert, or amateur or layman.

1028 Bunsen to Marsh, 29 December 1893.

1029 Grinnell: Othniel Charles Marsh, pp. 291–298. Quote on page 298.

1030 Thomas Henry Huxley: American Addresses. With a Lecture on the Study of Biology, London 1877, pp. 31–70.

1031 Bowler; Morus: Making Modern Science, p. 373.

biology (around 1874).[1032] When Huxley came to America, he intensively studied the lineage of the horse contained within Marsh’s collection.[1033]

Marsh and Darwin had professional ties; four letters written by Darwin are preserved in the Marsh correspondence. In the first letter, dated January 25, 1873, Darwin thanked Marsh for some scientific papers he had sent him, a fairly commonplace practice within the scientific professional network. Darwin also remarked that it was the treasure-trove of the American West which provided the rich fossil-findings that formed the basis for the ascent of US-American paleontology (see chapter 5):

> I am much obliged to your kindness for having sent me a large supply of your various papers, which I shall be very glad to read, as I do not often see the American Journal of Science. I have, however, read with great interest notices in the American Naturalists & other Journals of your several remarkable discoveries. The richness of the western parts of the U. States in fossils seems quite unparalleled.[1034]

When Marsh travelled to England in 1878, he was invited to visit Darwin at his home, as evidenced by a letter and a postcard sent to Marsh in his hotel in London.[1035] Marsh accepted the invitation and spent a most enjoyable time with the Darwins, as evidenced by a letter Darwin sent on July 14, thanking him for the pleasant visit and sending him two photographs of himself: “I send the [?] two Photographs, & feel honoured by your wish to have Them. I enjoyed most Thoroughly [?] your short visit here.”[1036]

In 1880 Marsh’s reconstruction of the lineage of the ancient horse and his publication on ancient birds with teeth were seen by many scientists, including Charles Darwin, as hard proof for the theory of evolution. Darwin knew of Marsh’s discoveries and descriptions and thanked the professor for these contributions to his theory in a personal letter:

> I received some time ago your very kind note of July 28th, & yesterday the magnificent volume. I have looked with renewed admiration at the plates, & will

1032 For more information on the history of the discovery and description of the toothed birds by Marsh and his associates see: Schuchert; LeVene: O. C. Marsh, pp. 425–444.

1033 Schuchert; LeVene: O. C. Marsh, pp. 230–234.

1034 Charles Darwin, Beckenham to Othniel Charles Marsh, New Haven, CT 25 January 1873, MS 343, Series I. Correspondence, Box 8, Folder 327.

1035 Charles Darwin, Beckenham to Othniel Charles Marsh, London, 2 January 1878; Charles Darwin, Beckenham to Othniel Charles Marsh, London, 5 January 1878, MS 343, Series I. Correspondence, Box 8, Folder 327.

1036 Charles Darwin, Beckenham to Othniel Charles Marsh, London, 14 January 1878, MS 343, Series I. Correspondence, Box 8, Folder 327.

> soon read the text. Your work on these old birds and on the many fossil animals of N. America has afforded the best support to the theory of evolution, which has appeared within the last 20 years. The general appearance of the copy which you have sent me is worthy of its contents, and I can say nothing stronger than this.[1037]

Among all the prominent advocates for Darwin's theory, Thomas Henry Huxley is remembered as the fiercest and most controversial. His frequent and public debates earned him the nickname "Darwin's Bulldog," and the Bulldog was coming to America. When Huxley travelled through North America in 1876, he held some scientific lectures and visited many museums und collections. He and Marsh became friends. Bowler says that the work of Marsh and Huxley is of particular importance for the discourse on the concept of progress within the debate on evolution:

> A description of the process by which the later nineteenth century naturalists laid the foundation of our modern knowledge of the development of life would take a book in itself, much of which would be irrelevant to our general theme of progressionism. The following section will study just two workers who made notable contributions to the field as examples: O. C. Marsh and T. H. Huxley. Both were staunch Darwinians who collaborated on a number of occasions, and both used the theory as the foundation of their attempted reconstructions. But they adopted opposing views on the relationship of Darwinism to progressionism. Marsh followed what was at first the more popular line, concluding that the fossil record was fully in agreement with Darwin's assumption that in the long run natural selection must give rise to progress. His 'law of brain growth' may be counted as a paleontological equivalent of the more philosophical progressionism of writers such as Spencer and Haeckel. Huxley, on the other hand, was suspicious of the fossil evidence for progression, holding that the concept was almost incapable of meaningful definition and that it was pointless to connect the new evolutionary theory with such an outdated notion. The contributions of these two naturalists provide an excellent illustration of how the new breed of paleontologist tried to come to grips with the complexities of the evolutionary process, while at the same time showing the two extremes of opinion on whether there was an underlying tendency toward progress. Marsh represents the typical late nineteenth century attachment to

---

1037 Charles Darwin, Beckenham to Othniel Charles Marsh, New Haven, CT 31 August 1880, quoted after a reproduction of the letter in: Schuchert; LeVene: O. C. Marsh, p. 246. Note that for some unknown reason the letter is not part of the O. C. Marsh correspondence in the archive of Yale's Sterling Memorial Library, but has been reproduced in LeVene's and Schuchert's Marsh biography.

> the belief that the universe is at bottom a progressive system (even if the details on the advance are not predesigned by the Creator). But Huxley points the way toward the increasing suspicion of biological progressionism that has grown up in the twentieth century as the earlier period's optimistic faith in general progress crumbled.[1038]

Jensen, who wrote an article about Huxley's journey to America, notes that Huxley's books were not being appreciated in the US prior to the 1870s. He quotes William Henry Appleton (1814–1899), the publisher of Huxley's works in the US, telling Huxley that the clergy and a few conservative scientists, first of all Louis Agassiz, agitated against the publications and the theory of evolution in general:

> The fact is that while your books met with thorough appreciation from the few earnest progressive thinkers they were shamefully abused by the majority of the papers in which they were noticed. Agassiz the scientific autocrat of this Continent led off in his organ the 'Atlantic Monthly' about naturalists amusing themselves with tracing their genealogy, and the signal being thus given from the Cambridge watchtower the clergy echoed and re-echoed it in various notes from one end of the land to the other.[1039]

This changed during the late 1860s and early 1870s, and Huxley's publications were more widely read and appreciated in the United States. Furthermore, he had formed close professional ties with a few of the more progressive US-American scientists, including Marsh. In 1876 he decided to go to America to take a grand tour, to hold lectures, and to visit his sister Eliza, who had been living in the United States since 1850.[1040]

Huxley's son described his father's 1876-voyage to America.[1041] He quoted a Mr. Smalley, the London correspondent of the "New York Tribune," who was on board the same ship (the "Germanic") as Huxley and heard him exclaim "in the Old World the first things you see as you approach a great city are steeples; here you see, first, centres of intelligence."[1042] (Huxley had allegedly spotted the Western Union Telegraph Building and the building of the "Tribune" when the ship arrived in New York). Huxley's

1038 Bowler: Fossils and Progress, pp. 131–132.

1039 J. V. Jensen: Thomas Henry Huxley's Lecture Tour of the United States, 1876, in: Notes and Records of the Royal Society of London, vol. 42, no. 2 (Jul., 1988), pp. 181–195. Quote on page 182.

1040 Jensen: Thomas Henry Huxley's Lecture Tour of the United States, 1876, pp. 182–186.

1041 See Leonard Huxley: Life and Letters of Thomas Henry Huxley, vol. 2, 2nd ed., London 1908, pp. 201–214.

1042 Huxley: Life and Letters of Thomas Henry Huxley, vol. 2, p. 201.

first destination in the US was New Haven and Marsh's fossil collection (on August 9), which he had collected "with great labour and sometimes at risk of his scalp."[1043] Instead of wasting any more time by inspecting the buildings of Yale College, Huxley asked Marsh to show him his collection at once, and Marsh recalled this meeting in an obituary for Huxley published in the "American Journal for Science" in 1895. Huxley reported to his wife (who had accompanied him to America but was visiting Saratoga instead of inspecting fossils) that he had "seen some things which were worth all the journey across."[1044] About Marsh he wrote: "He is a wonderfully good fellow, full of fun and stories about his Western adventures, and the collection of fossils is the most wonderful thing I ever saw. I wish I could spare three weeks instead of one to study it."[1045]

As to the importance of Marsh's extensive collections, Leonard Huxley (1860–1933), Thomas Henry's son, later wrote:

> At each inquiry, whether he had a specimen to illustrate such and such a point or exemplify a transition from earlier and less specialized forms to later and more specialized ones, Professor Marsh would simply turn to his assistant and bid him fetch box number so and so, until Huxley turned upon him and said: 'I believe you are a magician; whatever I want, you just conjure it up.'[1046]

Huxley later (on August 17) wrote to Marsh:

> I really cannot say how much I enjoyed my visit to New Haven. My recollections are sorting themselves out by degrees and I find how rich my store is. The more I think of it the more clear it is that your great work is the settlement of the pedigree of the horse.[1047]

On the same day he wrote to Clarence King:

1043 Huxley: Life and Letters of Thomas Henry Huxley, vol. 2, p. 202.

1044 Huxley: Life and Letters of Thomas Henry Huxley, vol. 2, p. 204.

1045 Huxley: Life and Letters of Thomas Henry Huxley, vol. 2, pp. 204–205.

1046 Huxley: Life and Letters of Thomas Henry Huxley, vol. 2, p. 203. Note that LeVene and Schuchert also quote this passage, but mistakenly attribute it to: Leonard Huxley: Life and Letters of Thomas Henry Huxley, vol. 1, London 1901, p. 495. See: Schuchert; LeVene: O. C. Marsh, pp. 235–236.

1047 Thomas Henry Huxley, Newport, RI, to Othniel Charles Marsh, New Haven, CT, 07 August 1876, MS 343, Series I. Correspondence, Box 18, Folder 709. Note that an abbreviated transcription of this letter is also included in the biography Leonard Huxley wrote, see: Huxley: Life and Letters of Thomas Henry Huxley, vol. 2, p. 205.

> I am disposed to think that whether we regard the abundance of material, the number of complete skeletons of the various species, of the extent of geological time covered by the collection, which I had the good fortune to see at New Haven, there is no collection of fossil vertebrates in existence which can be compared with it. I say this without forgetting Montmartre, Siwalik, or Pikermi – and I think that I am quite safe in adding that no collection which has been hitherto formed approaches that made by Professor Marsh, in the completeness of the chain of evidence by which certain existing mammals are connected with their older tertiary ancestry. It is of the highest importance to the progress of Biological Science that the publication of this evidence, accompanied by illustrations of such fullness as to enable palaeontologists to form their own judgment as to its value, should take place without delay.[1048]

In a speech given in Baltimore he comments on the status and importance of education in the United States:

> I cannot say that I am in the slightest degree impressed by your bigness or your material resources, as such. Size is not grandeur, territory does not make a nation. The great issue, about which hangs a true sublimity, and the terror of overhanging fate, is, what are you going to do with all these things?... The one condition of success, your sole safeguard, is the moral worth and intellectual clearness of the individual citizen. Education cannot give these, but it can cherish them and bring them to the front in whatever station of society they are to be found, and the universities ought to be and may be, the fortresses of higher life of the nation.[1049]

Later, and as part of a series of lectures on evolution held in New York on September 18, 20, and 22, he concluded that thanks to scientific endeavors such as Marsh’s collection the theory of evolution would rest

> upon exactly as secure a foundation as the Copernican theory of the motions of the heavenly bodies did at the time of its promulgation. Its logical basis is of precisely the same character – the coincidence of the observed facts with theoretical requirements.[1050]

1048 Huxley: Life and Letters of Thomas Henry Huxley, vol. 2, pp. 205–206.

1049 Huxley: Life and Letters of Thomas Henry Huxley, vol. 2, p. 209.

1050 Huxley: Life and Letters of Thomas Henry Huxley, vol. 2, p. 213.

The friendship and enthusiasm were mutual. Marsh even dedicated his never-completed autobiography to Huxley:

> To the memory of Thomas Henry Huxley, 'guide, philosopher and friend,' who made me promise to write this narrative, and often chided me for delay in so doing, this volume is dedicated as a token of friendship, and acknowledgement of many kindnesses in word and deed.[1051]

One of the first things Marsh describes in the autobiography is how he found the remains of extinct horses on his first trip to the West at Antelope Station. He named the specimen "Equus pavulus."[1052] These findings were of the greatest importance as proof for Darwin's theory, and reconstructions of the ancestry of the horse can be found to this day in many schoolbooks. In his autobiography he writes that the birds with teeth he found in 1870 "turned out [to be] one of the most important discoveries ever made."[1053]

Furthermore, Leonard Huxley quotes Marsh writing the following about Thomas Henry Huxley:

> How kind Huxley was to everyone who could claim his friendship, I have good cause to know. Of the many instances which occur to me, one will suffice. One evening in London at a grand annual reception of the Royal Academy, where celebrities of every rank were present, Huxley said to me, 'When I was in America, you showed me every extinct animal that I had read about, or even dreamt of. Now, if there is a single living lion in all Great Britain that you wish to see, I will show him to you in five minutes.' He kept his promise, and before the reception was over, I had met many of the most noted men in England, and from that evening, I can date a large number of acquaintances, who have made my subsequent visits to that country an ever-increasing pleasure.[1054]

Leonard Huxley writes that he obtained the statement on page 6 of Marsh's "recollections." It is very unfortunate these recollections seem to be lost. Huxley did not refer to Marsh's autobiography, for it contains no such passages, and it is unlikely that Leonard Huxley had read them when he wrote his book.

1051 MS 343, reel 26, frame 264, autobiography, p. 1.

1052 MS 343, reel 26, frame 266, autobiography, p. 3.

1053 MS 343, reel 26, frame 269, autobiography, chapter 4, p. 1.

1054 Huxley: Life and Letters of Thomas Henry Huxley, vol. 2, pp. 249–250.

After his return to England, Huxley wrote to Marsh to let his American friend know that the knowledge generated across the Atlantic would now enrich his lectures held in the Old World:

> I blew your trumpet the other day at the London Institution in a lecture about the Horse question [.] I did not know then that you had got another step back as I see you have by the note to my last lecture [,] which Youmans has just sent me. I must thank you very heartily for the pains you have taken over the woodcuts of the lectures. It is a great improvement to have the patterns of the grinders. I have promised to give a lecture at the Royal Institution on the 21st January next [,] and I am thinking of discoursing on the Birds with teeth. Have you anything new to tell on that subject? I have implicit faith in the inexhaustibility of the contents of those boxes.[1055]

In the same letter, but on a more personal note, he writes:

> Our voyage home was not so successful as that out. The weather was cold & I got a chill which laid me up for several days. In fact I was not well for some weeks after my return. But I am vigorous again now [.] Pray remember me kindly to all New Haven friends. My wife joins with me in kindest regards and good wishes for the new year [:] 'Tell him we expect to see him next year.'

It is noteworthy that Huxley reports the regards of his wife, for it underlines the personal friendship the Huxleys shared with Marsh. In later years Leonard and Henrietta Huxley (1825–1915), Thomas Henry's wife, wrote to Marsh, illustrating that the friendship of Huxley and Marsh transcended a purely professional working relationship. Marsh had become a friend to the Huxley family.

Marsh replied on January 12, 1877. Note that this is one of very few letters written by Marsh and preserved in the Marsh papers, and it might well be a rough draft he later amended, sending the enhanced version to Huxley. The letter preserved in the Marsh papers contains many corrections and crossed out lines. Marsh sent mostly professional tips for Huxley's next lectures. Of all the professional courtesies, the following line stands out:

> I am very sorry I did not show [?] you the account of my Eohippus in time for your lecture, I had him 'corralled' in the basement of our Museum when you

1055 Thomas Henry Huxley, London to Othniel Charles Marsh, New Haven, CT, 27 December 1876, MS 343, Series I. Correspondence, Box 18, Folder 709. For a full transcription of the letter see: Huxley: Life and Letters of Thomas Henry Huxley, vol. 2, pp. 213–214.

> were there, but he was so covered with Eocene mud that I did not know him from Orohippus.[1056]

When Marsh came to England in 1878, Huxley wrote another very personal letter to him:

> Welcome to England! I am delighted to hear of your arrival – but the news has only just reached me [,] as I have been away since Saturday with my wife & sick daughter who are at the seaside. A great deal has happened to us in the last six or seven weeks. My eldest daughter married [,] & then a week after an invasion of diphtheria [,] which struck down my eldest son, my youngest daughter [,] & my eldest remaining daughter altogether. Two of the cases were light [,] but my poor Madge suffered terribly [,] & for some ten days we were in sickening anxiety about her. She is slowly gaining strength now [,] & I hope there is no more cause for alarm, but my household is all to pieces, the Lares & Penates gone [,] & painters & disinfectors in their places. You will certainly have to run down to Margate and see my wife — or never expect forgiveness in this world.[1057]

Three letters written by Henrietta Huxley, the wife of Thomas Henry Huxley, are preserved in the Marsh papers. They tell a more personal story of the friendship between Marsh and the Huxleys. In the first letter, written on May 22, 1881, Henrietta thanks Marsh for a photograph of his home: "It was so very kind of you to send me a photograph of your house a charming one it is. Now I can picture you occupying it; but where is the lady of the house?"[1058] Once more it becomes evident that Marsh's bachelor lifestyle was rubbing some of his more family-minded contemporaries the wrong way (see chapter 3. 3). The next letter of Henrietta Huxley was written on August 26, 1895. Marsh had written a letter of condolence after the death of Thomas Henry Huxley on June 29, 1895. It reads as very emotional and affected by the recent loss, which must have depressed Henrietta greatly:

> Dear Professor Marsh,
> Most heartily do I thank you for your kind expression of sympathy with me. For

1056 Othniel Charles Marsh, New Haven, CT, to Thomas Henry Huxley, London, 12 January 1877, MS 343, Series I. Correspondence, Box 18, Folder 709.

1057 Thomas Henry Huxley, London to Othniel Charles Marsh, London, 24 June 1878, MS 343, Series I. Correspondence, Box 18, Folder 709. For a full transcription of the letter, see: Huxley: Life and Letters of Thomas Henry Huxley, vol. 2, pp. 248–249.

1058 Henrietta Huxley, London, to Othniel Charles Marsh, New Haven, CT 22 May 1881, MS 343, Series I. Correspondence, Box 18, Folder 707.

> [?] my great & everlasting loss – but also I thank you for the tribute of admiration. For [?] my dear husband – If anything could help me in my dire distress [?], it is the expression of love & reverence for him sent me by all who knew him. [...] Death always comes too soon! I feel thoroughly shattered. The four month strain of anxiety & its fatal ending have been almost more than I could bear.[1059]

She then asks Marsh for any letters written by her late husband and urges him to send them to Leonard Huxley, who would write a book about his father's life: "Our eldest son Leonard is to write his father's life. Perhaps you may have some of his letters." Even though Marsh had heeded Leonard Huxley's request for letters (see below) he had written to his father, Henrietta asked for more in a letter addressed "[t]o the President & Faculty of Yale University."[1060] It was written on April 17, 1899, and bemoans the death of Marsh: "I have received with much regret your announcement of the death of our old friend Professor O. C. Marsh [...]." In a postscript she asks for any letters written by Thomas Henry Huxley to be send to her son for his biographical project: "For any letters from my husband Thomas Henry Huxley to Prof. Marsh, I should be most grateful, as they would be useful for a 'The Life.'"

Leonard Huxley visited the US during the winter of 1889, and also paid a visit to Marsh, as evidenced by a letter, written on December 28, 1889, informing Marsh that he must postpone his visit.[1061] That the postponement was no problem and he had indeed visited Marsh on a later date is evidenced by a letter Leonard Huxley wrote to Marsh after his return from America. In the same letter he conveys more of the friendship his parents shared with marsh:

> They [Leonard's parents] were very glad to have such authentic news of you as I cd. [?] bring, & reviewed many recollections of their visit to New Haven, from the humming birds down to the story of 'Adam's branding.'[1062]

The last letter sent by Leonard Huxley to Marsh and archived in the Marsh papers was written on March 26, 1896. Huxley had started collecting material for his father's

1059 Henrietta Huxley, Eastbourne, to Othniel Charles Marsh, New Haven, CT 26 August 1895, MS 343, Series I. Correspondence, Box 18, Folder 707.

1060 Henrietta Huxley, Eastbourne, to the president & faculty of Yale University, New Haven, CT 17 April 1899, MS 343, Series I. Correspondence, Box 18, Folder 707.

1061 Leonard Huxley, New York, to Othniel Charles Marsh, New Haven, CT 28 December 1889, MS 343, Series I. Correspondence, Box 18, Folder 708.

1062 Leonard Huxley, Godalming, to Othniel Charles Marsh, New Haven, CT 18 February 1890, MS 343, Series I. Correspondence, Box 18, Folder 708.

biography and Marsh had sent him some of the letters Harriet Huxley had requested from him (see above):

> In a press of much writing I have put off acknowledging your batch of letters or [?] the all too short sketch of my fathers, but my thanks are none the less hearty for the delay. Having got so much from you, my appetite is naturally whetted for more: & so, someday I wonder if you could find leisure to jot down a few more reminiscences of him? Coming from such a source, they would be of double value to me. They would be authentic in the highest degree, by virtue of knowledge, sympathy & opportunity. I should make no apology for this bold suggestion: a biographer should be above apologising: but with you treat it as a suggestion & let it wait for its due season to ripen? I always measure the distance of that very pleasant visit I paid you at Newhaven [sic!] by the growth of my second boy [Noel Trevenen, 1889–1914] who was born in my absence that holiday. He is getting a big fellow, but the memory of that visit remains very fresh.[1063]

Of the seven letters of correspondence between Huxley and Marsh to survive in the Marsh papers, Leonard Huxley used four in the biography he wrote, this being proof that Marsh did send him the letters he was asking for, and that Leonard Huxley later returned them to him.

Jensen judges that Huxley's journey to America and his lectures had been very successful:

> Huxley was pleased with his American tour. Admirers there were in great good numbers. For example, a former captain in the Royal Artillery, a resident in New York for ten years, left a note at Huxley's New York hotel: 'You have created a profound impression, the interest in which will certainly wax rather than wane from the moment of your departure'. Autograph seekers pursued him during his tour, and citizens in various cities, such as Springfield, Mass., Providence, R. I., Buffalo and Troy, New York, requested that he give a lecture. Indeed, at his departure he had attracted considerable attention.[1064]

Thaddeus Stanton, a friend of Marsh's (see chapter 5. 4. 2.), wrote a letter to Marsh, telling him he had attended one of Huxley's lectures in America and that Huxley had made numerous references to Marsh and his work, indicating that this form of in-

---

1063 Leonard Huxley, Godalming, to Othniel Charles Marsh, New Haven, CT 26 March 1896, MS 343, Series I. Correspondence, Box 18, Folder 708.

1064 Jensen: Thomas Henry Huxley's Lecture Tour of the United States, 1876, p. 192.

ternational and very prominent appreciation of the US American would be greatly cherished and also showing that Marsh's scientific achievements had found further international recognition:

> Since my return I rec [?] that Prof. Huxley has been with you, and has complimented you in many ways, all of which is gratifying your friends as it must be to you. I have only seen one of his lectures, but in that he made frequent reference to you and your work. I am glad to see you vindicated (if, indeed, any vindication were necessary) by such good authority.[1065]

In a letter he wrote to Clarence King, Huxley underlined the immense importance of Marsh's contributions as evidence for Darwin's theory:

> In accordance with your wish, I very willingly put into writing the substance of the opinion as to the importance of Professor Marsh's collection of fossils which I expressed to you yesterday. As you are aware, I devoted four or five days to the examination of this collection, and was enabled by Prof. Marsh's kindness to obtain a fair conception of the whole.[1066]

Grinnell writes that Huxley gave tribute to Marsh in 1881, unfortunately failing to cite where he read the following passage:

> The discovery of the toothed birds of the cretaceous formation of N. America, by Prof. Marsh, completed the series of transitional forms between birds and reptiles, and removed Mr. Darwin's proposition that, 'many animal forms of life have been utterly lost, through which the early progenitors of birds were formerly connected with the early progenitors of the other vertebrate classes,' from the region of hypothesis to that of demonstrable fact.[1067]

His discoveries soon made Marsh world-famous: "The careful and methodical distribution of his writings to scientific centers throughout the world gave him eminence in practically every country."[1068] The discovery of the toothed birds made Marsh part of the transatlantic discussion on the evolutionary theory, and gained him the attention

1065 Thaddeus Stanton, Cheyenne, Wyoming Territory, to Othniel Charles Marsh, New Haven, CT 12 December 1876, MS 343, Series I. Correspondence, Box 30, Folder 1303.

1066 Thomas Henry Huxley, Newport, RI to Clarence King, location unknown, 19 August 1876, quoted after: Grinnell: Othniel Charles Marsh, p. 296.

1067 Grinnell: Othniel Charles Marsh, p. 301.

1068 Grinnell: Othniel Charles Marsh, p. 299.

and respect of Huxley and Darwin. Still, according to Grinnell, in the end the spectacular dinosaurs brought more fame and glory to Marsh.[1069]

In 1880 Marsh's "Odontornithes" was published by the USGS and while the publication further established the reputation of US-American science (and scientists) in Europe, it would also cause great trouble for Marsh and the USGS in general (see chapter 6. 5.):

> Widely regarded as the 'missing' link in Darwin's evolutionary series, Marsh's discovery of toothed birds in America was a dramatic accomplishment that called forth great praise from European scientists. When Huxley came to America, he went almost immediately to Yale to see Marsh's collections, and Darwin himself, besieged by critics, added his praise and thanks.[1070]

In 1879 Herbert Spencer, another strong advocate for the theory of evolution, contacted Marsh, thanking him for the copy of a paper he had written about "The Vertebrae of Recent Birds,"[1071] because it supported "the doctrine of Evolution by giving the interpretation of anomalies."[1072] Later Marsh sent him a copy of his "Odontornithes," as evidenced by a letter of appreciation written by Spencer in 1880.[1073] When Spencer visited the US in 1882, he was personally invited by Marsh to inspect his fossil collection:

> Thank you very much for your cordial letter. I shall, of course, be delighted to have the opportunity of seeing your magnificent collection of fossels [sic!] from the Rocky Mountains. You must excuse me, however, if I refrain from committing myself to more than a visit of inspection; for the reason that my state of health, especially at present, is such as to render needful that complete command of my own time and mode of living which is impracticable save when staying at a hotel. Social excitements are bad for me; and I shall have very much to restrict myself in respect to them while making my tour of the States.[1074]

1069 Grinnell: Othniel Charles Marsh, pp. 300–303.

1070 Goetzmann: Exploration and Empire, pp. 460–461.

1071 Othniel Charles Marsh: The Vertebrae of Recent Birds, in: The American Journal of Science, ser. 3, vol. 17, no. 100, (Apr. 1879), pp. 266–269.

1072 Herbert Spencer, London to Othniel Charles Marsh, New Haven, CT, 24 March 1879, MS 343, Series I. Correspondence, Box 30, Folder 1295.

1073 Herbert Spencer, London to Othniel Charles Marsh, New Haven, CT, 5 [?] August 1880, MS 343, Series I. Correspondence, Box 30, Folder 1295.

1074 Herbert Spencer, London to Othniel Charles Marsh, New Haven, CT, 9 June 1882, MS 343, Series I. Correspondence, Box 30, Folder 1295.

Note that Spencer was possibly the strongest, definitely the most popular proponent of "Social Darwinism," and that this theory "was equally, if not more, influential than the more naturalistic Darwinism, especially because it appealed to America's abiding faith in progress and because it did not directly challenge special creation."[1075] Robert Bannister analyzes how Spencer adapted Darwinian principles like natural selection to human societies and coined the phrase "survival of the fittest." He suggested that it was a natural aspect of human societies that superior individuals would dominate the inferior ones. According to Spencer this also was true for inter-societal conflicts; according to this logic it was only natural that superior "races" would wage war on inferior ones, that the stronger or more adaptive "races" would prosper and the inferior ones would eventually die out. It almost goes without saying that in Spencer's mind the "white race" was the most advanced, progressive, and superior of all.[1076] Bannister also describes in great detail how Social Darwinism, which he calls – perhaps more accurately – "Spencerianism," took root in the United States, and attests that "Spencerianism" was a "minority cause from the 1860s to the 1880s" but "provided an important link between midcentury American liberalism and the New Liberalism of the progressive era."[1077]

Herbert Spencer came to the US and dined at Marsh's home on October 21, 1882.[1078] Shortly after Spencer's visit, the teaching of the Darwinian theory of evolution was heavily criticized by president Noah Thomas Porter (1811–1892) of Yale, who proposed that institutions like Yale had an obligation to their supposedly Christian foundations and that the theory of evolution would hardly agree with the Holy Scripture. The debate, it seems, fizzled out and the president did not restrict the use of certain textbooks, as he had proposed.[1079]

After Spencer's tour of the US, he and Marsh stayed in contact and Marsh sent him a copy of his "Dinocerata" via his personal friend Henry Woodward (see chapter 3. 3.). Again, he values Marsh's work only as proof for the theory of evolution:

> I received two days ago through Mr. Woodward a copy of your magnificent second volume of the Dinocerata. Many thanks for it I congratulate you upon the completion of another step in your great undertaking. I wish it had come to me at a time when its contents might have been utilized in dealing with biological generalizations; but even as it is, if I live long enough to complete other

1075 Hamlin: From Eve to Evolution, pp. 6–7.

1076 Robert C. Bannister: Social Darwinism. Science and Myth in Anglo-American Social thought, Philadelphia 1978, pp. 34–36. Also see: Bowler: Evolution, pp. 220–223.

1077 Bannister: Social Darwinism, pp. 57–163. Quote on page 61.

1078 Schuchert; LeVene: O. C. Marsh, p. 244.

1079 Schuchert; LeVene: O. C. Marsh, pp. 244–245.

> works, and revise the Biology,[1080] I may have an opportunity of utilizing the striking evidences of evolution it contains.[1081]

The last surviving letter Spencer wrote to Marsh is a request for help with an article Spencer was about to write. Note that he wrote a nearly identical letter to Cope, asking Marsh's rival for the same favor (see below). The proposed article was, true to Spencer's other works, of a rather racist nature:

> A correspondent draws my attention to the fact that a phenomenon parallel to that which I have narrated in the recent essay on 'The Inadequacy of Natural Selection' concerning the Quagga, has been observed in the United States, when white women have borne children to negroes. Here is the passage: 'The children of white women by a white father had been repeatedly observed to show traces of black blood in cases when the woman had previous connection with a negro.' I should like to be able to give something like scientific verification of this. I wrote by this same post to Dr Youmans & to prof. Cope on the matter but can you yourself tell me anything about it, or can you tell me of any physiologist in the Southern States, who is likely to have personal knowledge?[1082]

Note that Cope received a word-for-word identical copy of this letter, with the exception that Spencer told him that he had written to Youmans and Marsh as well. Also, the letter to Marsh contains a postscript in which Spencer thanks Marsh again for his numerous contributions to the theory of evolution:

> This letter ought rather to be a letter of thanks for many pieces of information I have received from you during past years, in the shape of monographs on your marvelous series of extinct creatures. But instead I am asking for more information! Nevertheless pray accept my thanks for the large amount I have already received.

Despite Cope's stance on evolution, he had working ties with scientists promoting Darwinian evolution and with proponents of atheism, such as Haeckel. One letter of Haeckel's survives in the Haverford College's archives. The letter suggests that scien-

---

1080 Herbert Spencer: The Principles of Biology, vol. 1, London 1864.

1081 Herbert Spencer, London to Othniel Charles Marsh, New Haven, CT, 1 April 1885, MS 343, Series I. Correspondence, Box 30, Folder 1295.

1082 Herbert Spencer, London to Othniel Charles Marsh, New Haven, CT, 8 March 1893, MS 343, Series I. Correspondence, Box 30, Folder 1295.

tific monographies or other publications were exchanged between Cope and Haeckel.[1083] Another surviving letter was sent in 1882 by Francis Darwin (1848–1925), son of Charles Darwin. He asks Cope whether he owned any letters of his father's, so that he may make copies of them. Presumably, Francis was working on a biography about his famous father, or was simply trying to preserve the memories of the prominent biologist.[1084] It is unknown whether Charles Darwin was a correspondent of Cope's, since no letters of Charles Darwin can be found in the Cope Collections at Haverford, or in the letters his wife and daughter had preserved.[1085] Spencer, too, wrote to Cope and asked him for some help with an article, an almost exact copy of a request he had sent Marsh, written on the same day.[1086]

Cope was a Neo-Lamarckist, who did not subscribe to the Darwinian theory of evolution, most likely due to his deep religious believes. This in turn influenced how he interpreted his research:

> Perhaps the only major paleontologist of this later generation to start out from a partly linear view of development was the American Edward Drinker Cope, whose paper 'On the origin of genera' (1868) suggested that the generic forms represent fixed and divinely preordained hierarchies along which a series of different lines corresponding to the different species advance. [...] As his paleontological studies expanded, however, he rapidly abandoned this idea in favor of a Lamarckian theory of adaption in which the inheritance of acquired characteristics became the chief driving force of evolution. Thus, although he refused to accept natural selection, Cope's superficial conception of the history of life became close to that of the Darwinists and he was able to make notable advances in the field of reconstructing the hidden evolutionary steps.[1087]

In an article about Cope's thoughts on evolution Bowler even suggests that there was a link between Cope's and Marsh's concepts of evolution and the “Bone Wars.” He writes that their conflict

---

1083 Ernst Haeckel, Jena, to Edward Drinker Cope, Philadelphia, 4 December 1894, HC.MC-956.

1084 Francis Darwin, England, location unknown to Edward Drinker Cope, Philadelphia, 19 December 1882, HC.MC-956.

1085 The number of letters and other correspondence of Cope's is very limited, for it appears that much of it was lost or deliberately destroyed: “I believe that Cope's wife and his daughter, Julia Collins, both of whom had access to all his papers at his death in 1897, probably destroyed documents which may have represented their husband and father in a bad light before they presented to the American Museum of Natural History the corpus of materials which Osborn used [for his biography of Cope].” Davidson: The Bone Sharp, p. 3.

1086 Herbert Spencer, London to Edward Drinker Cope, Philadelphia, 8 March 1893, HC.MC-956.

1087 Bowler: Fossils and Progress, p. 130.

> was symbolic of an even deeper difference of opinion at the theoretical level: Marsh was a staunch supporter of Darwin's theory of evolution by natural selection, while Cope had become a leading proponent of the neo-Lamarckian mechanism of the inheritance of acquired characteristics.[1088]

In addition, Cope did coin an evolutionary concept called "Cope's Rule," or "Cope's Law." He suggested that a huge body caused or at least encouraged extinction. He supposed that bigger animals needed larger quantities of food and thus had a harder time adapting when food was sparse.[1089] Cope also coined the "Law of the Unspecialized," claiming that early lifeforms lacked in specialization and that their descendants had adapted to fill their respective ecological niche. Today many examples prove the opposite of "Cope's Rule," which therefore cannot be considered to be universally true.[1090] The falsification of "Cope's Rule" shows how experiments and the accumulation of data can be used by paleontologists to support or disprove a hypothesis. The study of entire evolutionary family trees via the observation of fossils is a labor-intensive undertaking, though it is often worth the effort.[1091]

Another example of the influence Marsh's findings had on the international discourse on evolution is the correspondence between Marsh and Ernst Haeckel, one of the foremost champions of evolution in Germany. Note that the initial reaction to the Darwinian theory of evolution in Germany was very positive, and it was adapted immediately by most Germany scientists; also "[s]ome German scientists, of whom Ernst Haeckel was the most active, were political radicals who saw Darwin's rejection of design as a weapon in their fight against conservatism."[1092] But with the implications of racist theories in the mix, Haeckel "has been identified controversially as a key influence on the later development of fascism and Nazism."[1093]

In the Marsh papers two of the letters preserved were written by Haeckel, whom Jensen calls an "outspoken agnostic and supporter of Darwinism" and who he attests "was a close friend and devotee of Huxley."[1094] Both letters were written in German.

1088 Peter J. Bowler: Edward Drinker Cope and the Changing Structure of Evolutionary Theory, in: Isis vol. 68, vo. 2 (Jun. 1977), pp. 249–265. Quote on page 249.

1089 Derek Turner: Paleontology. A Philosophical Introduction, Cambridge 2011, pp. 66–67.

1090 Turner: Paleontology, pp. 100–111.

1091 Derek Turner: Beyond Detective Work: Empirical Testing in Paleontology, in: David Sepkoski; Michael Ruse (eds.): The Paleobiological Revolution. Essays on the Growth of Modern Paleontology, Chicago 2009, pp. 201–214, DOI: 10.7208/chicago/9780226748597.001.0001.

1092 Bowler: Evolution, p. 187.

1093 Bowler: Evolution, p. 294.

1094 Jensen: Thomas Henry Huxley's Lecture Tour of the United States, 1876, p. 181.

Note that when Haeckel wrote to Cope he used English,[1095] indicating that Haeckel knew Marsh would understand a letter written entirely in German. The first letter is dated December 18, 1885.[1096] Haeckel thanks Marsh for sending him his own "grand paleontological works" ("prachtvollen palaeontologischen Arbeiten"). He then writes that Marsh and his discoveries were referenced, and praised, so often in Jena that Marsh would become one of the most well-known naturalists to students and professors of Jena alike.[1097] Only one other letter of Heackel to Marsh survives in the Marsh papers, written to the day ten years after the first one, on December 18, 1895.[1098] Haeckel tells Marsh he had received two plates from him, depicting "admirable restorations of fossils vertebrates" ("bewunderungswürdigen Restaurationen von fossilen Vertebraten"), which he would hang in his auditorium. He further writes that he admires Marsh for a talent most other scientists would condemn, namely his ability to close gaps in the fossil records by manner of educated guessing, presenting the fragmentary record as a comprehensive one.[1099]

The "Bone Wars" and the distinct personalities of Marsh and Cope came into play concerning their take on evolution: Cope was troubled by Darwin's theory of evolution, for it seemed to inherit a spirit of chaos and atheism which was very disconcerting for the devoted Quaker. Cope envisioned an alternative to Darwin's theory, harking back to Lamarck's notions and deemphasizing the importance of natural selection. This Neo-Lamarckian theory instead promoted that evolution was driven by the organism's reactions to its environment and that evolution was guided by some (divine) intelligence.[1100] Cope also believed that evolution was progressive; lower sim-

---

1095 Cope spoke German. It is unclear if Haeckel knew this, or if he wrote to him in English to avoid potentially insulting a man he probably knew less well than Marsh. As to Cope's mastery of the German language, at least in 1863 he wrote to his brother: "There were also some Germans from up the Rhine, into whose company I went and tried my skill at talking. I have been talking German ever since and can get along very well. I do not try French because I mix the two–but this horrid Netherlandish I cannot get." Osborn: Cope, p. 116.

1096 Ernst Haeckel, Jena, to Othniel Charles Marsh, New Haven, CT 18 December 1885, MS 343, Series I. Correspondence, Box 13, Folder 561.

1097 "Schon lange habe ich Ihnen meinen herzlichen Dank sagen wollen für die gütige Zusendung Ihrer prachtvollen palaeontologischen Arbeiten […]. Ihr Name und Ihre grossartigen Entdeckungen werden hier in Jena so oft genannt und gerühmt, dass Sie bei meinen Collegen und studenten einer der best bekannten grossen Naturforscher der Gegenwart sind." Haeckel to Marsh, 18 December 1885.

1098 Ernst Haeckel, Jena, to Othniel Charles Marsh, New Haven, CT 18 December 1895, MS 343, Series I. Correspondence, Box 13, Folder 561.

1099 "Ich danke Ihnen dafür um so mehr, als ich an Ihren grossartigen palaeontologischen Arbeiten gerade das sehr bewundere, was viele Andere tadeln: Ihr Talent, aus vielen einzelnen & unvollständigen Bruchstücken wieder ein natürliches Ganzes herzustellen, und die vielen Lücken der palaeontolog. Urkunden mit kritischer Phantasie befriedigend auszufüllen." Haeckel to Marsh, 18 December 1895.

1100 For more on the history and scientific contents of Neo-Lamarckism, see: Bowler: Evolution, pp. 236–244.

ple lifeforms would develop into higher and more complex organisms, modern men taking the top spot in the divine improvement campaign. Osborn, once a student and now a friend of Cope's, seconded some of the Quaker's ideas about evolution, especially concerning its progressive nature. In the 1880s Osborn began to concern himself with the evolution of horse teeth, which he assumed would illustrate the progressiveness of evolution. Osborn supposed that mammal teeth would best demonstrate an evolutionary favoring of more complex and specialized forms. This set Osborn on a collision course with Marsh, who possessed the most extensive collection of fossil horses. According to Osborn, Marsh was squandering this precious resource by assigning each new and slight variation of a horse's tooth to a new type of horse genus, while Osborn believed the variations to be an indication of an organism's reactions to its environment. This disagreement furthered Osborn's antipathy for Marsh and consolidated his alliance with Cope. In 1888 both Osborn and Marsh attended the annual meeting of the BAAS in Bath in England and were to give a short lecture on fossil teeth. Marsh, undiplomatic as usual, gave a very long presentation, forcing Osborn to shorten his lecture quite a bit.[1101]

7.3 Conclusion

The Nashville-based "Daily American" wrote that Marsh was "a bold discoverer, whose services to the science of paleontology in our western wildernesses have been more heroic and not less important than those of Cuvier, in his researches in Parisian suburbs."[1102]

The US-American contributions to the theory of evolution exemplify how the science of paleontology grew to be the worldwide forerunner in the last quarter of the nineteenth century. Fossils were found in the West, on terrains providing the best condition for the preservation, and later excavation, of bones. This was, at the time, a uniquely American feature, not matched anywhere in the world, though arguably Mongolian, Chinese, and some African regions would provide equally promising features in the beginning of the twentieth century. These fossils became part of a scientific debate, hotly fought and of great consequence for all societies and religions, highlighting another reason why US-American paleontology was perceived to deliver the best and most remarkable scientific results at the end of the nineteenth century.

1101 Regal: Henry Fairfield Osborn, pp. 63–66.

1102 Quoted after: Schuchert; LeVene: O. C. Marsh, pp. 242–243.

8

Transatlantic Knowledge Transfer – Paleontology and the Evolution of US Higher Education

"As a developing nation, the United States was fortunate to have acquired contributions from many sources and cultures; those from the German speaking countries of Europe have been considerable and varied. None, however, has been more significant to the civic and economic development of the United States than those brought from Germany into the sphere of education."[1103]

Angela Schwarz writes that at no other point in time (up till then) had science been conducted more internationally than during the second half of the nineteenth century, and that the professionalization of science and the improvements in transportation and communication contributed fundamentally to this scientific exchange:

> Basically, science – and the concept of Humboldtian science with its emphasis on a degree of empiricism and precision characteristic of nineteenth century science in particular – is or it is thought to be (and was at the time constructed as) independent from national affiliations.[1104]

Over the course of the nineteenth century, it became apparent that the United States' educational system, which was built upon imparting values and moral training, had become out of sync with the Zeitgeist. A generation of young US scientists who returned home after having studied in Germany dearly missed the laboratories which they had discovered overseas. Furthermore, big business did not demand morally trained scholars, but practical, utilitarian knowledge. Said new generation of scientists then imported the German system of laboratories and freedom in education and scientific publication back to the US through an ever-increasing number of scientific journals.[1105]

Additionally, scientific research and education became connected with popular culture during the nineteenth century. Museum exhibitions enabled students to study specimens, provided investors with profits, and made natural history accessible to the public. Many of these museums were directed by professional naturalists, assuring both high scientific standards and providing job opportunities to naturalists. The efforts of some US-American scientists to associate natural sciences with the divine were, at least during the first half of the century, part of the effort to popularize science and make it more socially acceptable, culminating in a natural theology of

---

1103 Daniel Fallon: German Influences on American Education, in: Frank Trommler; Elliott Shore (eds.): The German-American Encounter. Conflict and Cooperation between Two Cultures, 1800–2000, New York 2001, pp. 77–87. Quote on page 77.

1104 Angela Schwarz: Intersecting Anglo-German Networks in Popular Science and their Functions in the Late Nineteenth Century, in: Heather Ellis; Ulrike Kirchberger (eds.): Anglo-German Scholarly Networks in the Long Nineteenth Century, Leiden 2014, pp. 65–83, https://doi.org/10.1163/9789004253117_004. Quote on page 67.

1105 Struik: Yankee Science in the Making, pp. 337–349.

sorts. This emphasis on natural theology was one of the reasons why US-American science still limbed behind its European counterpart. After the Civil War, the US government's support for sciences was part of the effort to rebuild the nation. Religious implications were deemphasized, original research and its publication promoted. Still, most US-American colleges lacked specialized personnel trained to teach one scientific discipline instead of a mixture of subjects simply summarized as "science." Research laboratories and specialists were to be employed if the reform-efforts of the 1870s and 1880s were to be successfully implemented. Job opportunities for college graduates had to be created.[1106]

This chapter details how a supposedly uniquely German system of higher education developed in the US during the eighteenth and nineteenth centuries. It will analyze how, and if, the "Humboldtian" reforms were implemented in Germany, and how they then resonated internationally. It will also explore how the US-American system of higher education was reformed during the nineteenth century, and whether the "German University" became the role model for those reform plans. The final two subchapters will focus on the reform of the natural sciences, especially on paleontology, and on the impact said restructurings had on public education in the United States.

8.1 The Emergence of the Modern "German University"

In his book "Einsamkeit und Freiheit" Hartmut Schelsky examines the German university system and its university reforms. He begins his study in the medieval period when German universities were founded by the nobility and not by the church, as was the case in places like France or Italy. During this period, more and more freedoms were won through a lengthy struggle with members of the aristocracy, who were mainly interested in universities as places for the training of civil servants. Schelsky proclaims that universities then developed into places of social and intellectual freedom, relatively independent from the church and local aristocracy. In this system one could peruse scholarly ambitions in relative freedom and fairly independently from the otherwise all-present estate-based society.[1107]

1106 Keith R. Benson: From Museum Research to Laboratory Research. The Transformation of Natural History into Academic Biology, in: Roland Rainger et al. (eds.): The American Development of Biology, paperback ed., New Brunswick, NJ 1991, pp. 49–83.

1107 Helmut Schelsky: Einsamkeit und Freiheit. Idee und Gestalt der deutschen Universität und ihrer Reformen, Reinbek 1963, pp. 13–20.

Still, until the second half of the eighteenth-century universities largely remained places of vocational training. Research was conducted, if it was conducted at all, in the context of the scholarly societies which were founded all around the German-speaking area during the eighteenth century. In the second half of the century aspirations arose to make science more practical and utilitarian. This promoted specialization, slowly stimulating the differentiation of the scientific disciplines. Those disciplines were to be taught not at universities but at specialized training institutions probably best described as technical colleges ("Fachhochschulen"). Schelsky attests that this greater differentiation of scientific disciplines and their distribution among disparate institutions almost spelled the end for German universities, since the opinion that excessive erudition and the tendency to remain in the ivory tower of scholarly pursuits were obstacles in the progress of the enlightenment was widely held among German scholars.[1108]

In her study of the relationship between nationalism and science Crawford makes an interesting statement concerning the influence nationalism has on the development of scientific disciplines:

> The most basic, durable, and generalizable construct for analyzing scientific development is the *disciple*. As a basic unit of social and cognitive organization in the sciences, in its modern form it goes back to the mid-nineteenth century; furthermore, its geographic spread is strongly linked to the creation of national scientific enterprises in Europe and America. The story of how those enterprises developed their autonomy is often one of the emergence of disciplines, viewed conceptually, socially, and politically. Pride, power, and national prejudice have been seen as important elements in the drive towards autonomy, as illustrated by the wording of some titles of national disciplinary histories – 'the physicists' (in America), 'the now mighty theoretical physicists' (in Germany), and 'the kaiser's chemists.'[1109]

She later adds that nationalism in science is not an aberration but a "historical phenomenon linked in a certain stage of sociopolitical and economic development."[1110] She further attests that "[i]ts contours appear most clearly in Europe and North America in the high industrial age of the late nineteenth century." Crawford further suggests that science also contributes greatly to the unity of a nation, creates a "high culture," that scientific disciplines are "Kulturträger" (bearers of culture), infusing scientific

1108 Schelsky: Einsamkeit und Freiheit, pp. 31–44.

1109 Elisabeth Crawford: Nationalism and Internationalism in Science, 1880–1939. Four Studies of the Nobel Population, Cambridge 1992, p. 19.

1110 Crawford: Nationalism and Internationalism in Science, p. 31.

knowledge and values into the cultural life of a nation;" and that it was furthermore "generating the discoveries that would make citizenry in general identify with and be proud of its scientists."[1111]

Wolfgang Nitsch et al. suggest that state and society were one unit in Prussia during the beginning of the nineteenth century (in contrast to France and Great Britain). Although a middle class existed, it constituted no counterbalance to the aristocracy. A societal elite had developed, consisting of high-ranking civil servants. This class was neither directly dependent on the sovereign, nor a part of the middle class. It sought to introduce reformations, which were of public interest, through top-down decisions. After Prussia had been defeated by France (Revolutionary and Napoleonic), she lost much of her martial prowess and had to reinvent herself culturally. In German regions under French administration / occupation universities were shut down.[1112] The French had begun to do away with their own outmoded university system a while before. But the Prussian Kingdom sought to become a "culture state" ("Kulturstaat") instead of a "power state" ("Machtstaat"), and facilitated the creation of a new kind of university, freer and more independent from state and aristocracy.[1113] The founding of new universities and the reformation of the German system of higher education can therefore be understood as a form of resistance to the French restructuring, and as a sign that the educated middle class was gaining in strength.[1114]

The establishment of the Berlin University in 1809/10 was in some ways a countermove to the tendency for a utilitarian conduct of science and higher education, which meant parting with the eighteenth-century conception of science being mostly conducted in technical colleges and academies. Conceptions of science changed at this time and the goal of science was no longer to fathom the divine plan but to learn more about an objective reality. Furthermore, it was believed that an educated member of society would also be a moral one. To educate oneself one would best be independent from the material necessities of every-day life. This was part of a transition of Prussian society, which aimed to give the individual more freedom from the state. Highly educated and therefore supposedly moral individuals were to become model citizens

1111 Crawford: Nationalism and Internationalism in Science, p. 32.

1112 Between 1794 and 1818 the number of German universities was almost halved, many universities decreased dramatically in size. See: Irmtraut Scheele: Grundzüge der institutionellen Entwicklung der biologischen Disziplinen an den deutschen Hochschulen seit dem 18. Jahrhundert, in: Gert Schubring (ed.): "Einsamkeit und Freiheit" neu besichtigt: Universitätsreformen und Disziplinenbildung in Preußen als Modell für Wissenschaftspolitik im Europa des 19. Jahrhunderts, Stuttgart 1991, pp. 144–154, see p. 146.

1113 Wolfgang Nitsch et al.: Hochschule in der Demokratie. Kritische Beiträge zur Erbschaft und Reform der deutschen Universität, Berlin 1965, pp. 6–18.

1114 Charles E. McClelland: State, Society, and University in Germany 1700–1914, Cambridge 1980, pp. 93–105.

of the "culture state," an idealistic notion never fully implemented in reality.[1115] René König writes that the Prussian shift from a utilitarian one to a more theoretical conduct of science was, at least in part, motivated by the urge to differ from the French system. At that time Napoleon strived to replace the old French university system with a completely utilitarian academy system. König also adds that, even though planning for a new university in Berlin had begun as early as 1800, the treaties of Tilsit (1807) gave the final push for the foundation of the Berlin University. The treaty brought great territorial losses for Prussia, and Prussia lost all her universities except for those in Frankfurt (Oder) and Königsberg (now Kaliningrad, Russia).[1116] Or to put it in other words:

> By mid-century Europe had already shown what had to be done to advance science as the times demanded. Some nations learned faster than others, and the German states fastest of all. The German response may have owed something to the rivalry of small states, within each of which a sovereign could spend money without consulting taxpayers. A lot still went for royal frills. But the armies of Napoleonic France had knocked it into the crowned heads of Germany, especially that of Prussia, that to survive they needed the power of science, or at least of the national pride it helped foster.[1117]

Wilhelm von Humboldt (1767–1835), who was one of the founding members of Berlin University, thought that science as a whole was worthwhile and should not be broken down into specialized disciplines, that the world had to be studied as a whole. Knowledge had to be attained through study and experiment, not just be memorized and learned by heart from some old tome. The unity of teaching and research ("Einheit von Forschung und Lehre") was, in Humboldt's mind, the absolute prerequisite for autonomous scholarly pursuits. Still, science could be of practical use to society, and education would produce socially responsible members of society.[1118] Humboldt also believed that the professors at the old universities lacked didactical skills, and oftentimes scientific skills as well. He believed this led to the crudeness, immorality,

1115 Nitsch et al.: Hochschule in der Demokratie, pp. 242–262.

1116 René König: Vom Wesen der Deutschen Universität, Berlin 1935, pp. 34–40, 53–61.

1117 Bruce: The Launching of Modern American Science 1846–1876, p. 7.

1118 Schelsky: Einsamkeit und Freiheit, pp. 82–90. Also see: Nitsch et al.: Hochschule in der Demokratie, pp. 18–27. For more detail on Humboldt's ideas of creating responsible individuals, who were not just to be trained and educated but who should be led to self-determination, see: Clemens Menze: Die Bildungsreform Wilhelm von Humboldts, Hannover 1975, pp. 18–26.

and immaturity he perceived in contemporary students and was to be remedied by a reform of higher education and the establishment of a modern university at Berlin.[1119]

According to Johann Gottlieb Fichte (1762–1814), another important educational reformer of the time and a founding member of the Berlin University, scholars should be the teachers to all of humanity, guiding humanity (back) to a path of self-knowledge and self-understanding, onto a path of true and objective enlightenment and cognition. In his mind it was up to universities to train scientists who could carry the acquired knowledge into society and benefit it that way.[1120]

Other participants who contributed significantly to the education reform were Friedrich Wilhelm Joseph Ritter von Schelling (1775–1854) and Friedrich Schleiermacher (1768–1834). Schelling wanted to create a university completely focused on science, which did away with utilitarianism and obsolete traditions. His ideal university would be committed to all scientific disciplines and constituted an institution where professors would have a holistic approach to science. Schleiermacher, in a similar vein, thought that the utilitarianism most states pushed for impeded the very human need to educate oneself. He, however, was willing to accept the usefulness of practical knowledge as it was taught at academies.[1121] Schleiermacher's willingness to compromise with the demands the Prussian state expressed to make the science at Berlin University more utilitarian diluted Humboldt's ideals, and is often cited as one of the reasons why said ideals were never really implemented in practice.[1122]

In accordance with Humboldt's ideals there were to be flat hierarchies at the university. Students and professors would work together and it would be highly motivating to teach in tandem with research (this is the "Einheit von Forschung und Lehre" mentioned above). As both research and teaching should be free from and unrestricted by outside influences ("Freiheit der Lehre"), it would be best if the students isolated themselves from the rest of society to conduct their scholarly pursuits as independently and unaffectedly as possible.[1123]

Initially Berlin University was led by the considerations and ideals of Humboldt and the other reformers. But soon the state stepped in, anxious that the university had become a hotbed for revolutionary ideas and fearing that professors were indoctrinating students in subversive and antireligious behavior. Soon, professors were censored and reprimanded while the state sought to turn the university into a training facility

---

1119 Menze: Die Bildungsreform Wilhelm von Humboldts, pp. 280–288.

1120 König: Vom Wesen der Deutschen Universität, pp. 68–97.

1121 Menze: Die Bildungsreform Wilhelm von Humboldts, pp. 288–303.

1122 Rüdiger vom Bruch: Die Gründung der Berliner Universität, in: Rainer C. Schwinges (ed.): Humboldt International. Der Export des Deutschen Universitätsmodells im 19. und 20. Jahrhundert, Basel 2001, pp. 53–73.

1123 Schelsky: Einsamkeit und Freiheit, pp. 91–101.

for civil servants and the clergy.[1124] Nonetheless, after his return from Paris in 1827, Alexander von Humboldt (1769–1859), brother to Wilhelm, managed to establish a curriculum focused entirely on natural sciences, and empiricism once again triumphed over philosophy.[1125]

As to the prevalence of natural sciences at Germany universities: at first Heidelberg was the forerunner. Beginning in 1817 physics courses were taught, beginning in 1818 courses in mineralogy and chemistry. The period between 1850 and 1870 was the most exceptional and productive concerning the teaching of natural sciences at Heidelberg. Many professorships were established, Bunsen and Kirchhoff were appointed (Marsh learned from both, see chapter 3. 2.), and practical research was conducted in modern laboratories, acquired thanks to increased government funding. In contrast, Göttingen was of no interest to Marsh because a professorship for geology and paleontology was only established there in 1870, five years after his return to the United States. Munich, on the other hand, was one of the first universities to teach paleontology as a distinct scientific discipline, beginning with the appointment of Andreas Wagner (1797–1861) in 1843. Marsh had contact with more than one paleontologist from Munich, including Zittel, Baur, and Schlosser (see chapter 6.).[1126] As mentioned above, natural sciences at Berlin really took off after 1827. The following appointment of the Rose brothers as professors at the University of Berlin is part of the rise of natural sciences at Berlin, and both were teachers of Marsh's.[1127] Baumgarten provides a ranking of the German universities during the nineteenth century, the University of Berlin taking first place, Munich second, Leipzig third, Bonn fourth and Heidelberg fifth (Breslau, another university Marsh attended, ranking eleventh). Baumgarten writes that many professional scientists had the goal to work at one of the preeminent

1124 This was the result of the Carlsbad Decrees ("Karlsbader Beschlüsse") of 1819, a conservative reaction to the liberal reforms and the growing sentiment for German unification. The reactionary policies that followed meant, among other things, that the state would exert more influence over the German universities. After 1819 the management of the universities was rarely of public interest, which in turn meant that research and scientific publication became more important to professors than any political agitation. See: McClelland: State, Society, and University in Germany, pp. 106–149, 162–189.

1125 Menze: Die Bildungsreform Wilhelm von Humboldts, pp. 405–431.

1126 Marita Baumgarten: Professoren und Universitäten im 19. Jahrhundert. Zur Sozialgeschichte deutscher Geistes- und Naturwissenschaften, Göttingen 1997, pp. 66–69, 73. She provides the dates when a professorial chair was established at a German university. Mineralogy was established at Berlin in 1810, at Göttingen in 1811, at Heidelberg in 1818, at Gießen in 1819, at Munich in1826, a second chair for mineralogy was established in Berlin in 1839. A professorship for paleontology was established at Munich in 1843, and at Heidelberg (as geology / paleontology) in 1870. Beside the professorship for paleontology, Munich had a chair for natural history between 1826 and 1869. A chair for geology was established at Göttingen in 1798, but discontinued in 1817, Munich had a chair for geology from 1853 to 1890, and Heidelberg established her professorship for geology as late as 1913. See Baumgarten: Professoren und Universitäten im 19. Jahrhundert, pp. 282–286.

1127 Baumgarten: Professoren und Universitäten im 19. Jahrhundert, pp. 75–79.

institutions, while lower ranking universities were at best seen as an entry point to a scholarly career.[1128]

Between 1819 and 1866 the German states increased the sponsoring of universities and education in general.[1129] Still, enrollment at German universities hardly increased between the 1830s and 1860s, but then doubled between 1870 and 1914, while in some regions enrollment increased even more. According to McClelland, this increase was the result of the improvement of the financial situation taking place in Prussia and Germany as a whole during these years, further strengthening the educated middle class.[1130]

Now that professors could focus on their scientific activities and were less dependent on the practical use of their sciences, and because universities were less dependent on the sponsorship of influential individuals, transregional exchange between universities was facilitated. Professors could seek employment at another university more easily. Now that professional reputation gained importance, it became the deciding factor for the career of many a scholar, where before political contacts and the good will of influential benefactors were of greater consequence. These developments went hand in hand with the specialization of the scholars. The growing influence of the states on the education system came with an increase in bureaucracy, which meant that throughout the German states the governance of the universities became more homogeneous. A uniquely German system of higher education developed, and it was dominated by the Northern German / Prussian model.[1131] Baumgarten adds that once professorships were no longer allotted due to personal connections and kinship but due to individual talents, the quality of the German system of higher education increased. Beginning in the 1870s, a real job market for professors developed. She writes that for the better part of the nineteenth century Berlin University was considered to be the preeminent university in Germany, if not the world.[1132] Being situated in a major city facilitated the university's collaboration with other scientific and bureaucratic organs, as well as the other Prussian universities. In 1871, when the German national

---

1128 Baumgarten: Professoren und Universitäten im 19. Jahrhundert, p. 272.

1129 McClelland: State, Society, and University in Germany, pp. 203–217.

1130 McClelland: State, Society, and University in Germany, pp. 239–258.

1131 Roy Steven Turner: German Science, German Universities: Historiographical Perspectives from the 1980s, in: Gert Schubring (ed.): “Einsamkeit und Freiheit” neu besichtigt: Universitätsreformen und Disziplinenbildung in Preußen als Modell für Wissenschaftspolitik im Europa des 19. Jahrhunderts, Stuttgart 1991, pp. 24–36.

1132 Baumgarten: Professoren und Universitäten im 19. Jahrhundert, pp. 11–29, 147–159. Veysey adds that the preeminence of the Germany university was an accepted fact in the US: “Occasional Americans had been studying in Germany since 1816, and by the fifties considerable interest had developed concerning Continental universities, the German then being without doubt pre-eminent in the world.” See: Veysey: The Emergence of the American University, p. 10.

state was formed, the Berlin University became the flagship of the, now unified, German system of higher education.[1133]

Sylvia Paletschek questions the idea that the founding of the Berlin University was perceived as a revolutionary break with the old system. She instead argues that this assessment was made during the twentieth century and in hindsight because it was then understood to have shaped the identity and mentality of the "German University" as no other ideas or reforms had.[1134] This judgement goes well with Rüdiger vom Bruch's estimation that Humboldt's ideals were mostly evoked in later times of crisis within the German system of higher education.[1135]

Paletschek further argues that the founding of Berlin University did not constitute a turning point because most of the reform was a continuation of reforms initiated at the end of the eighteenth century in other German universities, such as Göttingen. It was, however, a countermove to the extreme specialization and practice-oriented modus operandi of the French academies. Merely the fact that the ideals of Humboldt were somewhat implemented might have been a unique selling point of the Berlin University.[1136]

At the end of the eighteenth and the beginning of the nineteenth century, the French conduct of science was gladly imported to Great Britain, the French scientific institutions and structures less so. During the nineteenth-century educational reform in the United Kingdom the new German approach to science was employed. The German model of higher education was much more specialized than the broad British approach, contributing to the diversification and genesis of many distinct scientific disciplines and to the professionalization of science (in contrast to the "hobby-scientists" of the eighteenth and early nineteenth centuries, who "knew everything," see

1133 Sylvia Paletschek: Verbreitete sich ein ‚Humboldt'sches Modell' an den deutschen Universitäten im 19. Jahrhundert?, in: Rainer C. Schwinges (ed.): Humboldt International. Der Export des Deutschen Universitätsmodells im 19. und 20. Jahrhundert, Basel 2001, pp. 75–104, see pp. 81–83.

1134 "Seit Beginn des 20. Jahrhunderts ist das Schlagwort der Humboldt'schen Universität für das deutsche Universitätswesen identitätsstiftend und mentalitätsprägend schlechthin." See: Paletschek: Verbreitete sich ein ‚Humboldt'sches Modell' an den deutschen Universitäten im 19. Jahrhundert?, pp. 75–77. Quote on page 75.

1135 Rüdiger vom Bruch writes that the German university system gained its world renown thanks to the reforms of the nineteenth century, which are today, at least in name, mostly attributed to Humboldt. He further notes that the ideals of Humboldt and company were never really implemented, and that during the course of the century many of the reforms were adjusted and rolled back. Still, the ideas and ideals of the so-called Humboldtian reforms were evoked again and again to this day, especially after phases of national and educational crisis, namely after World War Two. See: Rüdiger vom Bruch: Langsamer Abschied von Humboldt? Etappen deutscher Universitätsgeschichte 1810–1945, in: Mitchell G. Ash (ed.): Mythos Humboldt? Vergangenheit und Zukunft der deutschen Universität, Vienna 1999, pp. 29–57.

1136 Paletschek: Verbreitete sich ein ‚Humboldt'sches Modell' an den deutschen Universitäten im 19. Jahrhundert?, pp. 87–97.

chapter 2. 5.).[1137] The German scientific penchant for independent thought ("Freiheit der Wissenschaft"), however, was viewed skeptically and with reservation. It was suggested that too much freedom could inspire students to become overly critical, or even worse, turn them into revolutionaries. Nonetheless, the overall reform of the British educational system in 1870 drew much upon the German model: "Anyone wishing to pursue a career in science was forced to look to Germany for a training."[1138]

John Gascoigne writes about the importance of science for "elite" culture, the high society of the British Empire at the beginning of the nineteenth century. The circumstances he describes also apply to colonial America in the latter half of the eighteenth century and the young independent United States a few years later.

"Elsewhere in the British Empire, too, science formed an important part of the elite culture and its cultivation was both a mark of gentility and an espousal of the goal of improvement which formed an important part of the justification of empire. The growth of local scientific societies was one index of the increasing local consolidation of colonial elites. Such local scientific bodies provided an opportunity for rational amusement, social and political networking and, it was hoped, the promotion of the goals of improvement."[1139]

8.2 The German Influence on the Education Reform in the United States

Due to the perceived "civilizing" aspect of the universities on the North American continent, the history of higher education was of great consequence to the US-American self-conception during the long nineteenth century. The Puritan founding of Harvard

---

1137 David Allen adds that, beginning in the 1870s, British who had studied in Germany imported the German conception of higher education to the UK. Consequently, British scientists were becoming increasingly "professionalized," just like in the US. And just as in the US, scientific publication boomed, the Journal of Physiology and the Annals of Botany were founded in 1878, and 1887, respectively. The language with which science was conveyed professionalized as well and soon it was very hard for the laymen to grasp the meaning of scientific publications. Contrast this with the situation at the beginning of the nineteenth century when geology had become a national pastime (see chapter 2. 5.), and it seems that science had become cold, detached, and very unromantic. See: Allen: The Naturalist in Britain, pp. 163–166.

1138 John R. Davis: Higher Education Reform and the German Model: A Victorian Discourse, in: Heather Ellis; Ulrike Kirchberger (eds.): Anglo-German Scholarly Networks in the Long Nineteenth Century, Leiden 2014, pp. 39–62, https://doi.org/10.1163/9789004253117_004. Quote on page 57.

1139 John Gascoigne: Science and the British Empire from its Beginnings to 1850, in: Brett M. Brennett; Joseph M. Hodge (eds.): Science and Empire. Knowledge and Networks of Science across the British Empire 1800–1970, Basingstoke 2011, pp. 47–67. Quote on page 60.

in 1636 is recognized as being of the greatest importance to the history of education in the US, while international influences on the US-American university system are less well researched. Note that the first American university was founded in Santo Domingo on Hispaniola in 1538 (present-day Dominican Republic). The misconception that Harvard was the first American university goes hand in hand with the narrowing-down, or appropriation, of the term "American" as solely referring to the USA.[1140]

In the British colonies of North America, the college curricula encompassed spiritual education and taught in accordance with Christian morals; the colleges themselves were co-governed by civil and clerical authorities.[1141]

In the British colonies and the early republic alike, nature was viewed through the lens of natural theology and utilized to promote colonialism. Natural theology was still prevalent during the 1830s and 40s, when it was used to study nature, putting emphasis on scripture and the divine.[1142]

Bernhard Cohen examines the political philosophy of the young republic and analyzes the writings of the Founding Fathers. He states that it was only natural that the Founding Fathers took a scientific approach to political ideas, writing towards the end of the Age of Enlightenment. He points out that culture and society do not only influence the reception and conduct of science but, vice versa, scientific terms seep into everyday life and expressions, terms like "getting to the heart of the matter," or (for a more modern example) "quantum leap." [1143]

After the War of Independence, a new and genuinely US-American college system was developed. In the minds of the founders of the United States "the survival of republics depended on the *virtue* of their citizens."[1144] Meaning the moral education of students – molding them into virtuous citizens – was perceived to be the most important mission of the colleges. This goes hand in hand with the establishment of numerous new colleges after US independence and soon the old colonial colleges were joined by newly founded institutions, many of which were established as explicitly republican colleges.[1145]

Thomas Jefferson's endeavors to modernize the US-American educational system were somewhat similar to Humboldt's reforms. In 1819 Jefferson's University of Vir-

---

1140 Anja Werner: Striving for the Top: Reevaluating Regional and Transatlantic influences in the History of U.S. Higher Education, in: Laurenz Volkmann (ed.): Education and the USA, Heidelberg 2011, pp. 87–103, see pp. 87–89.

1141 Roger L. Geiger: The History of American Higher Education. Learning and Culture from the Founding to World War II, Princeton, NJ 2015, pp. 15–18, https://doi.org/10.1515/9781400852055. Also see: Joseph Ben-David: American Higher Education. Directions Old and New, New York 1972, pp. 11–16.

1142 Shapiro: Science Education, pp. 320–332.

1143 Cohen: Science and the Founding Fathers, pp. 20–21, 25.

1144 Geiger: The History of American Higher Education, p. 90.

1145 Geiger: The History of American Higher Education, pp. 89–122.

ginia deviated much from the, then still prevalent, colonial system of education, becoming the first truly modern university in the US. Furthermore, Jefferson sought to reform the school system so that the educational path from elementary school to university was clearly structured, with all the steps on the path interlocking. Many of his reforms were met with the same resistance by the state which had met Humboldt's reforms in Prussia (see above). Jefferson, however, was more practically minded in his approach to science than Humboldt. Jefferson, furthermore, wanted to educate all children, regardless of their parents' social class, to make education available to everyone and to turn a new generation into good and moral citizens. Humboldt was less egalitarian. Still, both Jefferson's and Humboldt's reforms aimed at the development, strengthening, and preservation of their respective nation, with the caveat that Jefferson's reforms were aimed towards the talented male white youth. Girls and enslaved males were still barred from higher education, and most enslaved people were denied any education at all.[1146] Jefferson tried to establish various modern schools for the conduct of the natural sciences, intellectual ancestors to modern faculties. Students should be able to freely choose which of the schools they wanted to attend. Though Jefferson's visions were implemented at the University of Virginia, they did not expand to other colleges, as they were perceived as too radical and too expensive. Only in the 1860s and 70s were some of the new ideas implemented in other places and blended with reforms inspired by the German example.[1147] Ben-David even judges that "[u]ntil about the 1870s, German universities were virtually the only institutions in the world in which a student could obtain training in how to do scientific or scholarly research."[1148]

Rudolph, who writes about the history of US-American colleges and universities, and seems to admire Jefferson almost beyond the reasonable, adds that the War of 1812 had emphatically imprinted on US-American national identity and also influenced the early Jeffersonian efforts to educational reform. Another early reformer was George Ticknor (1791–1871), one of the first US Americans to study at Göttingen, who sought to reform the education at Harvard following the German example, with limited success.[1149]

1146 "Für Jefferson und Humboldt war die Reform der schulischen und universitären Erziehung das Kernstück zum Aufbau und zur Kräftigung und Erhaltung ihrer Nationen." See: Jurgen Herbst: Thomas Jefferson und Wilhelm vom Humboldt: Universitäts- und Schulgründer, in: Rainer C. Schwinges (ed.): Humboldt International. Der Export des Deutschen Universitätsmodells im 19. und 20. Jahrhundert, Basel 2001, pp. 273–287. Quote on page 278.

1147 Frederick Rudolph: The American College and University, New York 1962, pp. 125–128.

1148 Joseph Ben-David: Centers of Learning. Britain, France, Germany, United States, New York 1977, p. 22.

1149 Rudolph: The American College and University, pp. 110–125.

Kohlstedt adds that while some US-American scientific discoveries of the early nineteenth century were acclaimed in Europe for their inventiveness, most US-American scholars were thought to lack "real" scientific investigation. Their inability to produce new theories was ascribed to the lack of modern European educational techniques and laboratories. After the War of 1812 and with American nationalism on the rise, national scientific progress was linked to the perceived success of the American ideals, manifested in material and cultural advancement (for an in-depth look at the link between patriotism and science in the early days of the republic see chapter 2. 6.).[1150]

When a new university was to be established in New York in 1832, a German-inspired, utilitarian approach to science was to be realized. But this practical approach to science was viewed by most students and professors as inferior to the classic holistic system of US-American higher education, and the university was transformed into a college. This is but another example showing how almost all efforts to reform the US-American educational system proved to be fruitless till after the conclusion of the Civil War.[1151]

US-American science, much like European science, became increasingly specialized during the nineteenth century: in 1802 Benjamin Silliman Sr. became professor for chemistry, natural history, and mineralogy at Yale. Upon his retirement in 1858, his chair was subdivided into chemistry, Silliman's son becoming his successor, and natural history, granted to James Dwight Dana. When a chair for botany was created in 1864, Dana became professor of geology.[1152]

> Since it must have been evident to anyone looking over their journals that scientists did not characteristically concern themselves directly with practical things, scientists were careful to stress that the moral and religious aspects of science were as valuable to society as was its practical utility [...]. Add to that a public which believed firmly that nature was the creation of an omnipotent and benevolent God, the study of God's works could be represented as a duty, not merely an idle pastime.[1153]

1150 Sally Gregory Kohlstedt: The Formation of the American Scientific Community. The American Association for the Advancement of Science 1848–1860, Urbana, IL 1976, p. 2.

1151 Rudolph: The American College and University, pp. 128–135.

1152 Daniels: American Science in the Age of Jackson, p. 35. Bruce notes that "Silliman persuaded Yale in 1804 to send him to Europe with ten thousand dollars for the purchase of books and apparatus." Bruce: The Launching of Modern American Science 1846–1876, p. 14.

1153 Daniels: American Science in the Age of Jackson, p. 48.

With those words Nathan Reingold summarizes the historical research on the professionalization of science in the US. He outlines the seemingly inevitable replacement of scientific amateurs with full-time professionals, bringing with them a more complete understanding of the disciplines and a higher standard of training and conduct. On the other hand, the rise of professionalization can be read as the elite establishing a monopoly on scientific publication for themselves at the cost of discouraging the brilliant and promising scientific amateur, who dominated science at the beginning of the nineteenth century (see chapter 2. 5.). Because the struggle between the elite and the populous in general is a popular focus in the study of democracy, the subject of professionalization is especially prominent in American literature concerning the history of science. Reingold notes that the association with a scientific society became more desirable as the nineteenth century progressed, for with it came a certain social prestige. He differentiates between mere "members" of a learned society, who might attend local meetings, and "contributors," who regularly traveled the country, attending meetings and actively contributing to the journal of their society. "Practitioners" were yet another class. They practiced science for a living and were paid to do so. During the nineteenth century the "contributors" went almost extinct and were replaced by "practitioners." Reingold then introduces the smallest group of scholars, the "researchers," who undertook original research and contributed the most to the international scientific reputation of their nation. Reingold cites the number of "practitioners" as being 1,500 in the year 1860, 2,100 in 1870, at least 3,300 in 1880, 7,300 in 1890, and 14,200 in 1900; there was roughly one "researcher" for every ten "practitioners".[1154] Jürgen Herbst comes to a similar conclusion concerning the professionalization of US-American science and higher education:

> The rise of service and research inevitably lowered the prestige of teaching and of the undergraduate colleges that continued to cherish teaching as their central concern. This is not to say research specialists could or would not teach, but it is to say that a shift occurred in the priorities scholars assigned to their varied tasks and in the self-image they cultivated. While within the large universities some professors and departments remained faithful to their teaching, others committed themselves to research and graduate instruction. [...] Large research universities delegated much undergraduate teaching to graduate assistants and placed their professors in large lecture halls to speak before hundreds of students. As service and research rather than teaching became

1154 Nathan Reingold: Definitions and Speculations. The Professionalization of Science in American in the Nineteenth Century, in: Sanborn C. Brown; Alexandra Oleson (eds.): The Pursuit of Knowledge on the Early American Republic. American Scientific and Learned Societies from Colonial Times to the civil War, Baltimore, MD 1976, pp. 33–69.

> the professors' chief occupation their loyalties turned from their college and students to their specialty and their colleagues. [...] Institutional identification was temporary; commitment to their field remained permanent. [...] The scholar's peer group consisted not necessarily, not even customarily, of colleagues in college or university, but of colleagues in the profession. The scholar was above all a biologist, or engineer, or historian, he was a professor or teacher only secondarily.[1155]

Still, at the beginning of the nineteenth century the United States lacked specialized scientific institutions. There were few specialized and professional US-American scientists to be found before the Civil War. Higham supposes that specialization was at first perceived as very Un-American, for it violated the US-American values of individuality, self-reliance, and egalitarianism (an expert depended on a network of other specialists and engaged with subject matters too complicated for the laymen to understand, therefore participating in an almost secret art). This might be part of the reason why US Americans did not readily participate in scientific specialization during the first half of the century. This changed after the Civil War when higher education was reformed and modernized, remodeled after European examples, and the US-American attitude towards specialization was reversed in the second half of the nineteenth century. When the Ph.D. system was imported to the United States it became a certificate of scientific specialization (whereas in Europe it was also associated with the notion of a broad education).[1156]

Edward Shils also stresses that the US was severely lagging behind Europe in terms of education, especially concerning its universities. The university was not as established as an institution of higher learning, as it was in most European countries. In the first half of the nineteenth century most of the US-American scientific work was not done at a university, but in private and by individuals. "In any case, in no field, except perhaps historical studies, did the United States have clusters of eminent amateur scholars and scientists of the quality attained at the higher reaches in Europe."[1157] This slowly changed as the universities were remodeled after the German example.

1155 Jurgen Herbst: Diversification in American Higher Education, in: Konrad H. Jarausch (ed.): The Transformation of Higher Learning 1860–1930. Expansion, Diversification, Social Opening and Professionalization in England, Germany, Russia and the United States, Chicago 1983, p. 196–206. Quote on page 203.

1156 John Higham: The Matrix of Specialization, in: Alexandra Oleson; John Voss (Eds.): The Organization of Knowledge in Modern America, 1860–1920, Baltimore, MD 1979, pp. 3–18.

1157 Edward Shils: The Order of Learning in the United States. The Ascendancy of the University, in: Alexandra Oleson; John Voss (eds.): The Organization of Knowledge in Modern America, 1860–1920, Baltimore, MD 1979, pp. 19–47. Quote on page 21.

At the beginning of the century US-American science also depended on scientific literature published in Europe, which hampered and slowed science in the US:

> American scientists depended heavily on imported books and journals. Getting them took time and money. A chemist ordered the books of Berzelius and Rose one February; he was still waiting in November. He figured the cost at thirty-seven dollars, probably as much as he earned in a week or more. A few booksellers in the large port cities made a specialty of importing books, sending their own agents to Europe, even opening branches there. English books predominated, however, for lack of polyglot customers.[1158]

Though this situation improved towards the end of the century, it shows the importance of private contacts and of the exchange of information, books, and journals within the working network of all scientists. The interest in US-American scientific literature in Europe was also very limited before the second half of the century.[1159]

US-American science gained real traction only after the end of the Civil War but was still exclusive to a relatively small group of professional scientists and amateurs by 1876. About two thousand people were engaged in science in one way or the other, most of them situated in New England and the Mid-Atlantic States. Individuals who had previously called themselves *natural philosophers* were now beginning to identify as *scientists*. Theology and moral education were deemphasized at colleges and universities as the production of practical and theoretical knowledge gained a foothold in higher education. Kevles supposes that material prosperity was a fundamental prerequisite for the public appreciation of theoretical science. He emphasizes his point by describing the westward expansion and rapidly growing industrialization as a result of said prosperity, which in turn allowed for the growth of scientific institutions across the country:

> But these were the boom years of the post-Civil War period, when the golden spike was driven linking East and West in the first transcontinental railroad; when the advance of technology was combining with widespread economic opportunity to spur the economy ahead at a dizzying pace; when thousands were moving westward, drawn alike by cheap land and the increasing miles of

1158 Bruce: The Launching of Modern American Science 1846–1876, p. 11.

1159 Bruce writes that of 382 US-American books reprinted in England between 1833 and 43 only 9 were scientific, see: Bruce: The Launching of Modern American Science 1846–1876, p. 12.

> railroad track; when the factory system was pouring out a constantly swelling volume of goods; when people seemed unprecedentedly well off.[1160]

Bruce also notes concerning the effects the civil War had on the conduct of US science that

> For scientific institutions, the Civil War was not a total disaster. It broke up old patterns and let new ones form. Lincoln's first Congress in 1862, freed of Southern obstructionism and nerved for bold action by the crisis, ranks among the foremost half-dozen in American history for the weep and significance of its innovations. Among them were the Morrill Land Grant Act for the support of colleges and the acts establishing the Department of Agriculture and the National Academy of Sciences.[1161]

He later adds that:

> The Civil War had exalted the nation over the states in politics, but at its end the scientific community was moving from central authority toward a federalism of specialized fields. American scientific achievements in the century since, compared with those of more centralized science in other nations, suggest that the movement was not necessarily for the worse.[1162]

In the 1870s and 80s the devotion to pure, meaning abstract, science was considered to be clashing with the most American of enterprises: making money. Although academic scientists employed by universities and colleges earned much higher wages than clerics or common workers, they often had a hard time affording the proper (and expensive) lifestyle of the high society this academic elite was associated with. Marsh was very good at partaking in this lifestyle and securing the support of many important members of society, while not paying his assistants enough to also live the life of the high society (see chapter 6. 2.). Most scientists were removed from the troubles of financial necessity which plagued most American workers. They believed science ought to be valued for its own sake, even if no practical use would ever derive from the discoveries, and that science might even add to the "cultural stature" of the nation. The interests of the utilitarian government clashed often with those of theoretical sci-

1160 Daniel J. Kevles: The Physicists. The History of a Scientific Community in Modern American, reprint, Cambridge, MA 1987, pp. 3–13. Quote on page 8.

1161 Bruce: The Launching of Modern American Science 1846–1876, p. 287.

1162 Bruce: The Launching of Modern American Science 1846–1876, p. 305.

entists and the question arose by whom, how, and to what purpose public funding for science should be conducted.[1163]

Gilman Ostrander adds that reform-minded places like Yale (see below) were in search of role models for a comprehensive reformation of the US-American educational system. At first the reformers looked to Scotland, then France, but the latter was deemed too radical (in politics and in the restructuring of their system of higher education), so that at the beginning of the century Germany seemed like the only sensible option to emulate.[1164] "The prowess of German universities in advancing academic scholarship was increasingly evident [during the 1850s], as was the inadequacy of American efforts."[1165]

But still, no matter how impressed the returnees were with the German system, they still had a hard time implementing it at home. A good deal of scientists remained skeptical toward the foreign reforms, so that the German system was never wholly transplanted to the US, though it certainly inspired the evolution and gradual change of the US-American one.[1166]

The US-American system of higher education combined elements of the traditional British "Oxbridge" system with some elements of the German system, mainly in the form of graduate schools. By the middle of the nineteenth century there were almost no universities to be found in the US. Higher education was usually conducted at colleges, but by the 1890s this had changed entirely. Modern universities that promoted original research and granted doctorates had been established following the German example. The founding of Clark University in 1887 in Worcester, Massachusetts, was heavily inspired by the German role model. Maybe that is one of the reasons why Baur, as well as other German scientists, were employed there (see chapter 6. 2.).[1167] Also note that in 1890 a modern university was established in Chicago,[1168] and immediately employed Baur.

But how was this change brought about? Werner stresses that at first US Americans preferred to go to Göttingen, but that during the 1870s Leipzig, Munich, and Berlin became the preferred destinations for US-American students. Social networks played an immense part in this. The scholarly exile communities became somewhat

1163 Kevles: The Physicists, pp. 45–55.

1164 Gilman M. Ostrander: Republic of Letters. The American Intellectual Community, 1776–1865, Madison, WI 1999, pp. 20–28.

1165 Geiger: The History of American Higher Education, p. 316.

1166 Carl Diehl: Americans and German Scholarship 1770–1870, New Haven, CT 1978, pp. 50–52.

1167 Roy Steven Turner: Humboldt in North America? Reflections on the Research University and its Historians, in: Rainer C. Schwinges (ed.): Humboldt International. Der Export des Deutschen Universitätsmodells im 19. und 20. Jahrhundert, Basel 2001, pp. 289–311, see pp. 289–294.

1168 John R. Thelin: A History of American Higher Education, Baltimore, MD 2004, pp. 118–122.

of a microcosm, for students from all over the US met in these communities. Most of these students rounded out their education in Germany, but also circulated information about US universities and helped each other find employment upon their return home.[1169] Beginning in the 1850s, the US-American exile community in cities like Berlin and Göttingen was large and well-organized enough that students could manage their day-to-day lives without ever really coming into contact with the German population. All social life could take place within the boundaries of the "colony." Many US-Americans learned just enough German to be able to follow the lectures, but never really spoke German fluently.[1170] Pochmann writes that "by 1850 there was no German university that did not have its American colony." He adds that "from 1820 to 1830, an average of 5 [US-American] students were registered annually" at German universities. This number rose quickly, and by 1860 approximately 77 US-Americans had enrolled in German Universities, by 1880 this number rose to 173, by 1890 to 446 and in 1900 there were more than 900.[1171] Fallon writes that "[b]efore 1850, around 200 American students had visited German universities. By 1900, however, over 9,000 Americans had studied there."[1172]

Another factor motivating students to study in Germany was that at the beginning of the century students at US-American colleges were treated like minors who were to be trained in morals and discipline, could not choose their own field of study, etc. At German universities, on the other hand, students were treated more like the independent adults they really were.[1173]

Edwards Shils describes the situation of young US-American returnees and their effect on US higher education as follows:

> As the reflux of young men from the German universities began in earnest, complaints were heard that American universities did not conduct research, that they were reluctant to demand that professors undertake research, and that they did not give due reward, in terms of appointment and promotion, to past and prospective accomplishments in research.[1174]

1169 Werner: Striving for the Top, pp. 90, 94. Geiger: The History of American Higher Education, pp. 328–332.

1170 Diehl: Americans and German Scholarship, pp. 130–140.

1171 Pochmann: German Culture in America, p. 77.

1172 Fallon: German Influences on American Education, p. 83.

1173 Jurgen Herbst: The German Historical School in American Scholarship. A Study in the Transfer of Culture, Ithaca, NY 1965, pp. 23–38.

1174 Shils: The Order of Learning in the United States, p. 28.

Turner concurs and writes that the reformation of the US-American educational system following the German role model was furthered by young US scientists returning from Germany:

> American enthusiasm for German practices was driven mainly by the large numbers of American Students, most of them postgraduates, who went to school in Germany in the 19th century. Some 10,000 American students studied in German universities between 1815 and 1914, most of them between 1870 and 1895. Many returned exhilarated by the 'freedom' of the German university, flattered by the inordinate personal attention that they received from their German hosts, imbued with some Americanized version of the German devotion to *Wissenschaft*, and driven by a professional and discipline-based concept of the academic role quite different from the traditional American one. In some fields, German science and scholarship were transplanted holusbolus to the United States.[1175]

Compare this with Marsh's situation after his return from Europe. Yale had no paid professorship to offer and it was thanks to the privilege of having a rich and generous uncle that Marsh could begin his scientific career, focusing on research and the techniques he had acquired in Germany (see chapter 3. 2.). Indeed, Marsh was in a lot of ways very typical for an US-American student who went to Europe. He studied at Yale, focused on natural sciences, and visited more than one university in Europe.[1176]

1175 Turner: Humboldt in North America?, p. 292. Karl-Ernst Jeismann, who has analyzed the experiences of US-Americans visiting Prussian schools in the 1830s and 40s, writes that the US-Americans were very impressed by Prussian efficiency. They noted that the Prussian system prepared the children well for their professional or academic futures. Although the state interfered heavily with the schools, teachers seemed to be relatively free in structuring their lessons. And the lessons were not mere lectures, but dialogues between teacher and pupils, enabling the teacher to cater to the individual needs of the children. Furthermore, teachers were specially educated for their job, in contrast to the US where formal qualifications for teaching were less strict. See: Jeismann, Karl-Ernst: American Observations Concerning the Prussian Educational System in the Nineteenth Century, in: Geitz, Henry et al. (eds.): German Influences on Education in the United States to 1917, Washington DC 1995, pp. 21–41. Add to this that after the failed revolution of 1848 many German teachers immigrated to the US and brought with them their ideas about and experiences with the German school system. They played an imported part in the establishment of Kindergartens in the US. Many of these refugees were democratically minded idealists seeking to implement their democratic ideas like religious tolerance and mixed-gender classes in US-American schools. See: Karl-Heinz Günther: Interdependence between Democratic Pedagogy in Germany and the Development of Education in the United States in the Nineteenth Century, in: Henry Geitz et al. (eds.): German Influences on Education in the United States to 1917, Washington DC 1995, pp. 43–56.

1176 Diehl: Americans and German Scholarship, pp. 61–62. Also see: Adam; Lerg: Introductory Remarks, p. 300.

After their return from Germany, most aspiring scientists felt isolated because scientific knowledge was much less appreciated in their home country than it had been in Germany, or Britain, where they had studied:

> For these young men, returning to the United States in the years immediately following the Civil War meant reentering a life where the intellectual air was very thin. Those who found employment in colleges and universities felt isolated. The older generation of teachers was neither accustomed to nor interested in engaging in serious scholarly discussions at a local or national level.[1177]

Rainer Schwinges attests that the ideas of Humboldt and the other German reformers of higher education were admired more and implemented faster in the US than in Germany, presumably by those US-Americans who had studied at German universities and now returned to the US to reform the university system there.[1178]

In contrast, Fritz Ringer judges that the reformation of US-American higher education followed the German model only in theory, but not in practice, or that German universities had themselves strayed far from its original ideal:

> It has been generally held that German models played a significant role in the transformation of American higher education after 1865. Yet while certain conventional images of German learning were certainly invoked by academic reformers in the United States (and elsewhere) during the late nineteenth century, those images may have borne little resemblance to German realities.[1179]

Bruch writes that the Humboldtian ideals were never really implemented at any university, not even in Berlin. From the beginning the government sought to push the university towards a more utilitarian approach. He judges that at the end of the century research was no longer conducted at the universities, which focused almost entirely on education, instead specialized academies conducted all the groundbreaking research.[1180] Bruce adds that:

1177 Shils: The Order of Learning in the United States, pp. 34–35.

1178 Rainer C. Schwinges: Humboldt International. Der Export des deutschen Universitätsmodells. Eine Einführung, in: Rainer C. Schwinges (ed.): Humboldt International. Der Export des Deutschen Universitätsmodells im 19. und 20. Jahrhundert, Basel 2001, pp. 1–13.

1179 Fritz K. Ringer: The German Academic Community, in: Alexandra Oleson; John Voss (eds.): The Organization of Knowledge in Modern America, 1860–1920, Baltimore, MD 1979, pp. 409–429. Quote on page 409.

1180 Bruch: Langsamer Abschied von Humboldt? pp. 34–41.

> The frequency of admiring articles on German universities rose in the [eighteen] seventies and peaked in the eighties. As in antebellum times, American scientists viewed the German university through their own prism, ignoring its principles of *Bildung* and *Idealismus* in favor for its specialization and meticulously detailed research, which had in fact become more pronounced in Germany since the fifties.[1181]

According to Bruch it seems fair to say that the "German University" succeeds in the US despite and not because of the Humboldtian ideals. On the other hand, Fallon judges that "no German university ever succeeded in adhering so faithfully to Humboldt's ideals as the typical American research university."[1182]

Scholars and science professionals were very much revered in Germany. Here they were an integral part of the country's rapid modernization and industrialization during the nineteenth century, and education was the most prominent tool of upward social mobility. The "academic" became an integral part of the educated upper middle class. While the number of enrollments at German universities rose sharply during the last decades of the century, the number of full professors ("Ordinarien") grew much more slowly. To keep up with the rising number of students, instructors ("Privatdozenten") were hired. The "Privatdozenten" were research assistants working for the professors while being paid much less than a full employee of the university.[1183] Marsh incorporated this system in his laboratory at the Peabody Museum when he hired his German assistants as "Privatdozenten" (see chapter 6. 1.).

As for other scientific institutions, aside from colleges and universities, a similar trend can be found. Hunter Dupree provides an overview of the troubled history of the National Academy of Sciences, from its founding in 1863 as a rather exclusive organization (not all scientific branches were welcome) to its transformation into a more open and inclusive society during the 1870s.[1184]

As to the importance of international experience for German scientists, Baumgarten notes that only a small percentage of professionals working in German higher education had spent some time teaching abroad, Schlosser being one of the few that did. Though more natural scientists spend time abroad than those working in humanities, perhaps due to a lower language barrier. Foreigners were very seldom appointed to teach at German institutions of higher education. During the nineteenth century only

1181 Bruce: The Launching of Modern American Science 1846–1876, p. 336.

1182 Fallon: German Influences on American Education, p. 85.

1183 Ringer: The German Academic Community.

1184 A. Hunter Dupree: The National Academy of Sciences and the American Definition of Science, in: Alexandra Oleson; John Voss (eds.): The Organization of Knowledge in Modern America, 1860–1920, Baltimore, MD 1979, pp. 342–363.

2.3–6.3% of staff positions in higher education were filled by persons not born in Germany, with almost no US Americans among them.[1185]

The German influence on postgraduate education is undeniable. As to what degree, and if the education of undergraduates at the colleges was affected as well is a whole different matter and is hotly debated by historians of science.[1186]

However, the fundamental proposition of Veysey's judgement about the German influence on the US-American university rings true:

> Younger American scientists – born during and after the 1840's – obtained inspiration from a newly specific source: the German university. During the final quarter of the nineteenth century, few academic Americans who embraced the ideal of scientific research failed to acknowledge an intellectual debt to an explicitly German style of educational experience.[1187]

Still, Mitchell Ash warns that it would be a mistake to believe that natural sciences were conducted increasingly within an international network as time progressed. At the end of the nineteenth century nationalism was on the rise and the international Republic of Letters was about to be discontinued in central Europe. German, which had served as a kind of lingua franca of science during the course of the century was now replaced with the respective national languages. Science was very much affected by the mounting nationalistic sentiments.[1188] "With the rise of nationalism during the nineteenth century, national identity became an increasingly relevant factor in scholarly identity. [...] students were prone to be corrupted by national interests."[1189]

That the education reforms in the US were of consequence is evidenced by the fact that in the 1880s and 1890s less young US-American scientists felt the need to spend some time at a German university. During the first half of the century the US had next to no graduate schools. At the end of the century. the people who formally had to journey through Europe to receive a worthwhile education in science could now do so at home. As a consequence, the prestige of the German educational system declined rapidly in the US.[1190] Another indicator for the effectiveness of the US education reforms is that the enrollment rates in US-American institutions for higher education

---

1185 Baumgarten: Professoren und Universitäten im 19. Jahrhundert, pp. 240–243.

1186 Turner: Humboldt in North America?, pp. 298–302.

1187 Veysey: The Emergence of the American University, p. 126.

1188 Mitchell G. Ash; Jan Surman: The Nationalization of Scientific Knowledge in Nineteenth Century Central Europe: An Introduction, in: Mitchell G. Ash; Jan Surman (eds.): The Nationalization of Scientific Knowledge in the Habsburg Empire, 1848–1918, Basingstoke 2012, pp. 1–29.

1189 Adam; Lerg: Introductory Remarks, p. 301.

1190 Herbst: The German Historical School in American Scholarship, pp. 1–22.

rose continuously during the nineteenth century (beginning in c. 1870 more and more women received a higher education as well), and figuratively exploded in the 1920s. Keep in mind that this is not solely the achievement of the educational reform, but also a consequence of the changing basic conditions of the population. More and more jobs required a higher education and general prosperity was on the rise as well.[1191]

The reforms in American higher education were fully implemented and cemented in the very early twentieth century:

> In state universities across the country, new presidents, many of them Ph.D.'s, most of them progressives and products of the pro-science educational movement of the late nineteenth century, were taking office and making the promotion of research the order of these days of reform.[1192]

In contrast, Lawrence Veysey writes that toward the end of the nineteenth century increasingly more young US Americans were seeking employment in practical jobs and utilitarianism was king, which meant a loss in reputation for universities and professors. He writes that lawyers and the clergy had a better reputation and were better paid than professors, and that enrollment in a university was seen as a questionable decision, almost as a form of escapism. Furthermore, universities were viewed as atheistic, godless places.[1193]

8.3 Natural Sciences in the US-American System of Higher Education

While the previous subchapter described how the US-American system of higher education was slowly reformed and some German innovations were implemented by US-American returnees, this subchapter goes into more detail concerning the development of the natural sciences within the US-American system of higher education.

US-American scientific institutions had previously emulated the British model, especially the Royal Society, as is reflected in Cotton Mather's attempts to found a scientific society in Boston in the 1680s, or by Benjamin Franklin's desires to found such an institution in Philadelphia during the 1740s. Later, while the United Colonies

---

1191 Colin B. Burke: The Expansion of American Higher Education, in: Konrad H. Jarausch (ed.): The Transformation of Higher Learning 1860–1930. Expansion, Diversification, Social Opening and Professionalization in England, Germany, Russia and the United States, Chicago 1983, pp. 108–130.

1192 Kevles: The Physicists, p. 70.

1193 Veysey: The Emergence of the American University, pp. 1–18.

sought independence from Britain, they still emulated British associations for the advancement of science.[1194]

> The BAAS, especially, provided a model capable of filling observable needs for American science: a meeting place on the national level, a component of specialization, a source for research funds, and an institution capable of representing science to the public and to the national government.[1195]

During the eighteenth century, Philadelphia had been the uncontested center of culture and scholarship in the US. Here the Quaker-elite held education (and material wealth) in high honors. Medical science was of great prestige, in part due to its utilitarian applicability, which was important to the Quakers. The American Philosophical Society (APS) was founded in 1769 when various medical societies joined forces. Soon Philadelphia would become a center of natural sciences as well.[1196] The emergence of US-American scientific institutions drew on earlier trends, which had developed in the field of geology. Kohlstedt summarizes that the AAAS emerged out of the already nationally organized circle of geologists and naturalists: "The American Association, voted into existence as an expansion of the more specialized Association of American Geologists and Naturalists in 1847, provided a forum for discussion of the problems facing scientists."[1197]

The APS was the first American scientific society of note and was founded in Philadelphia, then the most populous city in the United States and a center of commerce and learning. The APS was very utilitarian in its conception. The "Transactions" of the APS was to become the first American scientific journal. In 1812 the Academy of Natural Sciences of Philadelphia (ANS) was founded and published a journal of its own, beginning in 1817. The APS recognized the leadership of the ANS in the field of natural history and transferred its fossil collection to the ANS in the 1850s. Silliman's "American Journal of Science and Arts," first published in 1818, and the "Proceedings" of the American Association for the Advancement of Science further enriched the US-American scientific landscape. In 1824 the ANS and the Franklin Institute (founded in Philadelphia in 1824) began to organize public lectures on science and its practical implications. In the first half of the century the "Journal for Science and Arts" surpassed the "Memoirs" of the AAAS and became the most important US-American scientific

1194 Kohlstedt: The Formation of the American Scientific Community, p. 27.

1195 Kohlstedt: The Formation of the American Scientific Community, p. 42.

1196 Ostrander: Republic of Letters, pp. 47–64.

1197 Kohlstedt: The Formation of the American Scientific Community, p. x. For a detailed account of how the establishment of geological societies contributed to the establishment of scientific societies in general see: Kohlstedt: The Formation of the American Scientific Community, pp. 59–77.

journal. The "American Mineralogical Journal," published between 1810 and 1814, was the first national outlet dedicated to the geosciences. John C. Greene summarizes the ante-bellum situation of US-American scientific societies and journals as follows:

> It appears, that the record of societies for promoting and knowledge in the period before 1860 was an honorable, though not brilliant, one. Deprived for the most part of government support, they could not hope to emulate the model of the academies on the continent of Europe. Scattered in urban centers, none of which could claim national preeminence, dependent for support on members for the most part of limited means, hampered by public attitudes that placed no high value on intellectual achievements per se, the were equally incapable of attaining the eminence and influence of the leading British societies.[1198]

As to the conduct of natural sciences in US higher education, it seems they were rather neglected during the first half of the nineteenth century. Most science lectures were held in the winter term when many students did not attend college but worked as schoolteachers.

The Morrill Act of 1862 made government funds for the establishment of new institutions for higher education available. New universities were founded mainly in the West and Midwest with the goal of teaching practical skills like engineering and agriculture, but not all universities stuck to limited curricula.[1199]

The standing of the natural sciences was further boosted when US-American expansionism pushed to the forefront of the nation's collective mind following the Civil War. With a renewed interest in conquering and exploiting the western parts of the continent (see chapter 5), science was to be employed for this endeavor now more than ever: "The work of the pioneers, both in advancing science and in popularizing it, combined with the richness of the American continent in making science an instrument for exploiting the great natural wealth of inland America."

Silliman Sr. was one of the most prevalent advocates for the teaching of natural sciences to this end. Together with his son and his son-in-law, James D. Dana, he began to transform Yale into a hot spot for science and experimentation, while also pushing for the acquisition of comprehensive specimen collections.[1200] To this end

---

1198 John C. Greene: Science, Learning and Utility. Patterns of Organization in the Early American Republic, in: Sanborn C. Brown; Alexandra Oleson (eds.): The Pursuit of Knowledge on the Early American Republic. American Scientific and Learned Societies from Colonial Times to the civil War, Baltimore, MD 1976, pp. 1–20. Quote on page 19.

1199 Thelin: A History of American Higher Education, pp. 75–83.

1200 Rudolph: The American College and University, pp. 222–228. Quote on page 223.

Geiger judges that "Yale appeared increasingly anachronistic as the academic revolution progressed. Reform sentiment among the faculty became difficult to ignore."[1201]

In 1869 Ezra Cornell (1807–1874) followed suit and established a true university. Students were allowed to choose their own fields of study and soon engineering and agriculture beat the classical courses, as far as enrollments were concerned. The Cornell concept proved to be a huge success as students enrolled in unprecedented and unexpected numbers. At the same time Yale College, too, evolved into a university, having established a Ph.D. program in 1861.[1202]

Rudolph attributes this development within the US-American educational landscape to an amalgamation of English traditions and German innovation, forming a uniquely US-American university:

> The elective principle was the instrument by which departments of knowledge were built, by which areas of scholarly interest were enlarged, and therefore it was the instrument that enabled colleges to become universities. In the end, it was the instrument, secular and democratic, that permitted the American university to enter into a vital partnership with the society of which it was a part. It transformed the English college in America by grafting upon it German ideals and in the process created the American university.[1203]

Herbst might agree with this conclusion and elaborates on the birth of the distinctively US-American university:

> In any society, culture may be said to consist of the interaction of its institutions and ideas. Thus the Americans who went to German universities to acquire the tools of scholarship brought home not only tools but ideas as well. When the ideas proved difficult to assimilate to American conditions, the scholars sought to modify or discard them, only to realize that their scholarly equipment, torn from its ideological setting, would no longer serve until a new context of ideas could be developed.[1204]

In an interesting allegory Veysey likens the evolution of the US-American university to immigration:

1201 Geiger: The History of American Higher Education, p. 334.

1202 Rudolph: The American College and University, pp. 266–269.

1203 Rudolph: The American College and University, p. 305.

1204 Herbst: The German Historical School in American Scholarship, p. 232. For further detail on the gradual change of the US-American system see: Thelin: A History of American Higher Education, pp. 87–90, 103–107.

> The Idea of the University, initially an alien concept, underwent a process not unlike that which affected the actual immigrants who arrived on American shores in the eighteenth century: one of assimilation to the New World environment, accompanied by profound internal tension and a mingled sense of gain and loss.[1205]

In 1846 and 47 Yale began establishing a scientific school. Two new professorships were established, the first for agricultural chemistry and animal and vegetable physiology, the other for practical chemistry. The latter was held by Silliman Jr., and in 1850 Dana succeeded Silliman Sr. as professor for mineralogy and geology. These professorships became the basis for the Yale Scientific School, which formed in 1854–55. In 1858 Joseph Earl Sheffield (1793–1882) donated $ 10,000 and one building to the conduct of science at Yale, and Marsh was one of the first graduate students to enroll at the newfound Sheffield scientific school in 1860 (see chapter 3. 1.).[1206]

At the end of the century the old-fashioned image of an elderly, all-knowing professor-type had become cliché. When more and more professors also bore a Ph.D., clear hierarchies became apparent within the group. These hierarchies became a convenient tool for the management of science. Furthermore, the specialization of science accelerated, and more disciplines were taught at US-American institutions of higher education. Professors, too, became specialists. At the same time the competitive pressure increased in line with the motto "Publish or Perish." Original research became more important than teaching and lectures. This meant that the academic freedom of professors increased immensely, and scientific publications flourished.[1207] Add to this that the requirements to start a career in science increased immensely. Beginning in the 1850s, scientists who had studied in Germany greatly increased their chances of a successful career in science, if not to say that, at least for a time, having studied in Germany became a requirement for attaining a professorship in science at a prestigious US university.[1208] In the context of this thesis, Cope and Marsh (but, for example, Osborn as well) represent the first generation of US professors to whom this is applicable. The older generation, take Leidy for example, had much more limited experience with foreign education.

Paleontologist Bob Bakker sums up the evolution of US-American paleontology towards the end of the nineteenth century as follows:

1205 Veysey: The Emergence of the American University, p. 439.

1206 Schuchert; LeVene: O. C. Marsh, pp. 43–44.

1207 Rudolph: The American College and University, pp. 394–416. Also see: Herbst: The German Historical School in American Scholarship, pp. 38–51.

1208 Diehl: Americans and German Scholarship, pp. 140–143.

> American museums were erecting dinosaur skeletons as fast as American shipyards erected new steel battleships to protect the fledgling star-spangled empire. Europe viewed both developments with mixed admiration and alarm. For a century, Old World scientists had been digging and studying dinosaurs, but no one had found Jurassic giants nearly as complete as the ones that tumbled out of almost two dozen American quarries, starting in 1878. [...] America, which had had to import its scientific apparatus, and had sent its scholars to England and Germany for doctorates only decades earlier, now began to export scientific wealth.[1209]

As to the German influence on US-American paleontology, and the willingness to accept the new forerunner position US-American paleontologists earned during the nineteenth century, Bakker writes:

> Enter the Germans. No culture had a more illustrious nineteenth-century tradition of paleontological scholarship. A German, Hermann von Meyer, had first recognized the unity of all the great Mesozoic creatures we now call dDinosaurs. And German anatomists were acknowledged worldwide as the best in laboratory dissections and microscopy. In the early 1900s, Germany was a new and ambitious nation, and it was perhaps to be expected that a certain chauvinism should manifest itself in many different areas, including the scientific. It was not surprising, that German paleontologists didn't immediately accept the conclusions about the posture of dinosaurs advocated by Americans.[1210]

A letter written by Marsh in February of 1867 to Joseph Henry of the Smithsonian Institution illustrates how much young US-American institutions must have longed for recognition by their European peers, especially in Germany. Henry had written to Marsh in January 1867, inquiring about the standing of the Smithsonian among German scholars since Marsh had recently returned from Europe to America:

> Although you have doubtless had many opportunities of learning in what high estimation the Smithsonian Institution is held by men of learning in all parts of the world, I have thought it might interest you to know somewhat more fully how it is regarded at the present time in Europe, and especially in Germany, where an opinion on the subject is most likely to be correct. During my late residence of three years in Europe, most of which was spent at the German

1209 Robert Bakker: The Dinosaur Heresies. A Revolutionary View of Dinosaurs, Harlow 1987, pp. 201, 203.

1210 Bakker: The Dinosaur Heresies, p. 204.

> Universities, it was to me very gratifying to hear the Smithsonian so frequently spoken of with approval. I found indeed that the [Smithsonian] was very generally regarded as the fountain-head of science in America. This may, perhaps, have been owing in some measure to the fact that the Smithsonian is the great medium of intercourse between the scientific men of Europe and this country, but i[n] most instances the general object and aims of the institution were fully understood.[1211]

Concerning the role of paleontology in the changing US university system, Ronald Rainger supposes that

> Vertebrate paleontology also failed to gain a strong foothold in the changing system of American higher education. Many colleges and universities possessed collections that included extinct animals and offered courses in geology or biology that included paleontology. But few schools had the resources or the interest to finance field parties to travel to important deposits or to maintain large collections. With the creation of new departments and disciplines, the question of where or whether to include vertebrate paleontology was a problem in some academic institutions.[1212]

Paleontology had few utilitarian aspects, but utilitarianism was emphasized in the reforms of US-American education. Furthermore, experimental biology now provided the best proof to document the changes in organisms, before that it was comparative anatomy which was employed by paleontologists to provide evidence for morphology. Experimenting under laboratory conditions with easily manipulated organisms was easier and much cheaper than field work and extensive collections. Maybe the future of paleontology was public, not academic, education. The best examples of that are Carnegie's *diplodocus* and Osborn's museum in New York; both could provide, or find, the necessary financial means to support the science.[1213]

> Vertebrate paleontology was not able to compete effectively with the new emphasis in biology. Many colleges maintained their collections, and around the turn of the century a number of new college museums were established. Yet it is questionable to what extent the collections in those museums were used by students. Courses on biology and geology included material on fossil ver-

---

1211 Othniel Charles Marsh, New Haven, CT to Joseph Henry, Washington, DC, 20 February 1867, MS 343, Series I. Correspondence, Box 16, Folder 646.

1212 Rainger: An Agenda for Antiquity, pp. 18–19.

1213 Rainger: An Agenda for Antiquity, pp. 19–23.

> tebrates, but few students chose or were encouraged to choose that subject as a field of specialization [...] vertebrate paleontology became a marginal subject at institutions of American higher education [...]. Large public museums equipped with financial support and dedicated to public education became the most important centers for vertebrate paleontology.[1214]

> In late nineteenth-century America, the study of paleontology flourished. After the Civil War, a multitude of geological and geographical surveys of the western states and territories provided new occupational opportunities that enabled scientists to uncover a wealth of fossil material. Work by vertebrate paleontologists Joseph Leidy, O. C. Marsh, and Edward Drinker Cope yielded a spate of discoveries that brought them international fame and placed America at the forefront of that field of science. [...] Paleontology, it has been claimed, was among the first disciplines in which American scientists made empirical and theoretical contributions that were not merely derivative of European science. Yet despite the efflorescence of research in that field, paleontology remained largely peripheral to the developments occurring in American biology. Institutionally, vertebrate and invertebrate paleontology generally had no place in the new centers, established for biology.[1215]

LeVene and Schuchert describe in short Marsh's innovations in the field of paleontology, which added greatly to the conduct of this science in the US and were later emulated all around the globe. First of all, Marsh was a great collector and acquired fossils in such volumes that he could often reconstruct a very accurate picture of the extinct animal, while before most often only very fragmentary material could be observed.[1216] Furthermore, he gave very detailed instruction to his working teams in the field on how to excavate and preserve fossils for transportation back to the east coast. Until then bones were dug out very carelessly, it was common practice to "drive the pick under the bone, pull it up, rake all the pieces together and throw them into a sack, in the hope the preparators at the laboratory would be able to fit the puzzle together."[1217] In accordance with Marsh's instructions fossils were now to be wrapped in plaster-soaked strips of cloth, so that they would not further deteriorate on their journey to the east. Then they would be unwrapped in the laboratory while shellac or glue would be poured into the fine cracks and fractures in the fossilized bones, then reset into plaster until hardened. Marsh is credited as the inventor of this technique,

1214 Rainger: An Agenda for Antiquity, pp. 20–21.

1215 Rainger: Vertebrate Paleontology as Biology, p. 219.

1216 Schuchert; LeVene: Marsh, p. 169.

1217 Schuchert; LeVene: Marsh, p. 171.

but LeVene and Schuchert speculate that this medical technique was first used by Williston, who had studied medicine before he worked as a collector for Marsh.[1218] Davidson and Everhart write, however, that plaster had been used to preserve fossils long before Marsh did it, and that LeVene and Schuchert might have embellished Marsh's innovation in the field of fossil preservation greatly. Still, they conclude:

> The tools, techniques and packaging of specimens improved through the 1870s, especially with the use of plaster and burlap jackets to enclose and protect fragile specimens. While Cope and Marsh certainly provided the overarching instructions on how to collect, the men in the field were actually the ones who developed the best practices and eventually perfected the nearly universal plaster and burlap methods that are still in use today.[1219]

In his short description of the US-American learned societies as information systems, Hunter Dupree points out that minerals and biological specimens served as nonverbal transmitters of information, greatly advancing the conduct of natural history in the US. The same can be said about the fossils that were transported en masse from the west to the east during the second half of the nineteenth century. This way, laymen without scientific education could participate, further contributing to the sciences associated with natural history.[1220]

8.4 Paleontology and Public Education

"Science has always had a public face and still has it now. If only to defend their own turf, scientists have always cultivated an audience beyond the immediate one made up of their own peers and fellow researchers. [...] We now think of audiences as having an active rather than just a passive relationship to the production of scientific knowledge. Not only does the way in which the scientist chose to present his or her different audiences and the context within which that work is presented have important consequences for the way science is understood, but audiences themselves

---

1218 Schuchert; LeVene: Marsh, p. 174.

1219 Jane Pierce Davidson; Michael J. Everhart: Scattered and Shattered. A Brief History of the Early Methods of Digging, Preserving and Transporting Kansas Fossils, in: Transactions of the Kansas Academy of Science, vol. 120, no. 3–4 (Sep. 2017), pp. 247–258, https://doi.org/10.1660/062.120.0416. Quote on page 257.

1220 A. Hunter Dupree: The National Pattern of American Learned Societies, 1769–1863, in: Sanborn C. Brown; Alexandra Oleson (eds.): The Pursuit of Knowledge on the Early American Republic. American Scientific and Learned Societies from Colonial Times to the civil War, Baltimore, MD 1976, pp. 21–32.

actively interpret and redefine the knowledge as they go along. From this perspective, studying popular science does indeed engage with the actual content of science and the process of knowledge making."[1221]

Bowler and Morus demonstrate the importance of public engagement with science. In the case of paleontology, the most obvious example of public engagement is the exhibition of fossils, first and foremost dinosaur skeletons.

> The artist Charles Wilson Peale's Philadelphia Museum at the beginning of the nineteenth century catered to the interests of an American public already fascinated by the curious and the fantastic. Peale's museum featured natural historical curiosities such as the bones of a mastodon unearthed in New York State, his own historical paintings, antiquarian curiosities, and new mechanical inventions and contrivances. Even the showman P. T. Barnum's extravagant exhibitions of the exotic played on his public's fascination with science.[1222]

International influences were felt in the field of public education. Of the greatest interest for this thesis are the museums, some of which, like the Peabody Museum at Yale, were part of the system of higher education.

There was a growing interest in popular scientific lectures, drawing huge crowds beginning in the 1830s:

> Diffusion of science would, it seemed in the 1830s, facilitate American material and political aspirations. [...] The new enthusiasm for the scientific lectures probably aided men of science indirectly in gaining founding for 'useful' scientific projects such as the state geological surveys and the U.S. Coast Survey and Depot of Charts and Instruments.[1223]

Scientific knowledge was also circulated in popular journals and magazines, greatly widening the audience of people being exposed to scientific theories during the nineteenth century; this also led to the transformation of some of those scientific revelations into science fiction.[1224]

Kohlstedt describes some negative consequences of the newfound public appreciation of science in the United States:

1221 Bowler; Morus: Making Modern Science, pp. 367–368.

1222 Bowler; Morus: Making Modern Science, p. 374.

1223 Kohlstedt: The Formation of the American Scientific Community, pp. 9–10.

1224 Bowler; Morus: Making Modern Science, pp. 379–384.

> Yet popular involvement had negative results as well. Lectures were time-consuming and brought few direct results for science itself. The alternative of leaving the responsibility to less qualified men raised a fearsome specter – quackery. The need to provide scientific spokesmen in a rampant democracy became an important factor in the shaping of national scientific organization.[1225]

This constitutes a possible explanation for why paleontology and the publics most beloved subject thereof – the dinosaurs – were being recognized as an up-and-coming science. It seems that a scientific discipline had to satisfy the public's demand for entertainment, especially in a democratic society.

> Better science depended on financial support for individual research projects. Whether in a democracy or an autocracy, demonstration of the value of science was essential to securing support; as long as men of science lacked individual wealth or endowed institutions, the necessity of persuading the power of the purse remained. As the proposals for geological surveys indicate, scientists responded to and thus encouraged utilitarian expectations.[1226]

In an article on the practice of displaying prehistory at the AMNH, Rieppel writes that "American natural history museums developed into hybrid institutions that sought very different, at times conflicting, goals: scientific research, public education, and popular entertainment."[1227]

Sally Kohlstedt writes that the nineteenth century was somewhat of a Golden Age for museums and public exhibitions. She states that German immigrants (and other Europeans, like the all-influential Louis Agassiz) had a formative influence on the museum landscape in the US, and that many US-Americans were inspired by visiting European exhibitions and museums and implemented said inspiration upon their re-

1225 Kohlstedt: The Formation of the American Scientific Community, p. 16. Davidson and Everhart analyze a fascinating example of the amalgamation of scientific and biographical fact and fiction. They describe how Charles Hazelius Sternberg (1850–1948), one of the most popular and productive fossil hunters who collected fossils for Cope and Marsh, published fictionalized accounts of his experiences as a fossil collector. Because these accounts contained many verifiable facts alongside Sternberg's embellishments, they proved to be a good way to convey scientific fact and real experiences to a broad audience. See: Jane Pierce Davidson; Michael J. Everhart: Fictionalized Facts; "The Young Fossil Hunters" by Charles H. Sternberg, in: Transactions of the Kansas Academy of Science, vol. 117, no. 1–2 (Apr. 2014), pp. 41–48, DOI:10.1660/062.117.0106.

1226 Kohlstedt: The Formation of the American Scientific Community, p. 17.

1227 Lukas Rieppel: Bringing Dinosaurs Back to Life. Exhibiting Prehistory at the American Museum of Natural History, in: Isis, vol. 103, no. 3 (Sep. 2012), pp. 460–490, https://doi.org/10.1086/667969. Quote on page 460.

turn to the US, just like Marsh did with the Yale Museum (see chapter 3. 4.).[1228] During the nineteenth century, museums and collections became part of formal education. The previously rather limited collections of biological and paleontological specimens grew rapidly after the Civil War. They "became a fundamental tool for teaching natural history in undergraduate curricula in the 1860s. They predate the 'revolution' in higher education in the 1870s."[1229] Professors were expected to acquire relevant collections out of their own initiative and often out of their own pocket, thus Marsh's way of working is exemplary for the practice of establishing scientific collections in the US in the nineteenth century. When Louis Agassiz immigrated in 1846, he was one of the first implementers of the new method. The acquisition and exchange of preferably cohesive collections from all around the world became an important prerequisite for science education. The collections were (when open to the public) very visible sources of prestige. "Adding to the complexity of maintaining individual collections was the rapid growth of urban museums for art and for science in what has been billed a golden age of museums development." In 1873 the American Museum of Natural History had the largest museum building, rivaling the collections at Harvard and the Peabody Museum.[1230]

In contrast, Rainger postulates that Marsh and Leidy did collect a plethora of data, but that their scientific endeavors rarely contributed to the theoretical framework of biology or the theory of evolution. Cope, who also did just that with his Neo-Lamarckian theory, had no permanent ties to a scientific institution and thus did not contribute to the development of US-American higher education. Osborn's Department for Vertebrate Paleontology constitutes the exception to the rule and established paleontology as a zoological subdivision at the American Museum for Natural History. To be more precise: he employed the fossil record in morphology. Osborn established Princeton as a small but active research center and was part of the effort to transform Columbia College in New York into a full university. He then initiated the cooperation with the American Museum before moving on to that institution.[1231]

Schuchert and LeVene call the practice of displaying original fossils to the public a "sacrilege":

1228 Sally Gregory Kohlstedt: German Ideals and Practice in American Natural History Museums, in: Henry Geitz et al. (eds.): German Influences on Education in the United States to 1917, Washington DC 1995, pp. 103–114.

1229 Sally Gregory Kohlstedt: Museums on Campus: A Tradition of Inquiry and Teaching, in: Roland Rainger et al. (eds.): The American Development of Biology, paperback ed., New Brunswick, NJ 1991, pp. 15–47. Quote on pages 28–29.

1230 Kohlstedt: Museums on Campus, pp. 28–29.

1231 Rainger: Vertebrate Paleontology as Biology.

> In the middle 'eighties, when Marsh had completed his monograph on the fossil mammals known as Dinocerata, he felt that the species *Dinoceras mirabile* was well enough known to justify the exhibition of the mounted skeleton of it in the Museum. Not, however, one made up of the original bones! At that time, no paleontologist would have committed the sacrilege of showing the bones themselves mounted in a lifelike attitude, or even of modeling the missing parts in plaster. No! The bones must be kept apart forever in drawers, or laid away on padded shelves, so that the paleontologist alone might handle them and inspect every joint in the most minute detail.[1232]

Still, the very first dinosaur skeleton reconstruction to be exhibited publicly was that of Leidy's *hadrosaurus* in Philadelphia (sec chapter 2. 6.).

At the dawn of the twentieth century two of the leading paleontological museums in the US were founded by millionaires: Carnegie built his own museum in Pittsburgh and J. P. Morgan was a patron of the AMNH in New York. Carnegie's dinosaur-enthusiasm led to the discovery of a new specimen of *diplodocus* (a sauropod first described by Marsh in 1878). At the time, the *diplodocus carnegii* was the biggest known dinosaur, and still is one of the longest known dinosaurs. *Diplodocus* became an international sensation and casts of the skeleton were exported to many museums around the world. The discovery was discussed in many newspapers around the world. Carnegie's dinosaurs became representatives of the USA.[1233]

> Natural history played an important role in bridging the gap between the academic and non-academic worlds as well as in forging middle-class culture. On a bigger scale, within the extremely competitive early twentieth-century international panorama, natural history became a powerful vehicle for exporting both cultural and economic models.[1234]

8.5 Conclusion

Roger Geiger, who writes a very comprehensive history of US-American higher education between the seventeenth century and the end of World War Two, concludes the following for the development of US higher education during the nineteenth century:

1232 Schuchert; LeVene: O. C. Marsh, p. 296.

1233 Parsons: Drawing out Leviathan, pp. 2–4.

1234 Tamborini: "If the Americans Can Do It, So Can We," p. 252.

> The first decades of the nineteenth century marked a low point in all the vital signs of American higher education. It had become ineffective in promoting culture, career, or knowledge. The assumption of social superiority associated with collegiate education was resented and contested by democratic elements, especially in the expanding western settlements, but even in Federalist New England [...] The 1820s witnessed fruitless attempts to introduce useful knowledge. [...] Thomas Jefferson realized his vision of a republican university in Virginia, but in practice it served largely to provide acculturation to planter's sons.[1235]

Geiger continues and writes that the pitiful conditions at US colleges and universities only slowly changed during the 1870s:

> In the mid-nineteenth century, rapid growth and tentative innovation were premonitions of [...] major challenges that after 1870 would transform American higher education: [...] The growth of scientific knowledge forced consideration of how it could be incorporated into the colleges. [...] The issue of teaching useful knowledge presented itself in ever-more pressing terms. [...] [C]ollegiate culture transcended alleged effects of the ossified classical course and passed into the hands of the students themselves.[1236]

He concludes:

> The academic revolution posited the systematic pursuit of new knowledge, embodied in research and graduate education, as a central mission of universities. The organization of academic disciplines provided a new knowledge base throughout higher education, rendering the fixed classical course obsolete.[1237]

Joseph Ben-David postulates that science is done regionally, and certain regions or nations become centers of scientific invention. According to Ben-David a certain level of national prosperity was required for the establishment of modern science: in the seventeenth century there was no public funding for the sciences. Aspiring scholars had to be rather wealthy to be able to afford the education, equipment, and leisure time to pursue scientific study. He furthermore postulates that in the course of human history the centers of scientific knowledge production shifted constantly due to the shifts in national prosperity. In the early seventeenth century Italy took the

1235 Geiger: The History of American Higher Education, p. 542.

1236 Geiger: The History of American Higher Education, p. 543.

1237 Geiger: The History of American Higher Education, p. 545.

lead in knowledge production. At the end of the century English scholars were in the vanguard. In the early nineteenth century France is said to have been the center of modern science, then after c. 1840 Germany. The United States has dominated the sciences since the 1920s.[1238]

At the eve of the twentieth century transnational scholarly exchange had become more formalized, students were exchanged via the academic institutions while during the nineteenth century this exchanged mostly happened due to the initiative, recourses, and endeavors of private individuals, like Marsh, Zittel, and Baur.[1239]

Increasingly, dinosaur skeletons were employed in the competition between nation-states, as showcased in 1906, when colossal fossilized bones were discovered in the German colony in East Africa (in a part of it, which is nowadays Tanzania). Immediately, the paleontologist Eberhard Fraas was informed of the find and after initial examinations of the findings a massive public funding campaign was undertaken. The excavations were funded by small and individual private donations. The funding of the expedition was advertised to be of great national importance. Nationalism, not appreciation of science, was the main motivating factor in donating to the project. Prominent members of German high society were brought on board for the funding. Finally, the Prussian state noticed the immense popularity of the excavations and partially funded the season of 1912, but most of the expeditions were sponsored by individual donations. It is noteworthy that the state's motivation to spend the money was not the advancement of science but to promote German nationalism. Between 1909 and 1913 more than 225 tons of fossils were excavated in the colony and sent to Berlin. Finally, in 1937 a huge dinosaur was mounted, reconstructed from the colonial fossils. In line with the resurgence of German nationalism in the Third Reich, this was celebrated as a huge accomplishment of German science and even a victory over US-American paleontologists, for now the world's largest dinosaur was displayed in a German museum. Indeed, German paleontologists thought they were overtaking their US-American colleagues, which in turn meant that US-American paleontology had taken the international lead at the beginning of the twentieth century.[1240]

Hopes were raised "to enhance the prestige of German science and the German Empire. Not only would German paleontology benefit from this excavation, but German science as a whole would undoubtedly receive a boost to its international reputation."[1241]

Tamborini summarizes the situation as follows:

1238 Ben-David: The Scientist's Role in Society, pp. 14–16.

1239 Adam; Lerg: Introductory Remarks, pp. 300.

1240 Tamborini: "If the Americans Can Do It, So Can We".

1241 Tamborini: "If the Americans Can Do It, So Can We", p. 238.

> The rivalry between Germany and the United States reached a climax during the first decades of the twentieth century as German industry and research institutes sought to attain both public and private funding in order to catch up with the United States. At the same time, though, the United States was seen as a model to be emulated.[1242]

Even Charles Schuchert attested in 1915:

> You can understand that our interest in what you have found is very deep, because we not only have a great many very wonderful specimens of dinosaurs here in America, but we were proud in the belief that for all time we could say that America had reared the largest of all animals. However, this honor may now go to Germany, and if it does, all hail to you![1243]

The discussion of the German colonial dinosaur in the context of nationalism and international competition exemplifies how far US-American paleontology had come. Now, at the dawn of a new century US Americans were no longer looking to Germany and its universities for instructions. Innovative paleontological research was conducted in the US, where the largest and most fearsome dinosaurs were discovered, much to the joy of many a patriot. And now the ambitious and relatively young German nation state wanted to prove to the world it was on top of the scientific game, and that if not Germany herself then at least her colonies would produce the remains of fabulous dinosaurs.[1244]

During the long nineteenth century first Prussia and then the united German national state had reformed its system of higher education in places like Berlin and Heidelberg. The innovations of the "German University" were – and are to this day – often summarized under the term "Humboldtian Ideals," even though the ideals of Humboldt and the other reformers were never truly and fully implemented at any university. Nevertheless, the focus on original research, specialized laboratories, science, and the nominal independence of scholars and institutions of higher education which were attributed to the "German University" were soon admired across the globe and emulated in Great Britain and the United States.

1242 Tamborini: "If the Americans Can Do It, So Can We", p. 239.

1243 Quoted after Tamborini: "If the Americans Can Do It, So Can We", p. 248.

1244 Crawford even writes that "[u]nification and the proclamation of the Kaiserreich in 1871 had only created the political shell that was the state; this now had to be filled with the material well-being and cultural coherence that would make for a nation." Part of this cultural cohesion would spring for a national German science and system of higher education, being part of a united national culture. See: Crawford: Nationalism and Internationalism in Science, p. 33.

In matters of higher education, the US was understood to be severely lagging behind Europe, and this injured nationalistic pride. Education was perceived as a "civilizing" factor, bringing the light of civilization to a dark and wild continent. It became a point of patriotic pride to catch up to Europe in terms of education. To this end many a US-American scholar sought to implement elements of the "German University," which seemed to be the most progressive and productive, into the US system of higher education. Many men who rose to prominence in academia after the Civil War had spent some time at European universities and now sought to transplant what they had experienced there to America. This effected the conduct of natural sciences the most. The discipline of paleontology is one of the most striking examples thereof. Marsh, Cope, and Osborn knew the German system of higher education firsthand and with this experience and generous funding by both government and private fortunes, they pushed US paleontology to the top of the sciences. While Cope and Marsh contributed greatly to the science of paleontology, and especially dinosaur paleontology, and in Marsh's case also to Darwinian evolution, Osborn revolutionized the public-education aspect and how skeletons were displayed to the public at museums.

The circle was complete, US-American know-how, scientific insights, and textbooks were now exported to Europe. And finally, at the closing of the nineteenth and the beginning of the twentieth century the events surrounding the excavation, exhibition, and international rise to fame of a skeleton of one truly US-American dinosaur should epitomize this development in international paleontology.

9

Conclusion: Transatlantic Dinosaurs

"From the 1900s onwards, dinosaurs were used to make paleontology visible internationally, to enhance the prestige of national paleontology, to attract visitors to the natural history museums, and consequently to convince both the political and the scientific community of the scientific value of paleontological research."[1245]

In December 1898 Andrew Carnegie (1835–1919) decided to buy a dinosaur. The Scottish American was one of the richest people alive, he had made his fortune in the steel industry and had now, in his autumn years, turned to philanthropy. Not unlike George Peabody four decades earlier, Carnegie donated his money to libraries and institutions of higher education. He was also a dinosaur enthusiast: when the New York Post proclaimed that "the most colossal animal ever [had been] taken from the earth's stratas,"[1246] Carnegie decided to purchase the animal, or rather to pay for its full exhumation, for at the time the skeleton had only been partially dug up. Still, it promised to be a most spectacular discovery which continued to make headlines, even though not yet fully unearthed. The New York Journal and Advertiser called it "the most stupendous thing ever alive."[1247] After Carnegie had obtained the femur of the animal, he funded an expedition to Wyoming to hunt for the rest of the impressive skeleton. Though it turned out they were too late since the dinosaur had already been taken by another excavation team, Carnegie's team found another gigantic skeleton c. 30 km away from the original excavation site. The skeleton belonged to a *diplodocus*; a genus of dinosaur first described by Marsh in 1878. Carnegie's *diplodocus* was found on July 4, 1899 and was immediately dubbed the "Star-Spangled Dinosaur." Furthermore, it was of impressive size and later described as a new species of diplodocus, aptly named *diplodocus carnegii*. Three years later King Edward VII of England (1841–1910) requested a cast of the "Star-Spangled Dinosaur" for the London Museum. Carnegie obliged and during the next decade various museums around the world received casts of the dinosaur, now called "Dippy."[1248] They were erected in Berlin (1908), Paris (1908), Vienna (1909), Bologna (1909), St. Petersburg (1910), Buenos Aires (1912), and Madrid (1913). Later, a cast was displayed in Mexico City (1930) and another one was sent to Munich (1932), but never mounted. Thus "Dippy" the "Star-Spangled Dinosaur" also became

1245 Tamborini: "If the Americans Can Do It, So Can We", p. 229.

1246 Quoted after Brent H. Breithaupt: The Discovery and Loss of the "Colossal" Brontosaurus Giganteus from the Fossil Fields of Wyoming (USA) and the Events that Led to the Discovery of Diplodocus Carnegii. The First Mounted Dinosaur on the Iberian Peninsula, in: Jornadas Internacionales sobre Paleontología de Dinosaurios y su Entorno, Sep. 2013, pp. 48–50. Quote on page 48.

1247 Quoted after Breithaupt: The Discovery and Loss of the "Colossal" Brontosaurus Giganteus from the Fossil Fields of Wyoming (USA) and the Events that Led to the Discovery of Diplodocus Carnegii, p. 48.

1248 Breithaupt: The Discovery and Loss of the "Colossal" Brontosaurus Giganteus from the Fossil Fields of Wyoming (USA) and the Events that Led to the Discovery of Diplodocus Carnegii.

the "first cosmopolitan Dinosaur."[1249] It is noteworthy that government officials and royalty requested casts of *diplodocus* for their respective museums of natural history, not the directors of the museums or other scientists.[1250] Carnegie had in a roundabout way, and thanks to his vast fortune, become a de-facto diplomat for his nation:

> Carnegie's main agenda in the early-twentieth-century was the promotion of world peace, and we ought to see the donation of Diplodocus in that light. Apart from filling his need for personal recognition – which certainly played a role – Carnegie perceived that the best way was to exert personal influence, and so pacify, the 'crowned heads of Europe'. Plying them with impressive dinosaurs, and so allowing them to curry favour with the public on their respective domestic fronts, was one of the means in that campaign.[1251]

This shows that dinosaurs were of great importance to the public interest, not only to science. But then in 1909 Gustav Tornier (1858–1938), a German paleontologist, published an article attacking the reconstruction of *diplodocus* at Carnegie's museum, where "Dippy" stood proudly upright with erected, straight legs. Tornier argued that the animal's posture must have been more "reptile-like," with legs splayed out sideways. J.W. Hollad, director of the Carnegie Museum in Pittsburgh and member of the expedition that found "Dippy," fired back arguing against the lizard-like posture.[1252] Jim Parsons comments concerning the feud:

> There is clearly a note of outraged national pride in Holland's pique at the perceived calumny on American paleontology issued by the Teutonic Tornier. Should an *American* dinosaur crawl at the command of a *German* scientist? No patriotic American could tolerate the insult! Holland's upright, tall-standing dinosaur was therefore a proud symbol of defiance of the Kaiser and his minions. Further, *Diplodocus* was *carnegii*; its eponym was a great captain of industry and, incidentally, the benefactor of the museum. Could Mr. Carnegie's own dinosaur be depicted as *crawling*? Surely no creature named after Mr. Carnegie could be permitted to creep on its belly![1253]

1249 Debus: Prehistoric Monsters, p. 133.

1250 "Requests came from Edward VII of England, the emperors of Germany and Austria, the president of France and the king of Italy." See: Parsons: Drawing Out Leviathan, p. 4.

1251 Ilja Nieuwland: The Colossal Stranger. Andrew Carnegie and Diplodocus Intrude European Culture, 1904–1912, in: Endeavour, vol. 34, no. 2 (Jun. 2010), pp. 61–68, DOI: 10.1016/j.endeavour.2010.04.001. Quote on page 66.

1252 Parsons: Drawing Out Leviathan, pp. 118–120.

1253 Parsons: Drawing Out Leviathan, pp. 120–121.

Note that Parsons makes a polemic argument against symbolically charging scientific arguments. He asserts that there is simply not enough direct evidence to interpret the dispute between scientists within their cultural background and to read cultural significance into their arguments. He argues that since the cultural reading of events cannot be proven by empirical evidence they should not be employed in the analysis of scientific processes. He seems to suggest that science is purely based on such empirical evidence.[1254] The author of this thesis disagrees. The history of US-American paleontology can be interpreted through the lens of cultural background and societal surroundings and history. In fact, any human endeavor must be, any study that lacks these elements is incomplete and lacks an understanding of how human societies function.

Parsons' mocking polemic comment should in fact not be dismissed. Carnegie's "Star-Spangled Dinosaur" is the quintessential US-American dinosaur and a symbol for the nation, as a contemporary newspaper article reveals. The article is titled "Mr. Carnegie's Imitation Dinosaur 'Makes a Hit' in England" and was published on June 4, 1905 in the New York Times. It is accompanied by a depiction of a man on a stage, presenting the *diplodocus* skeleton to a cheering crowd. Three portraits depict Lord Avebury, the curator of the British Museum of Natural History, Carnegie, and Holland. Representations of the two nations are also present in the form of John Bull and Uncle Sam, respectively. The upper right-hand corner is adorned with the Star-Spangled Banner and the Union Jack (see figure 6).

The article describes the unveiling of the "Dippy" cast at the Natural History Museum in London and the reaction of the attendees. The nationalistic importance of the cast is emphasized, but in this case not as part of a patriotic competition, but as a sign of friendship between the US and Great Britain:

> The British people can boast now that they own the biggest imitation skeleton in the whole wide world. For this distinction they are indebted to the munificence of Andrew Carnegie, who has just presented to their National Museum of Natural History a replica of the skeleton of that remarkable Dinosaurian reptile known as Diplodocus Carnegii. [...]
>
> The presentation was an interesting affair, the marked feature of which was the great gratitude displayed by the geologists and paleontologists of England for the magnificent gift that had come to the old British Museum from the youngest of the American museums.

1254 Parsons: Drawing Out Leviathan, pp. 121–125.

Figure 6: Mr. Carnegie's Imitation Dinosaur 'Makes a Hit' in England, in: R.W.W.: Mr. Carnegie's Imitation Dinosaur "Makes a Hit" in England, in: The New York Times, Jun. 4, 1905, p. 7.

> Speaking of the geologists Sir Archibald Geikie said that through Mr. Carnegie's kindness, the scientists of England were now for the first time able in their own country to study intelligently the animal life of which the Diplodocus was a type. This statement in various forms was repeated by all of the Britishers who made speeches, and to anyone who was able to see Mr. Carnegie's face while the Britishers were speaking it was abundantly evident that he was greatly pleased at the appreciation his imitation skeleton commanded. [...]
>
> And so it may be accepted as settled that, in its own mute and undemonstrative way, the great structure which Mr. Carnegie has erected in the British Museum of Natural History will help strengthen the ties of friendship between the great American Republic and the great British Empire.[1255]

Mitchell provides some more general observations about the practice of giving away dinosaur skeletons, adding more nuance to the subject of Mr. Carnegie's gift:

> The dinosaur emerged as an ideal object of philanthropic giving at the end of the nineteenth century for many reasons. [...] [T]he dinosaur was a highly visible gift [...], it was both monumental and monstrous, a surefire public at-

1255 R.W.W.: Mr. Carnegie's Imitation Dinosaur "Makes a Hit" in England, in: The New York Times, Jun. 4, 1905, p. 7.

> traction. [...] [T]his monstrous image came with associations of modernity and novelty (as a contribution to scientific progress and the filling in of the evolutionary record) as well as an aura of unimaginable antiquity. [...] [T]he American dinosaur had inherited the role pioneered by Jefferson's mastodon and could thus serve as an emblem of national pride. America's big bones were a demonstration of its 'natural constitution,' its virility, potency, and dominance in the Darwinian struggle among nations.[1256]

"Dippy" is the culmination of half a century of US-American paleontological research and progress. The skeleton was found in the West and is therefore inextricably linked to the "frontier myth," even though at the time of "Dippy's" discovery the "frontier" had been closed for eight years. This made the skeleton a truly US-American one. Furthermore, the first *diplodocus* was described by Marsh, one of the nation's foremost paleontologists. This meant that the excavation of the first *diplodocus* skeleton was, at least in parts, paid for by the money of a wealthy US-American philanthropist (George Peabody, whose money funded Marsh's expenses for decades after Peabody himself had died). Found on the Fourth of July and imbued with nationalistic significance, the casts of the skeleton were then sent around the world as a symbol for the rise of US-American science, but also as a symbol for the rise of the US as a nation. Then the "Star-Spangled Dinosaur" was attacked by a German and became a player in the struggle for the biggest dinosaur when Germans found one in their colonies, the *giraffatitan* in German East Africa, today's Tanzania. A chapter in the history of US-American science came to a close; a chapter characterized by the race to catch up to European science, by learning from and emulating European and mainly Germany's systems of higher education and research. A new chapter began, and it was characterized by nationalistic competition. Only shortly thereafter nationalistic sentiments would lead into the First World War.

The relationship between US paleontology and US nationalism can be summarized as follows: first paleontology was employed in the search for a national identity while the young republic struggled to further dissociate from Europe. Then efforts were made to make paleontology international, and transatlantic scientific networks thrived. Finally, at the end of the century US paleontology succumbed to nationalism. Science cannot truly be international as long as scientists think within national systems. Therefore, even though scientists in the nineteenth century proclaimed to work without borders, scientific conduct was still influenced by nationalistic limitations and scientific accomplishments of one nation were weighed against those of others.

The "true" meaning of fossils as the remains of usually extinct lifeforms of the past had been discovered and their supposed mythological origins had been discredited

1256 Mitchell: The Last Dinosaur Book, p. 156.

during the seventeenth, eighteenth and early nineteenth centuries, thanks to Steno, Cuvier, and others. This gave rise to new theories about life on earth in ages past, as well as the age of the planet itself. During this time most religious explanations for said phenomena were renounced in favor of more scientific ones, leading to the establishment of the science of natural history, and later geology, as well as paleontology. The gathering, collection, systematization, and empirical study of fossils constituted the basis for these sciences. The fossils were either studied in collections in private or displayed in gardens, cabinets of curiosities, and later in museums. Beginning in this age of scientific revolution, scholars communicated their findings within networks, in their entirety these became known as the Republic of Letters. During the latter half of the nineteenth century, the period of relevance for this study, science still was conducted within these international networks. Professional knowledge, publications, and fossils were exchanged, job-opportunities created, and personal friendships cultivated. At the same time scientific societies were established, published their journals, and circulated them internationally. During the late eighteenth and early nineteenth centuries, the most groundbreaking studies of fossils were conducted in France, the German Region, and Great Britain. While American fossils were studied in Europe, the young United States lacked scientific infrastructure and know-how to conduct original paleontological research during the early nineteenth century. In 1842 dinosaurs were recognized as a distinct clade and captured the public's imagination all around the world, furthering interest in and funding for the science of paleontology. At the same time US-American paleontological research began to flourish, mainly in Philadelphia and thanks to Joseph Leidy. When Cope and Marsh began their respective careers, US paleontology was still widely outclassed by its European counterparts. At the end of their lives and the nineteenth century the tables had turned. Fossils were employed by nationalism, which was on the rise at the end of the century.

Marsh's example echoes these broader developments: he journeyed through Europe, and especially the German Region to round out his US-American education at the most prestigious and innovative scientific institutions. He also established his own network of scientists, which profited him throughout his entire career. On his return to the States Marsh set himself up as a scientist and thanks to the generous funding of his uncle Peabody he held the first chair for paleontology in the US (without receiving a salary). He supervised the construction of a museum, paid for by his uncle's fortune, and acquired a fine and comprehensive fossil collection with this money and the money of the United States Geological Survey – a collection the likes of which the world had never seen. Both Marsh and Cope were extremely well-off thanks to their respective families and relatives, which helped immensely in the conduct of paleontology: "It takes wealth to acquire knowledge. Where would dinosaur paleontology be today had Marsh and Cope been required to earn their daily bread by the sweat of their brows?"[1257]

On the one hand, the Peabody-patronage greatly furthered paleontology, on the other hand it was a testament to the fact that only the rich could freely and successfully conduct science. Furthermore, science became a commodity, a product like everything else in capitalism. And, science cannot be conducted impartially if the interests of rich donors must be considered. Or to say it in Stephen Gould's words:

> Natural history is and has always been a beggar's game. Our work has never been funded by or for itself. We have always depended upon patrons, and upon other people's perceptions of the utility of our data ... many, but not all, of these partnerships have been honorable from our point of view, but we have never had the upper hand. Quite the contrary, our hand has always been out.[1258]

Marsh's relationship with the US government was ambivalent. He profited significantly from the government's money, for it was the government that funded the USGS. The USGS in turn paid for many of Marsh's expenses, such as some truly elaborate publications, professional bone hunters, and the laboratory workers who cleaned and reconstructed the fossils, and apparently also conducted a lot of the research Marsh claimed for himself. Furthermore, troopers employed by the government escorted Marsh's expeditions. On the other hand, when USGS funding ran dry and Marsh was encouraged to quit, the US government wanted back what it had paid for, and Marsh had to hand over the fossils that had been acquired by government money. This undoubtably caused Marsh some headache for he never really tracked which fossils had been acquired with his own money and which with the governments. Still, his involvement with government institutions demonstrates that the US government was willing to finance paleontology, at least as a byproduct of surveying the land.

The bedrock of Marsh's collection consisted of the most complete and spectacular of the massive vertebrates which had once roamed what is now the west of North America. Their discovery not only meant that US-American paleontology was given the tool for its eventual ascension to overshadow its European counterpart, but that locality and time also linked this chapter of US-American paleontology to one of the most enduring national myths of the United States: the "winning of the West." After a destructive Civil War, the nation looked west and expanded into hitherto unexploited and grandiose lands. Science served public and nationalistic interests when this uniquely American land was surveyed. Though government agencies pursued practical interests in equipping and paying for these expeditions, US-American paleontolo-

1257 Dodson: The Horned Dinosaurs, p. 10.

1258 Quoted after Tamborini: "If the Americans Can Do It, So Can We", p. 254.

gy profited greatly from them. Furthermore, they linked this branch of science to the grand national narrative of *Manifest Destiny*. Fossil-hunting became an adventurous and manly occupation, on the "frontier" US-American paleontology was "Americanized." As the lands, the Great Plains and the Rocky Mountains were understood as symbols of US-American identity and uniquely American features, so were the fossils found within this hallowed national space understood as symbols for the magnificence of the nation. Especially the awe-inspiring dinosaurs. It is no coincidence that at the closure of the nineteenth century when the US became a leading industrial power, *diplodocus*, the "Star-Spangled Dinosaur," captured hearts and minds all around the world. Here now was the proof that the United States had finally caught up to Europe, not only in terms of production and industry, but also in terms of science. This link to the great national narrative is one reason why this period is remembered as the "Heroic Age of Paleontology." Another reason is the "Bone Wars," lending a very personal and entertaining nuance to the history of US paleontology, which would otherwise hardly be remembered as fondly as it is today.

During the second half of the nineteenth century, dinosaurs had taken the place of Peal's *mastodon*. Now dinosaurs were symbols of a savage past and metaphorically linked to the struggle between modern human beings and nations for supremacy. The image of fighting dinosaur dominates artistical depictions of prehistoric life to this day. Paul Semonin, who writes about the link between US-American nationalism and what he calls "American Monsters," meaning extinct lifeforms, even writes that "the savagery of prehistoric nature today may be replacing the wild west as the symbol of violent nature in American culture."[1259]

Metaphorically, dinosaurs usually stand for aspects such as slow-moving, slow-thinking, overly-large, non-adaptive, and most of all for being doomed to extinction in the near future. In some cases, they can also symbolize savagery, ferociousness, brute strength. Usually dinosaurs represent a negative, obsolete image. They symbolized nationalism and the struggle between nations, although these concepts were not always seen as negative, and to this day some people believe them to be positive: "The link between dinosaurs and national power and colonialism – particularly for Britain and the United States – was expressed in a metaphor of global hegemony: 'Dinosaurs *ruled the earth* for 140 million years.'"[1260]

Mitchell adds to the metaphoric symbolism of dinosaurs:

> The dinosaur also stands for the fate of the human species within the world system of modern capitalism, especially the 'species anxieties' that are endemic to modernity, from decadence to disaster to uncontrollable eco-suicide. In

1259 Semonin: American Monster, pp. 392–411. Quote on page 409.

1260 Haste: Dinosaur as Metaphor. Quote on page 369.

> this respect, it is the true descendant of the dragons, those 'prodigies' whose appearance in traditional societies signified war, plague, natural disaster, or the wrath of God.[1261]

Perhaps the best example for the improved standing of US-American paleontology in the late nineteenth century manifested in a (back then) late-breaking and controversial international discourse, the Darwinian theory of evolution. Marsh's extensive fossil collections, the *odontornithes*, and first and foremost the ancestry of the horse delivered tenable proof for Darwin's theory and earned US-American paleontology scientific laurels.

Marsh's German assistants not only partook in said "Bone Wars," but also exemplify how Marsh's scientific network functioned. Marsh already had an excellent standing within the scientific community when Zittel came to America and met him. Then Marsh asked Zittel for young German scientists who were willing to come to New Haven and work with him. This was an expression of the reverence Marsh had for the German system of higher education, which he knew first-hand. But the experiences of his assistants, and first and foremost Baur's, give insight into the conditions to scientific conduct during this period. Baur was dependent on his employer in an all-encompassing manner; not only his future career depended on Marsh but also his and his family's financial well-being. The analysis of these correspondences, many of them previously unpublished, also gives insight into some of the more personal aspects of the scientific network and the friendships between the scientists.

But Marsh's efforts to implement German practices and know-how into his own research at Yale were also part of a wider effort to modernize the US-American system of higher education. During the first decades of the nineteenth century, German universities grew to be seen as the most innovative and practical in the world. Meanwhile the US-American system of higher education was still focused on the moral education of the students, not on research and science. Especially the "Humboldtian ideals" were to be emulated when the US system of higher education was to be reformed in the latter half of the nineteenth century. Most of the reformers were people like Marsh who had spent the last years of their own education in Europe, had risen to power since then, and were now trying to implement innovations they had witnessed in Germany at home. Again, science was closely linked to the emerging nationalism and did its part to lead the world into a devastating World War.

This thesis endeavored to answer the following questions, as stated in the introduction:

How did US-American paleontology develop as a scientific discipline, especially in exchange with German higher education and through international scientific net-

---

1261 Mitchell: The Last Dinosaur Book, pp. 67–68.

works and knowledge transfer? Why did US paleontology (and science in general) initially lag behind Europe? How did the situation of US-American science then change in the second half of the nineteenth century? Why did this change occur at that time? How could the US catch up to Europe so quickly, and even arguably overtake it? Where did the influences and inspirations for the change in US science and higher education come from? Who were the people who brought the change? How did they interact with each other? What made US-American paleontology, and dinosaur paleontology in particular, US-American? How did US paleontology and its findings shape US-American identity, the US-American national consciousness, and US nationalism? How, in turn, did US-American nationalism and self-understanding influence the way paleontology developed as a scientific discipline? Does paleontology serve to exemplify the broader changes happening within the US-American system of higher education?

The young United States, and previously the English / British colonies, lacked in educational logistics. Not enough scholars and institutions of higher education existed in North America before the 1860s to conduct original research on a large enough scale to keep up with European science. Young people seeking higher education almost necessarily needed to travel to Europe. In the US, the know-how, scientific networks, and scientific societies were not sufficiently developed to produce original paleontological knowledge. The continent provided raw data in the form of fossils, but there were not enough academics dedicated to describing and analyzing said data on the continent before the 1860s. Fossils were therefore brought to Europe and fueled European paleontology to an extent. This changed during the nineteenth century, once higher education and paleontology in the US began to expand and formalize.

US-American paleontology caught up in the latter half of the nineteenth century and overtook European paleontological institutes. The teaching of paleontology was first formalized in the US in 1866, when Marsh claimed the first professorship for this branch of science. Educational reform followed and paleontological subjects were increasingly taught at US-American colleges and universities. The USGS and other governmental institutions paid for paleontological excursions. These expeditions brought with them often complete and until then unheard-of fossils from the West. Not all expeditions were sponsored by the federal or state governments or by institutions of higher education, instead some were sponsored with private funds. People such as Cope and Marsh invested their own vast fortunes in the acquisition of fossils. New and complete fossils provided data for new and innovative research, which in turn meant that US-American research gained international renown. Furthermore, the products of paleontology were met with public interest. Especially the impressive dinosaurs spurred the public imagination. Prehistoric monsters, of which the most colossal and fiercest were now found in the American West, furthered the public's interest in paleontology. Another aspect, namely the proof Marsh's fossils provided for Darwin's theory of evolution, advanced the national and international renown of US paleontology, for they became part of a very controversial, timely, relevant, and public discourse.

Most stimuli for this new attitude towards science and towards education in general came from Germany. As mentioned above, many of the leading US scientists and educational reformers had spent at least some time at various European institutions of higher education. Especially the "German University," best exemplified by Berlin University (since 1949 aptly named "Humboldt-Universität") with its Humboldtian ideals and modern laboratories, was to be emulated in a young nation yearning to catch up in scientific and educational matters.

The change was brought on by people like Marsh, who himself had studied in Germany and had established a personal network with many leading scientists in Germany and Great Britain. He held contact via correspondence, exchanged information, scientific publications and in some cases fossils within this network. He even acquired two of his three German assistants through this network, when Zittel, back in Munich, arranged their employment with Marsh. Said assistants, or at least Baur who spent the remainder of his life in the US, also contributed to the change happing in US-American paleontology and higher education. He sought employment not only at Yale but also at Clark University and the University of Chicago. The German assistants also interacted within the above-mentioned scientific networks. But not only knowledge and friendly pleasantries were exchanged. The "Bone Wars" shaped the personal interactions of the paleontologists. Alliances were formed and Marsh's assistants became entangled in the conflict.

After the conclusion of the Civil War, the nation looked west. The exploration and "conquest" of the west of the continent constituted a formative part of US nationalism and national self-perception. Public and governmental interests in the exploration of this region granted funds for scientific expeditions, which, among other scientific insights, brought fossils back to the east and its centers of learning. This direct link to the "frontier" and the "frontier experience / myth" associated paleontology with this great national narrative. Improvements to the infrastructure played an important part as well. Ever growing telegraph and railroad networks furthered paleontological study in the West. Fossils were transported via rail; new discoveries could be transmitted almost instantaneously via telegraph. Government interest meant that the paleontological expeditions would be provided with military escorts. It was in the West and on the "frontier" that paleontology became truly American. It left behind its European roots and now fed directly from the American land, a treasure-trove of fossils.

Paleontology, and furthermore the spectacular dinosaurs it unearthed were now associated with the fate of the nation, a product of the land itself. This came at a time when US popular culture was established. For the first time dime novels, newspaper articles, shows, exhibitions, and other media promoted a nationwide culture. Regional differences between north and south, east and west lost importance (though they never really faded completely). Furthermore, paleontology and dinosaurs in particular were employed by rising nationalism, which gripped the world at the end of the nineteenth century. These most impressive products of paleontology could now be used to prove to the world that the United States of America as a relatively young nation also had a deep past, and a thriving scientific community. Indeed, US-American paleontology and its findings did not have to shy away from any of its counterparts. The grand land had provided US-American paleontology with the tools to overtake its European ancestors.

Next steps to build upon this thesis could be to delve even deeper into the histories of Marsh's assistants. Maybe their biographies could be supplied through research in family archives or other archives that were not accessible to the author of this thesis. Even Marsh's biography is at least out of date and rather biased. A modern and truly comprehensive biography of Marsh's could build upon LeVene's and Schuchert's work, just as Davidson's "Bone Sharp" built upon Osborn's "Cope: Master Naturalist." It could also prove fruitful to directly compare the development of US-American paleontology with the development of another scientific discipline. If direct parallels occur, broader theoretical conclusions concerning the development of scientific disciplines within the US-American context could be deducted. Finally, it was not the intention of this thesis to participate in science history. But a science history of paleontology could be written by going fossil by fossil, description by description, theory by theory, and by directly analyzing and comparing the scientific descriptions.

This thesis has added to the historical study of the development of US-American paleontology as a scientific discipline. The novel approach of this study lays in its focus on the professional networks of US-American and German paleontologists, and especially on the ways in which German sciences and the university system contributed to the advancement of US-American paleontology and higher education. To this end the perspectives of Marsh's German assistants, so-far neglected by scholars, were analyzed, using hitherto unedited sources, likely unknown to other scholars of the subject. In this respect this study also adds nuance to the analysis of the "Bone Wars" as it elaborates on the relationship between Osborn and these German scientists. In addition, the link between the rise of US nationalism and science was analyzed in tandem with the construction of US national identity on the "frontier." It also showed how the products of paleontology, especially the fearsome dinosaurs, were then employed to further the nation's standing against the ever-increasing international competition. This association between paleontology and nationalism is most vividly exemplified by the story of "Dippy:" a gift donated in the name of international peace, yet still a symbol for the prestige that US-American paleontology held by the beginning of the twentieth century.

The *suuwassea* family will soon find a new home either at the Hagenbeck Zoo or the Center of Natural History in Hamburg. Initially, the dinosaurs will be studied in detail at the German university department. There the American fossils will stand as a testament to the contributions dinosaur skeletons made to the advancement of paleontology as a scientific discipline, both in the US and in Germany. At the zoo they could stand next to their concrete counterparts, themselves a testament to the unbroken popularity of the extinct giants, beloved ever since their Victorian equivalents were erected at the Crystal Palace.

Bibliography

Archive Index

Academy of Natural Sciences, Drexel University (Collections #1 and #567).

American Museum of Natural History, Henry Fairfield Osborn Papers, Correspondence: Individuals.

American Museum of Natural History, PA 1/108, General Correspondence.

American Museum of Natural History, VPA 1/8, General Correspondence Bas–Bel, BAUR, GEORGE (Prof.) Univ. of Chicago 1884–1897, Folder 36.

American Museum of Natural History, VPA 1/84, General Correspondence Schau-Sci, SCHLOSSER, Dr. Max Paleo. Mus., Munich 1888–1925, Folder 50.

American Museum of Natural History New York, VPA 12, Correspondence loans and exchanges, Museums and Universities Aboard 1896-1934, Box 5, Folder 3.

American Philosophical Society Library, Franz Boas Papers (Mss B B61).

Haverford College Quaker & Special Collections Edward Drinker Cope Papers, HC.MC-956.

Library of Congress, Digital Collections, Ulysses S. Grant Papers: Series 2, Letterbooks, 1869–1877; Vols. 2 (cont.)-5.

Staatsbibliothek zu Berlin, La 1880 (13), Schlosser, Max.

Yale University, Sterling Memorial Library, Charles Schuchert Papers (MS 435).

Yale University, Sterling Memorial Library, Dana Family Papers (MS 164).

Yale University, Sterling Memorial Library, Othniel Charles Marsh papers (MS 343).

Primary Sources

Abel, Othenio: Vorzeitliche Tierreste im Deutschen Mythus, Brauchtum und Volksglauben, Jena 1939.

Advertisement text for the card game “BONE WARS: The Game of Ruthless Paleontology”, http://www.zygotegames.com/bw.html, as consulted online on April 10, 2020.

Agassiz, Louis; Gould, Augustus Addison: Principles of Zoology. Touching the Structure, Development, Distribution, and Natural Arrangement of the Races of Animals, Living and Extinct, with Numerous Illustrations. Part I, Comparative Physiology, for the Use of Schools and Colleges, Boston 1851.

Ballou, William Hosea: Marsh Hurls Azoic Facts at Cope, in: The New York Herald, 19 January 1890, p. 11.

Ballou, William Hosea: Scientists Wage Bitter Warfare, in: The New York Herald, 12 January 1890, pp. 10–11.

Ballou, William Hosea: Scott Writes to Marsh, in: The New York Herald, 22 January 1890, p. 5.

Ballou, William Hosea: Some Great American Scientists. VII. Edward Drinker Cope, in: The Chautauquan, vol. 50, no. 1 (Mar. 1908), pp. 100–117.

Ballou, William Hosea: Some More Nuts for Marsh to Crack, in: The New York Herald, 26 January 1890, p. 25.

Ballou, William Hosea: Volley for Volley in Great Scientific War, in: The New York Herald, 13 January 1890, p. 3.

Ballou, William Hosea: Widening that Geological Chasm, in: The New York Herald, 14 January 1890, p. 4.

Barbour, Erwin Hinckley: Notes on the Paleontological Laboratory of the United States Geological Survey under Professor Marsh, in: American Naturalist, vol. 24, no. 280 (Apr. 1890), pp. 388–400.

Baur, Georg: A Review of the Charges Against the Paleontological Department of the U. S. Geological Survey, and of the Defence made by Prof. O. C. Marsh, in: American Naturalist, vol. 24, no. 279 (Mar. 1890), pp. 298–304.

Baur, Georg: Der älteste Tarsus (Archegosaurus), in: Zoologischer Anzeiger, vol. 9, no. 216 (Feb. 22, 1886), pp. 104–106.

Baur, George: Prof. Marsh on Hallopus and Other Dinosaurs, in: The American Naturalist, vol. 24, no. 282 (Jun. 1890), pp. 569–571.

Baur, George: The Oldest Tarsus (Archegosaurus), in: The American Naturalist, vol. 20, no. 2 (Feb. 1886), pp. 173–174.

Crichton, Michael: Dragon Teeth, New York 2017.

Cody, William Frederick: An Autobiography of Buffalo Bill (Colonel W. F. Cody), New York 1920.

Cody, William Frederick: The Life of Hon. William F. Cody Known as Buffalo Bill the Famous Hunter, Scout and Guide. An Autobiography, Hartford, CT 1879.

Cope, Edward Drinker: Remarks on a New Large Enaliosaur, in: Proceedings of the Academy of Natural Sciences of Philadelphia 20 (Mar. 1868): 92–93.

Cope, Edward Drinker: The Fossil Reptiles of New Jersey (Continued), in: The American Naturalist, vol 3, no 2 (May 1869), pp. 84–91.

Cope, Edward Drinker: On Professor Marsh's Criticisms, in: The American Naturalist, vol. 7, no. 7 (Jul.1873), Appendix.

Cope, Edward Drinker; Kingsley, John Sterling: EDITORIAL, in: American Naturalist, vol. 24, no. 278 (Feb. 1890), pp. 158–160.

Darwin, Charles: On the Origin of Species by Means of Natural Selection, or the Preservation of Favoured Races in the Struggle for Life, London 1859.

Darwin, Charles: The Descent of Man, and Selection in Relation to Sex, London 1871.

Davis, Mark (director): Dinosaur Wars (film), in: American Experience (series), PBS 2011, http://www.pbs.org/wgbh/americanexperience/films/dinosaur/, as consulted online on April 10, 2020.

Farber, Sharon N.: The Last Thunder Horse West of the Mississippi, in: Isaac Asimov's Science Fiction Magazine, vol. 12, no. 11 (Nov. 1988), pp. 20–44.

Heigel, Karl Theodor von: Zum Andenken an Karl von Zittel. Rede in der öffentlichen Festsitzung der K. B. Akademie der Wissenschaften am 14. März 1904, Munich 1904.

Hilgard, Eugene Woldemar: Dr. Meyer and the South-Western Tertiary, in: Science, new ser., vol. 7, no. 152 (Jan. 1, 1886), p. 11.

Holmes, Oliver Wendell: To Christian Gottfried Ehrenberg, for his "Jubilaeum" at Berlin, November 5, 1868, in: Holmes, Oliver Wendell: The Complete Poetical Works of Oliver Wendell Holmes, Cambridge Edition, Boston, MA c. 1895, p. 206.

Holmes, Thom: Fossil Feud, Parsippany, NJ 1998.

Hulke, John Whitaker: Appendix to "Note on a Modified Form of Dinosaurian Ilium, Hitherto Reputed Scapula", in: The Quarterly Journal of the Geological Society of London, vol. 32 (1876), pp. 364–366.

Huxley, Thomas Henry: American Addresses. With a Lecture on the Study of Biology, London 1877.

Jackson, Charles T.: On the Collection of Geological Specimens and on Geological Surveys, in: The American Journal of Science and Arts, vol. 30, no. 1 (Jul. 1836), pp. 203–208.

Jaffe, Mark: The Gilded Dinosaur. The Fossil War Between E. D. Cope and O. C. Marsh and the Rise of American Science, New York 2000.

Jefferson, Thomas: Autobiography, in: The Avalon Project – Documents in Law, History and Diplomacy (URL: http://avalon.law.yale.edu/19th_century/jeffauto.asp, as consulted online on January 19, 2018).

Jefferson, Thomas: Notes on the State of Virginia, Philadelphia 1788.

Keppler, Joseph Ferdinand: Our Indian Policy, in: Frank Leslie's Illustrated Newspaper, vol. 41, no. 1042 (18. Sep. 1875), p. 1.

Kimmel, Elizabeth Cody: Dinosaur Bone Wars. Cope and Marsh's Fossil Feud, New York 2006.

Malthus, Thomas Robert: An Essay on the Principle of Population, as it Affects the Future Improvement of Society. With Remarks on the Speculations of Mr. Godwin, Mr. Condorcet, and Other Writers, London 1798.

Marsh, Othniel Charles: A Statement of Affairs at Red Cloud Agency. Made to the President of the United States, 1875.

Othniel Charles Marsh: Description of the Remains of a New Enaliosaurian (Eosaurus Acadianus) from the Coal Formation of Nova Scotia, in: The American Journal of Science, ser. 2, vol. 34, no. 100, (Jul. 1862), pp. 1–16.

Marsh, Othniel Charles: Dinocerata. A Monograph of an Extinct Order of Gigantic Mammals, Washington, DC 1886.

Marsh, Othniel Charles: Discovery of Cretaceous Mammalia, in: The American Journal of Science, ser. 3, vol. 38, no. 223, (Jul. 1889), pp. 81–92.

Marsh, Othniel Charles: Notice of a New Fossil Annelid (Helminthodes Antiquus) from the Lithographic Slates of Solenhofen, in: The American Journal of Science, ser. 2, vol. 38, no. 114, (Nov. 1864), p. 415.

Marsh, Othniel Charles: Odontornithes. A Monograph on the Extinct Toothed Birds of North America, with Thirty-Four Plates and Forty Woodcuts, Washington, DC 1880.

Marsh, Othniel Charles: Restoration of Brontops Robustus, from the Miocene of America, in: The American Journal of Science, ser. 3, vol. 37, no. 218 (Feb. 1889), pp. 163–165.

Marsh, Othniel Charles: The Gold of Nova Scotia, in: The American Journal of Science, ser. 2, vol. 32, no. 96, (Nov. 1861), pp. 395–400.

Marsh, Othniel Charles: The Vertebrae of Recent Birds, in: The American Journal of Science, ser. 3, vol. 17, no. 100, (Apr. 1879), pp. 266–269.

Meyer, Otto: Birds with Teeth, in: Popular Science Monthly, vol. 36, no. 3 (Jan. 1890), pp. 382–389.

Meyer, Otto: Contributions to the Eocene Paleontology of Alabama and Mississippi, in: Geological Survey of Alabama – Bulletin, no. 1 (1886), pp. 61–85.

Meyer, Otto: Giant Reptiles of a Past Age, in: Popular Science Monthly, vol. 34, no. 4 (Feb. 1889), pp. 466–473.

Meyer, Otto: Some Remarks on the Present State of Our Knowledge of the North American Eastern Tertiary, in: The American Geologist, vol. 2, no. 2 (Aug. 1888), pp. 88–94.

Meyer, Otto: Successional Relations of the Species in the French Old-Tertiary, in: The American Journal of Science, ser. 3, vol. 30, no. 176 (Aug. 1885), pp. 151–153.

Meyer, Otto: The Genealogy and the Age of the Species in the Southern Old-Tertiary, in: The American Journal of Science, ser. 3, vol. 29, no. 174 (Jun. 1885), pp. 457–468.

Meyer, Otto: The Genealogy and the Age of the Species in the Southern Old-Tertiary, in: The American Journal of Science, ser. 3, vol. 30, no. 175 (Jul. 1885), pp. 60–72.

Meyer, Otto: Untersuchungen über die Gesteine des Gotthardtunnels, Berlin 1878.

Nast, Thomas: Education. Is There No Middle Ground?, in: Harper's Weekly, vol. 23, no. 1183 (30. Aug. 1879), p. 696.

N.N.: An Old Grievance Aired, In: New York Times, 13 January 1890, p. 8.

N.N.: Aberdeen Daily News, 18 September 1890, p. 2.

N.N.: Dinoforschung am CeNak, https://hamburg.leibniz-lib.de/aktuelles/news/news-archiv/2020-newsarchiv/2020-04-09-news.html, as consulted online on Mai 14, 2020.

N.N.: Eintrag in das Doktorbuch der Philosophischen Fakultät (Phil. Fak. B 128 b), Universität Leipzig.

N.N.: Eintrag in die Matrikel (Rektor M 29), Universität Leipzig.

N.N.: Georg Baur, in: The American Naturalist, vol. 32, no. 381 (Sep. 1898), pp. 717–718.

N.N.: Idaho Daily Statesman, 25 January 1890, p. 2.

N.N.: New Haven Evening Register, 13 January 1890, p. 1.

N.N.: Notes, in: The American Naturalist, vol. 7, no. 6 (1873), p. 384.

N.N.: Philadelphia Inquirer, 14 January 1890, p. 2.

N.N.: Scientist at War, in: Philadelphia Inquirer, 13 January 1890, p. 2.

N.N.: Springfield Republican, 13 January 1890, p. 5.

N.N.: Stiftung Hagenbeck sichert versteinerte Dinos für den Tierpark, https://www.hamburg-magazin.de/artikel/stiftung-hagenbeck-dinosaurier-für-hamburg, as consulted online on Mai 14, 2020.

N.N.: The Scientists New President, in: Columbus Daily Enquirer, 22 September 1890, p. 6.

N.N.: The Row Among the Geologists, in: Boston Evening Transcript, 16 January 1890, p. 6.

N.N.: Tombstone Epitaph Prospector, 1 February 1890, p. 4.

N.N.: Urvogel Archaeopteryx und andere Fossilienstars, https://www.naturpark-altmuehltal.de/fossilienstars/, as consulted online on December 03, 2019.

N.N.: War Among the Scientists, in: Chicago Tribune, 16 January 1890, p. 4.

N.N.: Worcester Daily Spy 22 January 1890, p. 4.

N.N.: Zeitschrift der Deutschen Geologischen Gesellschaft, vol. 16, (1864), p. 363.

N.N.: Zeitschrift der Deutschen Geologischen Gesellschaft, vol. 17, (1865), pp. 13, 267–269.

Osborn, Henry Fairfield: Cope. Master Naturalist. The Life and Letters of Edward Drinker Cope, Princeton, NJ 1931.

Osborn, Henry Fairfield: A Review of the “Discovery of the Cretaceous Mammalia”, in: The American Naturalist, vol. 25, no. 295 (Jul. 1891), pp. 595–611.

Osborn, Henry Fairfield: Biographical Memoir of Joseph Leidy, 1823–1891, in: National Academy of Sciences. Biographical Memoirs, vol. 7, Washington, DC 1913, pp. 335–396.

Osborn, Henry Fairfield: Karl Alfred von Zittel, in: Science, new ser., vol. 19, no. 474 (Jan. 29, 1904), pp. 186–188.

Ottaviani, Jim: Bone Sharps, Cowboys, and Thunder Lizards: Edward Drinker Cope, Othniel Charles Marsh, and the Gilded Age of Paleontology, Ann Arbor 2005, MI.

Powell, John Wesley: Introduction to the Study of Indian Languages, with Words, Phrases, and Sentences to be Collected, Washington, DC 1877.

R.W.W.: Mr. Carnegie’s Imitation Dinosaur “Makes a Hit” in England, in: The New York Times, Jun. 4, 1905, p. 7.

Roemer, Ferdinand von: Die Kreidebildungen von Texas und ihre Organischen Einschlüsse, Bonn 1852.

Roemer, Ferdinand von: Die Silurische Fauna des Westlichen Tennessee. Eine Palaeontologische Monographie, Breslau 1860.

Roemer, Ferdinand von: Texas. Mit Besonderer Rücksicht auf Deutsche Auswanderung und die physischen Verhältnisse des Landes nach Eigener Beobachtung Geschildert, Bonn 1849.

Rothpletz, August: Gedächtnisrede auf Karl Alfred von Zittel, gehalten in der öffentlichen Sitzung der K. B. Akademie der Wissenschaften zu München zur Feier ihres 146. Stiftungstages am 15. März 1905, Munich 1905.

Schlosser, Max: Nachträge und Berichtigungen zu: die Nager des Europäischen Tertiärs. Palaeontographica 31. Band, in: Zoologischer Anzeiger, vol. 7, no. 182 (Dec. 1884), p. 639.

Schlosser, Max: Über neue Funde von Leptodon Graecus Gaudry und die Systematische Stellung dieses Säugethieres, in: Zoologischer Anzeiger, vol. 22, no. 597 (Sep. 18, 1899), pp. 378–380.

Schlosser, Max: Über neue Funde von Leptodon Graecus Gaudry und die Systematische Stellung dieses Säugethieres, in: Zoologischer Anzeiger, vol. 22, no. 598 (Oct. 2, 1899), pp. 385–387.

Scott, William B.; Osborn, Henry Fairfield: The Mammalia of the Uinta Formation, in: Transactions of the American Philosophical Society, new ser., vol. 16, no. 3 (1890), pp. 461–572.

Seeley, Harry Grovier: Note on the Pelvis of Ornithopsis, in: Quarterly Journal of the Geological Society of London, vol. 45 (1889), pp. 391–397.

Spencer, Herbert: The Principles of Biology, vol. 1, London 1864.

Turner, Frederick Jackson: The Significance of the Frontier in American History, in: Annual Report of the American Historical Association, 1894, pp. 119–227.

Wallace, Davis Rains: The Bonehunter's Revenge. Dinosaurs, Greed, and the Greatest Scientific Feud of the Gilded Age, New York 1999.

Whipple, Thomas King: Study out the Land. Essays by T.K. Whipple, Berkeley, CA 1943.

Zittel, Karl Alfred von: Das Wunderland am Yellowstone, Berlin 1885.

Zittel, Karl Alfred von: Geschichte der Geologie und Paläontologie bis Ende des 19. Jahrhunderts, Munich 1899.

Secondary Sources

Adam, Thomas; Lerg, Charlotte A.: Introductory Remarks. Diplomacy on Campus. The Political Dimensions of Academic Exchange in the North Atlantic, in: Journal of Transatlantic Studies, vol. 13, no. 4 (Dec. 2015), pp. 299–310, https://doi.org/10.1080/14794012.2015.1088327.

Adams, Frank Dawson: The Birth and Development of the Geological Sciences, New York 1954.

Aiton, Eric John: Leibniz. A Biography, Bristol 1985.

Aldrich, Michelle L.: Women in Paleontology in the United States 1840–1960, in: Earth Sciences History, vol. 1, no. 1 (1982), pp. 14–22.

Allen, David Elliston: The Naturalist in Britain. A Social History, 2nd corr. ed., Princeton, NJ 1994.

Andersen, Hanne et al.: The Cognitive Structure of Scientific Revolutions, Cambridge 2006, https://doi.org/10.1017/CBO9780511498404.

Appel, Toby A.: The Cuvier–Geoffroy Debate. French Biology in the Decades before Darwin, New York 1987.

Ash, Mitchell G.; Surman, Jan: The Nationalization of Scientific Knowledge in Nineteenth Century Central Europe: An Introduction, in: Ash, Mitchell G.; Surman, Jan (eds.): The Nationalization of Scientific Knowledge in the Habsburg Empire, 1848–1918, Basingstoke 2012, pp. 1–29.

Backenköhler, Dirk: CUVIERs langer Schatten. "Il n'y a point d'os humains fossiles", in: Hoßfeld, Uwe; Junker, Thomas (eds.): Die Entstehung biologischer Disziplinen II. Beiträge zur 10. Jahrestagung der DGGTB in Berlin 2001, Berlin 2002, pp. 133–147.

Bakker, Robert: The Dinosaur Heresies. A Revolutionary View of Dinosaurs, Harlow 1987.

Bannister, Robert C.: Social Darwinism. Science and Myth in Anglo-American Social thought, Philadelphia 1978.

Barnes, Barry: Scientific Knowledge and Sociological Theory, London 1974.

Baumgarten, Marita: Professoren und Universitäten im 19. Jahrhundert. Zur Sozialgeschichte deutscher Geistes- und Naturwissenschaften, Göttingen 1997.

Beecher, Charles Emerson: Othniel Charles Marsh, in: The American Journal of Science and Arts, ser. 4, vol. 7, no. 42 (Jun. 1899), pp. 403–428.

Beek, Alan Lena van: Riesen in der Literatur des Mittelalters. Diskursive Formationen im deutschen Sprachraum, Frankfurt/Main 2021, https://doi.org/10.25716/amad-85226.

Benson, Keith R.: From Museum Research to Laboratory Research. The Transformation of Natural History into Academic Biology, in: Rainger, Roland et al. (eds.): The American Development of Biology, paperback ed., New Brunswick, NJ 1991, pp. 49–83.

Berkhofer, Robert F.: The White Man's Indian. Images of the American Indian from Columbus to the Present, New York 1978.

Ben-David, Joseph: American Higher Education. Directions Old and New, New York 1972.

Ben-David, Joseph: Centers of Learning. Britain, France, Germany, United States, New York 1977.

Ben-David, Joseph: The Scientist's Role in Society. A Comparative Study, Englewood Cliffs, NJ 1971.

Bowler, Peter J.: Edward Drinker Cope and the Changing Structure of Evolutionary Theory, in: Isis vol. 68, vo. 2 (Jun. 1977), pp. 249–265.

Bowler, Peter J.: Evolution. The History of an Idea, 3rd compl. rev. and exp. ed., Berkeley, CA 2003.

Bowler, Peter J.: Fossils and Progress. Paleontology and the Idea of Progressive Evolution in the Nineteenth Century, New York 1976.

Bowler, Peter J.; Morus, Iwan Rhys: Making Modern Science. A Historical Survey, Chicago 2005.

Breithaupt, Brent H.: The Discovery and Loss of the "Colossal" Brontosaurus Giganteus from the Fossil Fields of Wyoming (USA) and the Events that Led to the Discovery of Diplodocus Carnegii. The First Mounted Dinosaur on the Iberian Peninsula, in: Jornadas Internacionales sobre Paleontología de Dinosaurios y su Entorno, Sep. 2013, pp. 48–50.

Bridger, Bobby: Buffalo Bill and Sitting Bull. Inventing the Wild West, Austin, TX 2002.

Brinkman, Paul D.: Paleontology, in: Montgomery, Georgia M., Largent, Mark A. (eds.): A Companion to the History of American Science, Chichester 2016, pp. 227–240, DOI:10.1002/9781119072218.

Browne, Janet: Corresponding Naturalists, in: Lightman, Bernard V.; Reidy, Michael S. (eds.): The Age of Scientific Naturalism. Tyndall and His Contemporaries, London 2014, pp. 157–170.

Bruce, Robert V.: The Launching of Modern American Science 1846–1876, New York 1987.

Bruch, Rüdiger vom: Die Gründung der Berliner Universität, in: Schwinges, Rainer C. (ed.): Humboldt International. Der Export des Deutschen Universitätsmodells im 19. und 20. Jahrhundert, Basel 2001, pp. 53–73.

Bruch, Rüdiger vom: Langsamer Abschied von Humboldt? Etappen deutscher Universitätsgeschichte 1810–1945, in: Ash, Mitchell G. (ed.): Mythos Humboldt? Vergangenheit und Zukunft der deutschen Universität, Vienna 1999.

Bryant, Paul: Nature Writing and the American Frontier, in: Bryant, Paul et al. (eds.): The Frontier Experience and the American Dream. Essays on American Literature, College Station, TX 1989, pp. 205–216.

Buffetaut, Eric: A Short History of Vertebrate Palaeontology, London 1987.

Buffetaut, Eric: From Giant Birds to X-Rays. Victor Lemoine (1837–97), Physician and Palaeontologist, in: Duffin, Christopher J. et al. (eds.): Geology and Medicine. Historical Connections, London 2017, pp. 115–131, https://doi.org/10.1144/SP452.

Buffetaut, Eric et al.: The Discovery of French Dinosaurs, in: Sarjeant, William A. S. (ed.): Vertebrate Fossils and the Evolution of Scientific Concepts. Writings in Tribute to Beverly Halstead, by Some of His Many Friends, Amsterdam 1995, pp. 159–180.

Burke, Colin B.: The Expansion of American Higher Education, in: Jarausch, Konrad H. (ed.): The Transformation of Higher Learning 1860–1930. Expansion, Diversification, Social Opening and Professionalization in England, Germany, Russia and the United States, Chicago 1983, pp. 108–130.

Cannon, Susan Faye: Science in Culture. The Early Victorian Period, Folkestone 1978.

Champlin, Peggy: Raphael Pumpelly. Gentleman Geologist of the Gilded Age, Tuscaloosa, AL 1994.

Cochran, Thomas Childs: Railroad Leaders, 1845–1890. The Business Mind in Action, Cambridge, MA 1953.

Cohen, I. Bernard: Franklin and Newton. An Inquiry into Speculative Newtonian Experimental Science and Franklin's Work in Electricity as an Example Thereof, Philadelphia 1956.

Cohen, I. Bernard: Revolution in Science, Cambridge, MA 1985.

Cohen, I. Bernard: Science and the Founding Fathers. Science in the Political Thought of Jefferson, Franklin, Adams, and Madison, New York 1995.

Cook, James W.: The Arts of Deception. Playing with Fraud in the Age of Barnum, Cambridge, MA 2001.

Cope, Edward Drinker et al. (eds.): In Memoriam John Adam Ryder, Philadelphia 1895.

Coward, John M.: The Newspaper Indian. Native American Identity in the Press, 1820–90, Urbana, IL 1999.

Crawford, Elisabeth: Nationalism and Internationalism in Science, 1880–1939. Four Studies of the Nobel Population, Cambridge 1992.

Daniels, George H.: American Science in the Age of Jackson, Tuscaloosa, AL 1994 (orig. publ. 1968).

Daston, Lorraine: The Empire of Observation, 1600–1800, in: Daston, Lorraine; Lunbeck, Elizabeth (eds.): Histories of Scientific Observation, Chicago 2011, pp. 81–113, https://doi.org/10.1007/s11191-012-9515-z.

Daston, Lorraine: The Sciences of the Archive, in: Osiris, 2nd ser., vol. 27 (2012), pp. 156–187.

Davidson, Jane Pierce: A History of Paleontology Illustration, Bloomington, IN 2008.

Davidson, Jane Pierce: Bonehead Mistakes: The Background in Scientific Literature and Illustrations for Edward Drinker Cope's First Restoration of Elasmosaurus Platyurus, in: Proceedings of the Academy of Natural Sciences of Philadelphia, vol. 152 (2002), pp. 215–240.

Davidson, Jane Pierce; Everhart, Michael J.: Fictionalized Facts; "The Young Fossil Hunters" by Charles H. Sternberg, in: Transactions of the Kansas Academy of Science, vol. 117, no. 1–2 (Apr. 2014), pp. 41–48, DOI:10.1660/062.117.0106.

Davidson, Jane Pierce: Misunderstood Marine Reptiles. Late Nineteenth-Century Artistic Reconstructions of Prehistoric Marine Life, in: Transactions of the Kansas Academy of Science, vol. 118, no. 1–2 (Apr. 2015), pp. 53–67, DOI:10.1660/062.118.0107.

Davidson, Jane Pierce: Patrons of Paleontology. How Government Support Shaped a Science, Bloomington, IN 2017.

Davidson, Jane Pierce; Everhart, Michael J.: Scattered and Shattered. A Brief History of the Early Methods of Digging, Preserving and Transporting Kansas Fossils, in: Transactions of the Kansas Academy of Science, vol. 120, no. 3–4 (Sep. 2017), pp. 247–258, https://doi.org/10.1660/062.120.0416.

Davidson, Jane Pierce: The Bone Sharp. The Life of Edward Drinker Cope, Philadelphia 1997.

Davidson, Jenny: Breeding. A Partial History of the Eighteenth Century, New York 2009.

Davis, John R.: Higher Education Reform and the German Model: A Victorian Discourse, in: Ellis, Heather; Kirchberger, Ulrike (eds.): Anglo-German Scholarly Networks in the Long Nineteenth Century, Leiden 2014, pp. 39–62, https://doi.org/10.1163/9789004253117_004.

Dean, Dennis R.: Gideon Mantell and the Discovery of Dinosaurs, Cambridge 1999.

Dean, Dennis R.: James Hutton and the History of Geology, Ithaca, NY 1992.

Debus, Allen A.: Prehistoric Monsters. The Real and Imagined Creatures of the Past That We Love to Fear, Jefferson, NC 2010.

Delair, Justin B.; Sarjeant, William A. S.: The Earliest Discoveries of Dinosaurs, in: Isis, vol. 66, no. 1 (Mar. 1975), pp. 4–25.

DeYoung, Ursula: A Vision of Modern Science. John Tyndall and the Role of the Scientist in Victorian Culture, New York 2011.

Diehl, Carl: Americans and German Scholarship 1770–1870, New Haven, CT 1978.

Dippie, Brian W.: Catlin and his Contemporaries. The Politics of Patronage, Lincoln, NE 1990.

Dippie, Brian W.: The Vanishing American. White Attitudes and U.S. Indian Policy, Middletown, CT 1982.

Dodson, Peter: The Horned Dinosaurs. A Natural History, Princeton, NJ 1996.

Drayton, Richard Harry: Nature's Government. Science, Imperial Britain, and the 'Improvement' of the World, New Haven, CT 2000.

Dupree, A. Hunter: Science and the Federal Government. A History of Policies and Activities, Baltimore, MD 1957.

Dupree, A. Hunter: The National Academy of Sciences and the American Definition of Science, in: Oleson, Alexandra; Voss, John (eds.): The Organization of Knowledge in Modern America, 1860–1920, Baltimore, MD 1979, pp. 342–363.

Dupree, A. Hunter: The National Pattern of American Learned Societies, 1769–1863, in: Brown, Sanborn C.; Oleson, Alexandra (eds.): The Pursuit of Knowledge on the Early American Republic. American Scientific and Learned Societies from Colonial Times to the civil War, Baltimore, MD 1976, pp. 21–32.

Ellis, Heather: Enlightened Networks. Anglo-German Collaboration in Classical Scholarship, in: Ellis, Heather; Kirchberger, Ulrike (eds.): Anglo-German Scholarly Networks in the Long Nineteenth Century, Leiden 2014, pp. 23–38, https://doi.org/10.1163/9789004253117_004.

Engelhardt, Dietrich von: Historisches Bewußtsein in der Naturwissenschaft. Von der Aufklärung bis zum Positivismus, Freiburg/Breisgau 1979.

Engelhardt, Wolf von: Die Entwicklung der Geologischen Ideen seit der Goethe-Zeit, in: Olsen, Karl Heinrich (ed.): Beiträge zur Geowissenschaft. Vorträge Anläßlich der Wwissenschaftlichen Vortragsveranstaltung und der Gauß-Gedenkfeier vom 30. 4. 1979, Göttingen 1979, pp. 1–23.

Fallon, Daniel: German Influences on American Education, in: Trommler, Frank; Shore, Elliott (eds.): The German-American Encounter. Conflict and Cooperation between Two Cultures, 1800–2000, New York 2001, pp. 77–87.

Fastovsky, David E.; Weishampel, David B.: Dinosaurs. A Concise Natural History, Cambridge 2009.

Fischer, Klaus: Die Neue Ordnung des Wissens. Experiment – Erfahrung – Beweis – Theorie, in: Dülmen, Richard van; Rauschenbach, Sina (eds.): Macht des Wissens. Die Entstehung der modernen Wissensgesellschaft, Cologne 2004, pp. 155–185.

Fisher, Irving N.: The O. C. Marsh Papers, in: The Yale University Library Gazette, vol. 46, no. 1 (Jul. 1971), pp. 35–40.

Foth, Christian; Rauhut, Oliver W.M.: Re-evaluation of the Haarlem Archaeopteryx and the radiation of maniraptoran theropod dinosaurs, in: BMC Evolutionary Biology, vol. 17, art. no. 236 (Dec. 2017), https://doi.org/10.1186/s12862-017-1076-y.

Foucault, Michel: Orders of Discourse. Inaugural Lecture Delivered at the Collège de France, transl. Rupert Swyer, in: Social Science Information, vol. 10, no. 2 (Apr. 1971), pp. 7–30.

Frantz, Klaus: Die Indianerreservationen in den USA. Aspekte der Territorialen Entwicklung und des Sozio-Ökonomischen Wandels, Stuttgart 1993.

Freeman, Michael: Victorians and the Prehistoric. Tracks to a Lost World, New Haven, CT 2004.

Frymer, Paul: Building an American Empire. The Era of Territorial and Political Expansion, Princeton, NJ 2017.

Gascoigne, John: Science and the British Empire from its Beginnings to 1850, in: Brennett, Brett M.; Hodge, Joseph M. (eds.): Science and Empire. Knowledge and Networks of Science across the British Empire 1800–1970, Basingstoke 2011, pp. 47–67.

Gascoigne, John: Science in the Service of Empire. Joseph Banks, the British State and the Uses of Science in the Age of Revolution, Cambridge 1998.

Geiger, Roger L.: The History of American Higher Education. Learning and Culture from the Founding to World War II, Princeton, NJ 2015, https://doi.org/10.1515/9781400852055.

Geinitz, Hanns Bruno: Othniel Charles Marsh. Professor der Paläontologie an Yale Univeristät, New-Haven, Conn. und Landes-Paläontolog für Wirbelthiere in den Vereinigten Staaten Nordamerikas, in: Leopoldina, vol. 35, no. 7 (Jul. 1899), pp. 122–124.

Gierl, Martin: Korrespondenzen, Disputationen, Zeitschriften. Wissensorganisation und die Entwicklung der gelehrten Medienrepublik zwischen 1670 und 1730, in: Dülmen, Richard van; Rauschenbach, Sina (eds.): Macht des Wissens. Die Entstehung der modernen Wissensgesellschaft, Cologne 2004, pp. 417–438.

Goetzmann, William H.: Exploration and Empire. The Explorer and the Scientist in the Winning of the American West, New York 1966.

Goldstein, Adam A.: Darwinism, in: Montgomery, Georgia M.; Largent, Mark A. (eds.): A Companion to the History of American Science, Chichester 2016, pp. 306–319, DOI:10.1002/9781119072218.

Goodman, Dena: The Republic of Letters. A Cultural History of the French Enlightenment, Ithaca, NY 1994.

Gorman, James; Horner, John R.: Digging Dinosaurs, New York 1988.

Gould, Stephen Jay: The Structure of Evolutionary Theory, Cambridge, MA 2002.

Gould, Stephen Jay: Time's Arrow, Time's Cycle. Myth and Metaphor in the Discovery of Geological Time, Cambridge, MA 1987.

Grafton, Anthony: Worlds Made by Words. Scholarship and Community in the Modern West, Cambridge, MA 2009.

Greene, John C.: American Science in the Age of Jefferson, Ames, IA 1984.

Greene, John C.: The Development of Mineralogy in Philadelphia, 1780–1820. A Summary, in: Schneer, Cecil J. (ed.): Toward a History of Geology. Proceedings of the New Hampshire Inter-Disciplinary Conference on the History of Geology, September 7–12, 1967, Cambridge, MA 1969, pp. 184–185.

Greene, John C.: Science, Learning and Utility. Patters of Organization in the Early American Republic, in: Brown, Sanborn C.; Oleson, Alexandra (eds.): The Pursuit of Knowledge on the Early American Republic. American Scientific and Learned Societies from Colonial Times to the civil War, Baltimore, MD 1976, pp. 1–20.

Greene, John C.; Burke, John G.: The Science of Minerals in the Age of Jefferson, Philadelphia 1978.

Greene, Mott T.: History of Geology, in: Osiris, 2nd ser., vol. 1 (1985), pp. 97–116.

Gregory, Joseph T.: North American Vertebrate Paleontology, 1776–1976, in: Schneer, Cecil J. (ed.): Two Hundred Years of Geology in America. Proceedings of the New Hampshire Bicentennial Conference on the History of Geology, Hanover, NH 1979, pp. 305–335.

Grene, Marjorie; Depew, David: The Philosophy of Biology. An Episodic History, Cambridge 2004.

Grinnell, George Bird: Othniel Charles Marsh. Paleontologist, in: Jordan, David Starr [ed.]: Leading American Men of Science, New York 1910, pp. 283–312.

Grinnell, George Bird: Sketch of Professor O. C. Marsh, in: Popular Science Monthly, vol. 13 (Sep. 1878), pp. 612–617.

Guntau, Martin; Laitko, Hubert: Entstehung und Wesen wissenschaftlicher Disziplinen, in: Guntau, Martin; Laitko, Hubert (eds.): Der Ursprung der modernen Wissenschaften. Studien zur Entstehung wissenschaftlicher Disziplinen, Berlin 1987, pp. 17–89.

Guntau, Martin: The Natural History of the Earth, in: Jardine, Nicholas et al. (eds.): Cultures of Natural History, Cambridge 1996, pp. 211–229.

Günther, Karl-Heinz: Interdependence between Democratic Pedagogy in Germany and the Development of Education in the United States in the Nineteenth Century, in: Geitz, Henry et al. (eds.): German Influences on Education in the United States to 1917, Washington DC 1995, pp. 43–56.

Guthrie, Stewart Elliott: Faces in the Clouds. A New Theory of Religion, New York 1993.

Hacking, Ian: Working in a New World. The Taxonomic Solution, in: Horwich, Paul (ed.): World Changes. Thomas Kuhn and the Nature of Science, Cambridge, MA 1993, pp. 275–310.

Halstead, Lambert Beverly; Sarjeant, William A. S: Scrotum Humanum Brookes – The Earliest Name for a Dinosaur?, in: Sarjeant, William A. S. (ed.): Vertebrate Fossils and the Evolution of Scientific Concepts. Writings in Tribute to Beverly Halstead, by Some of His Many Friends, Amsterdam 1995, pp. 219–222.

Hamlin, Kimberly A.: From Eve to Evolution. Darwin, Science, and Women's Rights in Gilded Age America. Chicago 2014, DOI:10.7208/chicago/9780226134758.001.0001.

Hannah, Matthew G.: Governmentality and the Mastery of Territory in Nineteenth-Century America, Cambridge 2000.

Haste, Helen: Dinosaur as Metaphor, in: Sarjeant, William A. S. (ed.): Vertebrate Fossils and the Evolution of Scientific Concepts. Writings in Tribute to Beverly Halstead, by Some of His Many Friends, Amsterdam 1995, pp. 359–380.

Hay, Oliver Parry: George Baur, in: Science, new ser., vol. 8, no. 185 (Jul. 15, 1898), pp. 68–71.

Heesen, Anke te; Spary, Emma C.: Sammeln als Wissen, in: Heesen, Anke te; Spary, Emma C. (eds.): Sammeln als Wissen. Das Sammeln und seine wissenschaftsgeschichtliche Bedeutung, Göttingen 2001, pp. 7–21.

Herbst, Jurgen: Diversification in American Higher Education, in: Jarausch, Konrad H. (ed.): The Transformation of Higher Learning 1860–1930. Expansion, Diversification, Social Opening and Professionalization in England, Germany, Russia and the United States, Chicago 1983, p. 196–206.

Herbst, Jurgen: The German Historical School in American Scholarship. A Study in the Transfer of Culture, Ithaca, NY 1965.

Herbst, Jurgen: Thomas Jefferson und Wilhelm vom Humboldt: Universitäts- und Schulgründer, in: Schwinges, Rainer C. (ed.): Humboldt International. Der Export des Deutschen Universitätsmodells im 19. und 20. Jahrhundert, Basel 2001, pp. 273–287.

Herbert, Sandra: Charles Darwin, Geologist, Ithaca, NY 2005.

Herman, Daniel Justin: Hunting and the American Imagination, Washington, DC 2001.

Higham, John: The Matrix of Specialization, in: Oleson, Alexandra; Voss, John (eds.): The Organization of Knowledge in Modern America, 1860–1920, Baltimore, MD 1979, pp. 3–18.

Hixson, Walter L.: American Settler Colonialism. A History, New York 2013.

Hochfelder, David: The Communications Revolution and Popular Culture, in: Barney, William L. (ed.): A Companion to 19th-century America, Malden, MA 2001, pp. 305–316.

Hölder, Helmut: Kurze Geschichte der Geologie und Paläontologie. Ein Lesebuch, Berlin 1989.

Horner, John R.: Dinosaurs at the Table, in: Sepkoski, David; Ruse, Michael (eds.): The Paleobiological Revolution. Essays on the Growth of Modern Paleontology, Chicago 2009, DOI: 10.7208/chicago/9780226748597.001.0001.

Hutchings, Kevin: Transatlantic Extinctions and the “Vanishing American”, in: Hutchings, Kevin; Miller, John (eds.): Transatlantic Literary Ecologies. Nature and Culture in the Nineteenth-Century Anglophone Atlantic World, London 2017, pp. 58–72.

Hung, Edwin H.-C.: Beyond Kuhn. Scientific Explanation, Theory Structure, Incommensurability and Physical Necessity, Aldershot 2006.

Huxley, Leonard: Life and Letters of Thomas Henry Huxley, vol. 1, London 1901.

Huxley, Leonard: Life and Letters of Thomas Henry Huxley, vol. 2, 2nd ed., London 1908.

Jensen, J. V.: Thomas Henry Huxley’s Lecture Tour of the United States, 1876, in: Notes and Records of the Royal Society of London, vol. 42, no. 2 (Jul., 1988), pp. 181–195.

Irmscher, Christoph: The Poetics of Natural History. From John Bartram to William James, New Brunswick, NJ 1999.

Johnson, Charles Willison: Truman Hemin[g]way Aldrich, in: The Nautilus, vol. 46, no. 1 (Jul. 1932), pp. 34–35.

Kanz, Kai Torsten: Von der BIOLOGIA zur Biologie. Zur Begriffsentwicklung und Disziplingenese vom 17. bis zum 20. Jahrhundert, in: Hoßfeld, Uwe; Junker, Thomas (eds.): Die Entstehung biologischer Disziplinen II. Beiträge zur 10. Jahrestagung der DGGTB in Berlin 2001, Berlin 2002, pp. 9–30.

Kevles, Daniel J.: The Physicists. The History of a Scientific Community in Modern American, reprint, Cambridge, MA 1987.

Knell, Simon J.: Museums, Fossils and the Cultural Revolution of Science. Mapping Change in the Politics of Knowledge in Early Nineteenth-Century Britain, in: Knell, Simon J. et al. (eds.): Museum Revolutions. How Museums Change and Are Changed, London 2007, pp. 28–47.

Knell, Simon J.: The Culture of English Geology, 1815–1851. A Science Revealed through Its Collecting, Aldershot 2000.

Knight, David M.: Voyaging in Strange Seas. The Great Revolution in Science, New Haven, CT 2014.

Knopf, Adolph: Charles Schuchert 1858–1942, Washington, DC 1952.

Knorr-Cetina, Karin D.: The Manufacture of Knowledge. An Essay on the Constructivist and Contextual Nature of Science, Oxford 1981.

Koerner, Lisbet: Daedalus Hyperboreus. Baltic Natural History and Mineralogy in the Enlightenment, in: Clark, William et al. (eds.): The Sciences in Enlightened Europe, Chicago 1999, pp. 389–422.

König, René: Vom Wesen der Deutschen Universität, Berlin 1935.

Kohlstedt, Sally Gregory: German Ideals and Practice in American Natural History Museums, in: Geitz, Henry et al. (eds.): German Influences on Education in the United States to 1917, Washington DC 1995, pp. 103–114.

Kohlstedt, Sally Gregory: Institutional History, in: Osiris, 2nd ser., vol. 1 (1985), pp. 17–36.

Kohlstedt, Sally Gregory: Museums on Campus: A Tradition of Inquiry and Teaching, in: Rainger, Roland et al. (eds.): The American Development of Biology, paperback ed., New Brunswick, NJ 1991, pp. 15–47.

Kohlstedt, Sally Gregory: The Formation of the American Scientific Community. The American Association for the Advancement of Science, 1848–60, Urbana, IL 1976.

Kortländer, Bernd: Begrenzung – Entgrenzung. Kultur- und Wissenschaftstransfer in Europa, in: Jordan, Lothar; Kortländer, Bernd (eds.): Nationale Grenzen und internationaler Austausch. Studien zum Kultur- und Wissenschaftstransfer in Europa, Tübingen 1995, pp. 1–19.

Krämer, Fabian: Why There Was No Centaur in Eighteenth-Century London. The Vulgar as a Cognitive Category in Enlightenment Europe, in: Greyerz, Kaspar von et al. (eds.): Wissenschaftsgeschichte und Geschichte des Wissens im Dialog – Connecting Science and Knowledge. Schauplätze der Forschung – Scenes of Research, Göttingen 2013, pp. 317–345, https://doi.org/10.14220/9783737001717.317.

Kuhn, Thomas S.: The Structure of Scientific Revolutions, Chicago 1962.

Kuukkanen, Jouni-Matti: Revolution as Evolution. The Concept of Evolution in Kuhn's Philosophy, in: Kindi, Vasso; Arabatzis, Theodore (eds.): Kuhn's The Structure of Scientific Revolutions Revisited, New York 2012, pp. 134–152.

Lanham, Url: The Bone Hunters. The Heroic Age of Paleontology in the American West, New York 1991 (orig. publ. 1973).

Latour, Bruno; Woolgar, Steve: Laboratory Life. The Construction of Scientific Facts, 2nd ed., Princeton, NJ 1986.

Latour, Bruno: Reassembling the Social. An Introduction to Actor-Network-Theory, Oxford 2005.

Laudan, Rachel: From Mineralogy to Geology. The Foundations of a Science, 1650–1830, Chicago 1987.

Lenk, Hans: Einführung in die Erkenntnistheorie. Interpretation – Interaktion – Intervention, Munich 1998.

Lepenies, Wolf: Das Ende der Naturgeschichte. Wandel kultureller Selbstverständlichkeiten in den Wissenschaften des 18. und 19. Jahrhunderts, Munich 1976.

Levin, David: Giants in the Earth. Science and the Occult in Cotton Mather's Letters to the Royal Society, in: The William and Mary Quarterly, vol. 45, no. 4 (Oct. 1988), pp. 751–770.

Limerick, Patricia Nelson: The Frontier in the Twentieth Century, in: Grossman, James R.: The Frontier in American Culture. An Exhibition at the Newberry Library, August 26, 1994 – January 7, 1995, Berkeley, CA 1994, pp. 67–102.

Lipking, Lawrence: What Galileo Saw. Imagining the Scientific Revolution, Ithaca, NY 2014.

Loveland, Jeff: Rhetoric and Natural History. Buffon in Polemical and Literary Context, Oxford 2001.

Maddra, Sam A.: Hostiles? The Lakota Ghost Dance and Buffalo Bill's Wild West, Norman, OK 2006.

Maienschein, Jane: Whitman at Chicago. Establishing a Chicago Style of Biology?, in: Rainger, Ronald et al. (eds.): The American Development of Biology, pbk. ed., New Brunswick, NJ 1991 (orig. publ. 1988), pp. 151–182.

Martin-Catherin, Nadja: The Making of 'Indians'. Sitting Bull, Native Agency, and American Culture, Trier 2015.

Mayor, Adrienne: The First Fossil Hunters. Paleontology in Greek and Roman Times, Princeton, NJ 2001.

Mayor, Adrienne: Fossil Legends of the First Americans, Princeton, NJ 2005.

Mayr, Helmut: Karl Alfred von Zittel zum 150jährigen Geburtstag (25.9.1839–5.1.1904), in: Mitteilungen der Bayerischen Staatssammlung für Paläontologie und histor. Geologie, vol. 29 (Dec. 1989), pp. 7–51.

Mayr, Helmut: Schlosser, Max, in: Neue Deutsche Biographie 23 (2007), pp. 107–108 (online version, URL: https://www.deutsche-biographie.de/gnd117330353.html#ndbcontent, as consulted online on January 19, 2018).

McClelland, Charles E.: State, Society, and University in Germany 1700–1914, Cambridge 1980.

McDermott, John D.: Red Cloud. Oglala Legend, Pierre, SD 2015.

McDermott, John D.: Red Cloud's War. The Bozeman Trail 1866 – 1868, vol. I+II, Norman, OK 2010.

McNenly, Linda Scarangella: Native Performers in Wild West Shows. From Buffalo Bill to Euro Disney, Norman, OK 2012.

Menand, Louis: The Metaphysical Club, New York 2001.

Menze, Clemens: Die Bildungsreform Wilhelm von Humboldts, Hannover 1975.

Merrill, George P.: The First One Hundred Years of American Geology, New Haven, CT 1924.

Miller, Charles Allen: Jefferson and Nature. An Interpretation, Baltimore, MD 1988.

Mitchell, William John Thomas: The Last Dinosaur Book. The Life and Times of a Cultural Icon, Chicago 1998.

Mohr, Hans: Subjektivität in den Naturwissenschaften, in: Radermacher, Hans (ed.): Aktuelle Probleme der Subjektivität, Bern 1983, pp. 75–90.

Moses, Lester George: Wild West Shows and the Image of American Indians. 1883 – 1933, Albuquerque, NM 1996.

Nash, Roderick Frazer: Wilderness and the American Mind, 5th ed., New Haven, CT 2014.

Nieuwland, Ilja: The Colossal Stranger. Andrew Carnegie and Diplodocus Intrude European Culture, 1904–1912, in: Endeavour, vol. 34, no. 2 (Jun. 2010), pp. 61–68, DOI: 10.1016/j.endeavour.2010.04.001.

Nitsch, Wolfgang et al.: Hochschule in der Demokratie. Kritische Beiträge zur Erbschaft und Reform der deutschen Universität, Berlin 1965.

Numbers, Ronald L.: Science and Religion, in: Osiris, 2nd ser., vol. 1 (1985), pp. 59–80.

Nyberg, Kenneth: Linnaeus's Apostles and the Globalization of Knowledge, 1729–1756, in: Manning, Patrick; Rood, Daniel (eds.): Global Scientific Practice in an Age of Revolutions, 1750–1850, Pittsburgh, PA 2016, pp. 73–89.

Ogilvie, Brian W.: The Science of Describing. Natural History in Renaissance Europe, pbk. ed., Chicago 2008 (orig. publ. 2006).

Olson, James C.: Red Cloud and the Sioux Problem, repr. ed. Lincoln, NE 1975.

Olson, Richard G.: Science Deified & Science Defied. The Historical Significance of Science in Western Culture, vol. 1: From the Bronze Age to the Beginnings of the Modern Era, ca. 3500 B.C. to ca. A.D. 1640, Berkeley, CA 1982.

Olson, Richard G.: Science Deified & Science Defied. The Historical Significance of Science in Western Culture, vol. 2: From the Early Modern Age through the Early Romantic Era, ca. 1640 to ca. 1820, Berkeley, CA 1990.

Oleson, Alexandra: Introduction. To Build a New Intellectual Order, in: Brown, Sanborn C.; Oleson, Alexandra (eds.): The Pursuit of Knowledge on the Early American Republic. American Scientific and Learned Societies from Colonial Times to the Civil War, Baltimore, MD 1976, pp. xv–xxv.

Ostrander, Gilman M.: Republic of Letters. The American Intellectual Community, 1776–1865, Madison, WI 1999.

Outram, Dorinda: Georges Cuvier. Vocation, Science and Authority in Post-Revolutionary France, Manchester 1984.

Pacholski, Andreas: Wahrheit in Gestalt. Sprachbedingungen der Wissenschaft. Die Ansätze T. S. Kuhns und M. Merleau-Pontys, Marburg 2009.

Paletschek, Sylvia: Verbreitete sich ein ‚Humboldt'sches Modell' an den deutschen Universitäten im 19. Jahrhundert?, in: Schwinges, Rainer C. (ed.): Humboldt International. Der Export des Deutschen Universitätsmodells im 19. und 20. Jahrhundert, Basel 2001, pp. 75–104.

Palmer, Katherine van Winkle: The Clairbornian Scaphopoda. Gastropoda and Dibranchiate Cephalopoda of the Southern United States, in: Bulletins of American Paleontology, vol. 7, no. 32 (Dec. 1937).

Palmer, Thomas Waverly: A Register of the Officers and Students of the University of Alabama, 1831–1901, Tuscaloosa, AL 1901.

Parsons, Keith M.: Drawing out Leviathan. Dinosaurs and the Science Wars, Bloomington, IN 2001.

Paul, Gregory S.: The Princeton Field Guide to Dinosaurs, Princeton, NJ 2010.

Pauly, Philip: Biologists and the Promise of American Life. From Meriwether Lewis to Alfred Kinsley, Princeton, NJ 2000.

Pinch, Trevor: Does Science Studies Undermine Science? Wittgenstein, Turing, and Polanyi as Precursors for Science Studies and the Science Wars, in: Labinger, Jay A.; Collins, Harry (eds.): The One Culture? A Conversation about Science, Chicago 2001, pp. 13–26.

Pochmann, Henry A.: German Culture in America. Philosophical and Literary Influences 1600–1900, Madison, WI 1957.

Pomeroy, Earl Spencer: In Search of the Golden West. The Tourist in Western America, 1st Bison Book pr., Lincoln, NE 1990.

Porter, Roy: The Making of Geology. Earth Science in Britain, 1660–1815, Cambridge 1977.

Rabbitt, Mary, C.: John Wesley Powell: Pioneer Statesman of Federal Science, in: The Colorado River Region and John Wesley Powell: A Collection of Papers Honoring Powell on the 100th Anniversary of His Exploration of the Colorado River, 1869–1969 (Geological Survey Professional Paper 669), Washington, DC 1969, pp. 1–21.

Rainger, Ronald: An Agenda for Antiquity. Henry Fairfield Osborn & Vertebrate Paleontology at the American Museum of Natural History, 1890–1935, Tuscaloosa, AL 1991.

Rainger, Ronald: Collectors and Entrepreneurs. Hatcher, Wortmann, and the Structure of American Vertebrate Paleontology Circa 1900, in: Earth Sciences History, vol. 9, no. 1 (1990), pp. 14–21.

Rainger, Ronald: Vertebrate Paleontology as Biology. Henry Fairfield Osborn and the American Museum of Natural History, in: Rainger, Ronald et al. (eds.): The American Development of Biology, pbk. ed., New Brunswick, NJ 1991 (orig. publ. 1988), pp. 219–256.

Rappaport, Rhoda: When Geologists Were Historians, 1665–1750, Ithaca, NY 1997.

Regal, Brian: Henry Fairfield Osborn. Race, and the Search for the Origins of Man, Aldershot 2002.

Reingold, Nathan: Definitions and Speculations. The Professionalization of Science in American in the Nineteenth Century, in: Brown, Sanborn C.; Oleson, Alexandra (eds.): The Pursuit of Knowledge on the Early American Republic. American Scientific and Learned Societies from Colonial Times to the civil War, Baltimore, MD 1976, pp. 33–69.

Reingold, Nathan: Science in Nineteenth-Century America. A Documentary History, New York 1964.

Rico, Monica: Nature's Noblemen. Transatlantic Masculinities and the Nineteenth-Century American West, New Haven, CT 2013, DOI:10.12987/yale/9780300136067.001.0001.

Rieppel, Lukas: Bringing Dinosaurs Back to Life. Exhibiting Prehistory at the American Museum of Natural History, in: Isis, vol. 103, no. 3 (Sep. 2012), pp. 460–490, https://doi.org/10.1086/667969.

Ringer, Fritz K.: The German Academic Community, in: Oleson, Alexandra; Voss, John (eds.): The Organization of Knowledge in Mordern America, 1860–1920, Baltimore, MD 1979, pp. 409–429.

Rudolph, Frederick: The American College and University, New York 1962.

Rudwick, Martin J. S.: Earth's Deep History. How It Was Discovered and Why It Matters, Chicago 2014, DOI: 10.7208/chicago/9780226204093.001.0001.

Rudwick, Martin J. S.: The Great Devonian Controversy. The Shaping of Scientific Knowledge among Gentlemanly Specialists, Chicago 1985.

Rudwick, Martin J. S.: Worlds before Adam. The Reconstruction of Geohistory in the Age of Reform, Chicago 2008, https://doi.org/10.4000/miranda.1393.

Rupke, Nicolaas A.: Richard Owen. Victorian Naturalist, New Haven, CT 1994.

Rupke, Nicolaas A.: The Great Chain of History. William Buckland and the English School of Geology (1814–1849), Oxford 1983.

Ruse, Michael: Darwin and Design. Does Evolution Have a Purpose? Cambridge, MA 2003.

Ruse, Michael: Darwinism and Its Discontents, Cambridge, MA 2006.

Rydell, Robert W.; Kroes, Rob: Buffalo Bill in Bologna. The Americanization of the World, 1869–1922, Chicago 2005.

Scheele, Irmtraut: Grundzüge der institutionellen Entwicklung der biologischen Disziplinen an den deutschen Hochschulen seit dem 18. Jahrhundert, in: Schubring, Gert (ed.): „Einsamkeit und Freiheit" neu besichtigt: Universitätsreformen und Disziplinenbildung in Preußen als Modell für Wissenschaftspolitik im Europa des 19. Jahrhunderts, Stuttgart 1991, pp. 144–154.

Schelsky, Helmut: Einsamkeit und Freiheit. Idee und Gestalt der deutschen Universität und ihrer Reformen, Reinbek 1963.

Scheidegger, Nicoline: Der Netzwerkbegriff zwischen einem Konzept für Handlungskoordination und einer Methode zur Untersuchung relationaler Phänomene, in: Kulin, Sabrina et al. (eds.): Soziale Netzwerkanalyse. Theorie, Methoden, Praxis, Münster 2012, pp. 41–51.

Schnurmann, Claudia: Frontiers, Landscape and Ideology in the 19th-century USA: the Yosemite, in: Ellis, Steven G.; Eßer, Raingard (eds.): Frontiers and the Writing of History, 1500–1850, Hannover 2006, pp. 275–299.

Schuchert, Charles: Biographical Memoir of Othniel Charles Marsh, 1831–1899, Washington, DC 1938.

Schuchert, Charles; LeVene, Clara Mae: O. C. Marsh. Pioneer in Paleontology, reprint, New York 1978 (orig. publ. 1940).

Schwarz, Angela: Intersecting Anglo-German Networks in Popular Science and their Functions in the Late Nineteenth Century, in: Ellis, Heather; Kirchberger, Ulrike (eds.): Anglo-German Scholarly Networks in the Long Nineteenth Century, Leiden 2014, pp. 65–83, https://doi.org/10.1163/9789004253117_005.

Schwinges, Rainer C.: Humboldt International. Der Export des deutschen Universitätsmodells. Eine Einführung, in: Schwinges, Rainer C. (ed.): Humboldt International. Der Export des Deutschen Universitätsmodells im 19. und 20. Jahrhundert, Basel 2001.

Secord, James A.: How Scientific Conversation Became Shop Talk, in: Fyfe, Aileen; Lightman, Bernard V. (eds.): Science in the Marketplace. Nineteenth-Century Sites and Experiences, Chicago 2007, pp. 23–59, DOI:10.1017/S0080440107000564.

Secord, James A.: Visions of Science. Books and Readers at the Dawn of the Victorian Age, Oxford 2014.

Selden, Paul A.; Nudds, John R.: Evolution of Fossil Ecosystems, 2nd ed., London 2012.

Sellers, James B.: History of the University of Alabama, vol. 1: 1818–1902, Tuscaloosa, AL 1953.

Semonin, Paul: American Monster. How the Nation's First Prehistoric Creature Became a Symbol of National Identity, New York 2000.

Sepkoski, David; Ruse, Michael: Introduction. Paleontology at the High Table, in: Sepkoski, David; Ruse, Michael (eds.): The Paleobiological Revolution. Essays on the Growth of Modern Paleontology, Chicago 2009, pp. 1–11, DOI: 10.7208/chicago/9780226748597.001.0001.

Shapiro, Adam R.: Science Education, in: Montgomery, Georgia M.; Largent, Mark A. (eds.): A Companion to the History of American Science, Chichester 2016, pp. 320–332, DOI:10.1002/9781119072218.

Shils, Edward: The Order of Learning in the United States. The Ascendancy of the University, in: Oleson, Alexandra; Voss, John (eds.): The Organization of Knowledge in Modern America, 1860–1920, Baltimore, MD 1979, pp. 19–47.

Shor, Elizabeth Noble: Fossils and Flies. The Life of a Complete Scientist. Samuel Wendell Williston (1851–1918), Norman, OK 1971.

Slotkin, Richard: Gunfighter Nation. The Myth of the Frontier in Twentieth-Century America, New York 1992.

Slotkin, Richard: Regeneration Through Violence. The Mythology of the American Frontier; 1600–1860, 4th ed., Middletown, CT 1987.

Slotkin, Richard: The Fatal Environment. The Myth of the Frontier in the Age of Industrialization 1800–1890, New York 1985.

Sommer, Marianne: History Within. The Science, Culture, and Politics of Bones, Organisms, and Molecules, Chicago 2016.

Spiegel, Sibylle: Buffalo Bill's Wild West in München. Eine Veranstaltung Von „Höherem Wissenschaftlichen Interesse", Gerolzhofen 2002.

Spiegel-Rösing, Ina-S.: Wissenschaftsentwicklung und Wissenschaftssteuerung. Einführung und Material zur Wissenschaftsforschung, Frankfurt/Main 1973.

Steward, Mart A.: 'Let Us Begin With the Weather'. Climate, Race and Cultural Distinctiveness in the American South, in: Teich, Mikuláš et al. (eds.): Nature and Society in Historical Context, Cambridge 1997, pp. 240–256.

Strasser, Bruno J.: Collecting Nature. Practices, Styles, and Narratives, in: Osiris, 2nd ser., vol. 27 (2012), pp. 303–340.

Strohmeier, Gerhard: Wild West Imagery. Landscape Perception in Nineteenth-Century America, in: Teich, Mikuláš et al. (eds.): Nature and Society in Historical Context, Cambridge 1997, pp. 257–273.

Stuart, Paul: The Indian Office. Growth and Development of an American Institution, 1865–1900, Ann Arbor, MI 1979.

Struik, Dirk J.: Yankee Science in the Making, Boston, MA 1948.

Tamborini, Marco: "If the Americans Can Do It, So Can We". How Dinosaur Bones Shaped German Paleontology, in: History of Science, vol. 54, no. 3 (Sep. 2016), pp. 225–256, https://doi.org/10.1177/0073275316671526.

Taquet, Philippe: Establishing the Paradigmatic Museum. Georges Cuvier's Cabinet d'Anatomie Comparée in Paris, in: Knell, Simon J. et al. (eds.): Museum Revolutions. How Museums Change and Are Changed, London 2007, pp. 3–14.

Thenius, Erich; Vávra, Norbert: Fossilien im Volksglauben und im Alltag. Bedeutung und Verwendung vorzeitlicher Tier- und Pflanzenreste von der Steinzeit bis heute, Frankfurt/Main 1996.

Thelin, John R.: A History of American Higher Education, Baltimore, MD 2004.

Thomas, David Hurst: Skull Wars. Kennewick Man, Archaeology, and the Battle for Native American Identity, New York 2000.

Thomson, Keith Stewart: The Legacy of the Mastodon: The Golden Age of Fossils in America, New Haven, CT 2008.

Turner, Derek: Beyond Detective Work: Empirical Testing in Paleontology, in: Sepkoski, David; Ruse, Michael (eds.): The Paleobiological Revolution. Essays on the Growth of Modern Paleontology, Chicago 2009, pp. 201–214, https://doi.org/10.7208/9780226748597-012.

Turner, Derek: Paleontology. A Philosophical Introduction, Cambridge 2011.

Turner, Roy Steven: German Science, German Universities: Historiographical Perspectives from the 1980s, in: Schubring, Gert (ed.): „Einsamkeit und Freiheit" neu besichtigt: Universitätsreformen und Disziplinenbildung in Preußen als Modell für Wissenschaftspolitik im Europa des 19. Jahrhunderts, Stuttgart 1991, pp. 24–36.

Turner, Roy Steven: Humboldt in North America? Reflections on the Research University and its Historians, in: Schwinges, Rainer C. (ed.): Humboldt International. Der Export des Deutschen Universitätsmodells im 19. und 20. Jahrhundert, Basel 2001, pp. 289–311.

Veysey, Laurence R.: The Emergence of the American University, Chicago 1965.

Vizenor, Gerald; Lee, Arthur Robert: Postindian Conversations, Lincoln, NE 1999.

Warren, Leonard: Joseph Leidy. The Last Man Who Knew Everything, New Haven, CT 1998.

Weingart, Peter: Die Einheit der Wissenschaft – Mythos und Wunder, in: Weingart, Peter (ed.): Grenzüberschreitungen in der Wissenschaft – Crossing Boundaries in Science, Baden-Baden 1995, pp. 11–28.

Weingart, Peter et al.: Nachrichten aus der Wissensgesellschaft. Analysen zur Veränderung der Wissenschaft, Weilerswist 2007.

Wendler, Philipp: Die "Knochenkriege", die U.S.-amerikanische Presse und atlantische Wissenschaften in der zweiten Hälfte des 19. Jahrhunderts, unpubl. Magister thesis, Hamburg 2011.

Werner, Anja: Striving for the Top: Reevaluating Regional and Transatlantic influences in the History of U.S. Higher Education, in: Volkmann, Laurenz (ed.): Education and the USA, Heidelberg 2011, pp. 87–103.

Wheeler, William Morton: George Baur's Life and Writings, in: The American Naturalist, vol. 33, no. 385 (Jan. 1899), pp. 15–30.

White, Richard: Frederick Jackson Turner and Buffalo Bill, in: Grossman, James R.: The Frontier in American Culture. An Exhibition at the Newberry Library, August 26, 1994 – January 7, 1995, Berkeley, CA 1994, pp. 7–66.

Wieser, Matthias: Das Netzwerk von Bruno Latour. Die Akteur-Netzwerk-Theorie zwischen Science & Technology Studies und poststrukturalistischer Soziologie, Bielefeld 2012.

Wilson, Leonard Gilchrist: Lyell in America. Transatlantic Geology, 1841–1853, Baltimore, MD 1998.

Wrobel, David M.: The End of American Exceptionalism. Frontier-Anxiety from the Old West to the New Deal, Lawrence, KS 1993.

Xu, Xing et al.: First Ceratopsid Dinosaur from China and its Biogeographical Implications, in: Chinese Science Bulletin, vol. 55, no. 16 (jun. 2010), pp. 1631–1635, DOI:10.1007/s11434-009-3614-5.

ABOUT THE AUTHOR

Philipp Wendler holds a degree in History, Old German Literature, and Political Science from Universität Hamburg. He worked as research assistant to Prof. Dr. Claudia Schnurmann and earned a doctorate in the Global History Department of Universität Hamburg. His research interests are US and North American history. Currently he works as project coordinator for the German Migration Research Network, funded by the German Academic Exchange Service (DAAD).